T0227013

Demystifying Embedded Systems Middleware

Dedication

In loving memory of my father, who gave me the inspiration to write this book before he passed away,

for the team at Elsevier, all of my family, friends, and colleagues that I am lucky enough to still have in my life today and who continue to inspire me ....

Demystifying Embedded Systems Middleware

Tammy Noergaard

ELSEVIER

AMSTERDAM • BOSTON • HEIDELBERG • LONDON
NEW YORK • OXFORD • PARIS • SAN DIEGO
SAN FRANCISCO • SINGAPORE • SYDNEY • TOKYO
Newnes is an imprint of Elsevier

Newnes

Newnes is an imprint of Elsevier
The Boulevard, Langford Lane, Kidlington, Oxford OX5 1GB, UK
Radarweg 29, PO Box 211, 1000 AE Amsterdam, The Netherlands

First edition 2011

Notice
No responsibility is assumed by the publisher for any injury and/or damage to persons
or property as a matter of products liability, negligence or otherwise, or from any use
or operation of any methods, products, instructions or ideas contained in the material
herein. Because of rapid advances in the medical sciences, in particular, independent
verification of diagnoses and drug dosages should be made

British Library Cataloguing in Publication Data
A catalogue record for this book is available from the British Library

Library of Congress Cataloging-in-Publication Data
A catalog record for this book is availabe from the Library of Congress

ISBN–13: 978-0-7506-8455-2

For information on all Newnes publications
visit our web site at books.elsevier.com

10 11 12 13 14   10 9 8 7 6 5 4 3 2 1

Working together to grow
libraries in developing countries

www.elsevier.com | www.bookaid.org | www.sabre.org

ELSEVIER   BOOK AID
            International   Sabre Foundation

Contents

About the Author

Tammy Noergaard is uniquely qualified to write about all aspects of embedded systems. Since beginning her career, she has wide experience in product development, system design and integration, operations, sales, marketing, and training. She has design experience using many hardware platforms, operating systems, middleware, and languages. She worked for Sony as a lead software engineer developing and testing embedded software for analog TVs, and also managed and trained new embedded engineers and programmers. The televisions she helped to develop in Japan and California were critically acclaimed and rated #1 in *Consumer Reports* magazines. She has consulted internationally for many years, for companies including Esmertec and WindRiver, and has been a guest lecturer in engineering classes at the University of California at Berkeley, Stanford University, as well as giving technical talks at the invitation of Aarhus University for professionals and students in Denmark. She has also given professional talks at the Embedded Internet Conference and the Java User's Group in San Jose over the years. Most recently, her experience has been utilized in Denmark to help insure the success of fellow team members and organizations in building best-in-class embedded systems.

CHAPTER 1

Demystifying Middleware in Embedded Systems

Chapter Points

- Middleware is introduced in reference to the Embedded Systems Model
- Outline why understanding middleware is important
- Identifying common types of middleware in the embedded space

---

1.1 What is the Middleware of an Embedded System?

With the increase in the types and profitability of complex, distributed embedded systems, an approach common in the industry is designing and customizing these types of embedded systems in some manner that is independent of the underlying low-level system software and hardware components. To successfully achieve desired results within cost, schedule, and complexity goals many engineering teams base their approach on architecting various higher-level *middleware* software components into their embedded systems designs.

Currently within the embedded systems industry, there is no formal consensus on how embedded systems middleware should be defined. Thus, until such time as there is a consensus, this book takes the pragmatic approach of defining what middleware is and how different types of middleware can be categorized. Simply put, middleware is an abstraction layer that acts as an intermediary. Middleware manages interactions between application software and the underlying system software layers, such as the operating system and device driver layers. Middleware also can manage interactions between multiple applications residing within the embedded device, as well as applications residing across networked devices.

Middleware is simply software, like any other, that in combination with the embedded hardware and other types of embedded software is a means to an end to achieving some combination of the desirable goals shown in Table 1.1.

Demystifying Embedded Systems Middleware. DOI: 10.1016/B978-0-7506-8455-2.00001-7

Table 1.1: Examples of Desirable Requirements for Middleware to Meet

Requirement	Description
Adaptive	Middleware that enables overlying middleware and/or embedded applications to adapt to changing availability of system resources
Flexibility and Scalability	Middleware that allows overlying middleware and/or embedded applications to be configurable and customizable in terms of functionality that can be scaled in or out depending on application requirements, over all device requirements, and underlying system software and hardware limitations
Security	Middleware that insures the overlying middleware and/or embedded applications (and the users using them) have authorized access to resources
Portability	The 'write-once', 'run-anywhere' mantra. Middleware that allows overlying middleware and/or embedded applications to run on different types of embedded devices with different underlying system software and hardware layers. To avoid requiring time-consuming and expensive rewrites of the application code, middleware can mask the differences in underlying layers within different types of embedded systems, programming languages, and even implementations of the same standard produced by different design teams
Connectivity and Inter-Communication	Middleware that provides overlying middleware and/or embedded applications the ability to transparently communicate with other applications on a remote device through some user-friendly, standardized interface. Essentially, communication interfaces abstracted to level of local procedure call or method invocation

As shown in Figure 1.1a, middleware resides in the system software layer of an embedded system and is any software that is not a device driver, an operating system kernel, or an application. Middleware components can exist within various permutations of a real-world software stack: such as directly over device drivers, residing above an operating system, tightly coupled with an operating system package from an off-the-shelf vendor, residing above other middleware components, or some combination of the above, for example.

Keep in mind that what determines if a piece of software is 'middleware' is by where it resides within the embedded system's architecture, and not only because of its inherent purpose within the system alone. For example, as shown in Figure 1.1b, embedded Java virtual machines (JVMs) are currently implemented in an embedded system in one of three ways: in the hardware, in the system software layer, or in the application layer. When a JVM is implemented within the system software layer and resides on an operating system kernel is an example *when* a JVM is classified as *middleware*.

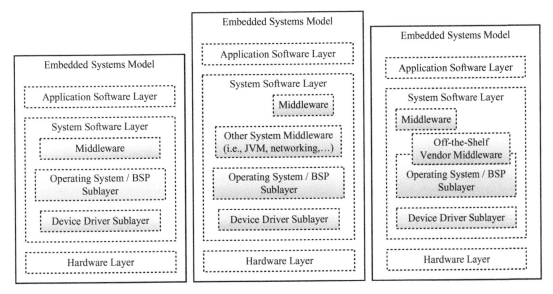

Figure 1.1a: Middleware and the Embedded Systems Model[1]

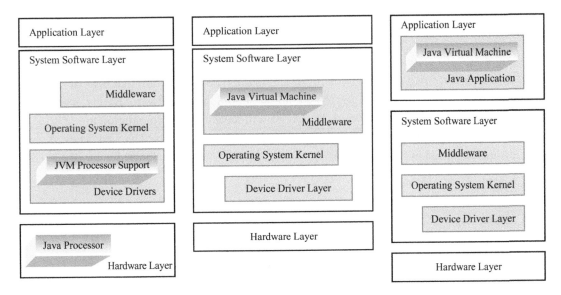

Figure 1.1b: Embedded JVMs in the Architecture[1]

Figure 1.1c shows a high-level block diagram of different types of middleware utilized in embedded devices today. Within the scope of this text, at the most general level, middleware is divided into two categories: ***core*** middleware and middleware that ***builds on*** these ***core*** components. Within each category, middleware can be further broken down into types, such as file systems, networking middleware, databases, and virtual machines to name a few. Open source and

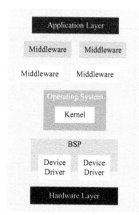

Figure 1.1c: Types of Middleware in Embedded Systems

real-world examples of these types of middleware will be used when possible throughout this book to demonstrate the technical concepts. Examples of building real-world designs based on these types of middleware will be provided, and the challenges and risks to be aware of when utilizing middleware in embedded systems will also be addressed in this text.

Core middleware is software that is most commonly found in embedded systems designs today that do incorporate a middleware layer, and is the type of software that is most commonly used as the foundation for more complex middleware software. By understanding the different types of core middleware, the reader will have a strong foundation to understanding and designing any middleware component successfully. The four types of core middleware discussed in this book are:

- **Chapter 4**.   Networking
- **Chapter 5**.   File systems
- **Chapter 6**.   Virtual machines
- **Chapter 7**.   Databases.

Middleware that builds on the core components varies widely from market to market and device to device. In general, this more complex type of middleware falls under some combination of the following:

- Message Oriented and Distributed Messaging, i.e.,
  - Message Oriented Middleware (MOM)
  - Message Queues
  - Java Messaging Service (JMS)
  - Message Brokers
  - Simple Object Access Protocol (SOAP)

- Distributed Transaction, i.e.,
  - Remote Procedure Call (RPC)
  - Remote Method Invocation (RMI)
  - Distributed Component Object Model (DCOM)
  - Distributed Computing Environment (DCE)
- Transaction Processing, i.e.,
  - Java Beans (TP) Monitor
- Object Request Brokers, i.e.,
  - Common Object Request Broker Object (CORBA)
  - Data Access Object (DAO) Frameworks
- Authentication and Security, i.e.,
  - Java Authentication and Authorization Support (JAAS)
- Integration Brokers.

At the highest level, these more complex types of middleware will be subcategorized and discussed under the following two chapters:

- **Chapter 3**.   Market-specific Complex Middleware
- **Chapter 8**.   Complex Messaging and Communication Middleware.

This book introduces the main concepts of different types of middleware and provides snapshots of open-source to help illustrate the main points. When introducing the fundamentals of various middleware components within the relative chapters, this book takes a multistep approach that includes:

- discussing the importance of understanding the standards, underlying hardware, and system software layers
- defining the purpose of the particular middleware component within the system, and examples of the APIs provided with a particular middleware component
- introducing middleware models and open-source software examples that would make understanding the middleware software architecture much simpler
- providing some examples of how overlying layers utilize various middleware components to apply some of what the reader has read.

The final chapter pulls it all together with pros and cons of utilizing the different types of middleware in embedded systems designs. As this book will demonstrate, there are several different types of embedded systems middleware on the market today, in addition to the countless homegrown solutions. Note that these embedded systems middleware solutions can be further categorized as other types of middleware depending on the field – such as being *proprietary* versus *open-source*, for example. In short, the key is for the reader to pick up on the high-level concepts and the patterns in embedded middleware software – and to recognize that these endless permutations of middleware solutions in the embedded space exist, because there is not 'one' solution that is perfect for all types of embedded designs.

1.2 How to Begin When Building a Complex Middleware-based Solution

For better or worse, successfully building an embedded system with middleware requires more than just solid technology alone. Engineers and programmers who recognize this wisdom from day one are most likely to reach production within quality standards, deadlines, and costs. In fact, the most common mistakes that kill complex embedded systems projects, especially those that utilize middleware components, are unrelated to the middleware technology itself. It is because team members did not recognize that successfully completing complex embedded designs requires:

- **Rule #1**: more than technology
- **Rule #2**: discipline in following development processes and best practices
- **Rule #3**: teamwork
- **Rule #4**: alignment behind leadership
- **Rule #5**: strong ethics and integrity among each and every team member.

So, what does this book mean by Rule 1 – that building an embedded system with middleware successfully requires more than just technology?

It means that many different influences, including technical, business-oriented, political, and social to name a few, will impact the process of architecting an embedded design and taking it to production. The architecture business cycle shown in Figure 1.2 shows a visualization

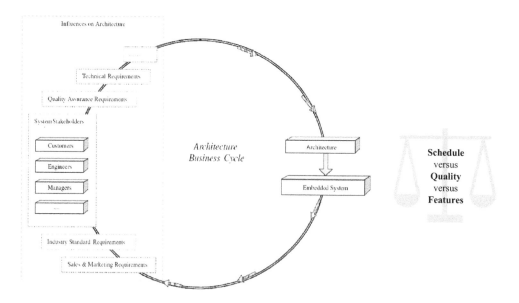

Figure 1.2: Architecture Business Cycle[2]

of this rule in which many different types of influences generate the requirements, the requirements in turn generate the embedded system's architecture, this architecture is then the basis for producing the device, and the resulting embedded system design in turn provides feedback for requirements and capabilities back to the team.

So, out of the architecture business cycle comes a reflection of what challenges real-world development teams building a complex middleware-based system face – balancing quality versus schedule versus features. This is where the other four rules stated at the start of this section come into play for insuring success. Ultimately, the options embedded teams have to choose from when targeting to successfully build a complex design are typically some combination of:

- **X** Option 1: Don't ship
- **X** Option 2: Blindly ship on time, with buggy features
- **X** Option 3: Pressure tired developers to work even longer hours
- **X** Option 4: Throw more resources at the project
- **X** Option 5: Let the schedule slip
- √ Option 6: Healthy Shipping Philosophy: '**Shipping a very *high-quality* system *on time*.**'

Not shipping unfortunately happens too often in the industry, and is obviously the option everyone on the team wants to avoid. 'No' products will ultimately lead to 'no' team, and in some cases 'no' company. So, moving on to the next option – why 'shipping a buggy product' is also to be avoided at all costs is because there are serious liabilities that would result if the organization is sued for a lot of money, and/or employees going to prison if anyone gets hurt as a result of the bugs in the deployed design (see Figure 1.3). When developers are forced to cut corners to meet the schedule relative to design options, are being forced to work overtime to the point of exhaustion, are undisciplined about using best practices when programming, code inspections, testing, and so on – this can then result in serious liabilities for the organization when what is deployed contains serious defects.

Option 3 – 'pressure tired developers to work even longer hours' – is also to be avoided. The key is to 'not' panic. Removing calm from an engineering team and pushing exhausted developers to work even longer overtime hours on a complex system that incorporates middleware software will only result in more serious problems. Tired, afraid, and/or stressed-out engineers and developers will result in mistakes being made during development, which in turn translates to additional costs and delays.

Negative influences on a project, whether financial, political, technical, and/or social in nature, have the unfortunate ability to negatively harm the cohesiveness of an ordinarily healthy team within a company – eventually leading to sustaining these stressed software teams as unprofitable in themselves. Within a team, even a single weak link, such as a team of exhausted and stressed-out engineers, will be debilitating for an entire project and even an

- **Breach of Contract**, i.e.,
  - ◦ if bug fixes stated in contract are not forthcoming in timely manner

- **Breach of Warranty and Implied Warranty**, i.e.,
  - ◦ delivering system without promised features

- **Strict and Negligence Liability**, i.e.,
  - ◦ bug causes damage to property
  - ◦ bug causes injury
  - ◦ bug causes death

- **Malpractice**, i.e.,
  - ◦ customer purchases defective product

- **Misrepresentation and Fraud**, i.e.,
  - ◦ product released and sold that doesn't meet advertised claims

- Based on the chapter "Legal Consequences of Defective Software" by Cem Kaner
Testing Computer Software. 1999

Figure 1.3: Why Not Blindly Ship? – Programming and Engineering Ethics Matter[3]

entire organization. This is because these types of problems radiate outwards influencing the entire environment, like waves (Figure 1.4).

The key here is to decrease the interruptions (see Figure 1.5) *and* stress for a development team during their most productive programming hours within a **normal** work week, so that there is more focus and fewer mistakes.

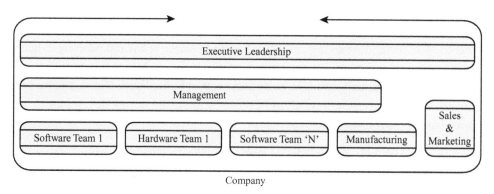

Figure 1.4: Problems Radiate and Impact Environment

"... developers imprisoned in noisy cubicles, those who had no defense against frequent interruptions, did poorly. How poorly? The numbers are breathtaking. The best quartile was 300% more productive than the lowest 25%. Yet privacy was the only difference between the groups.

Think about it – would you like 3× faster development?

It takes your developers 15 minutes, on average, to move from active perception of the office busyness to being totally and productively engaged in the cyberworld of coding. Yet a mere 11 minutes passes between interruptions for the average developer. Ever wonder why firmware costs so much? ..."

- Jack Ganssle. *A Boss's Quick-Start to Firmware Engineering.*
- DeMarco and Lister. *Peopleware.*

Figure 1.5: Real World Tidbit, Underpinnings of Software Productivity

Another approach in the industry to avoid a schedule from slipping has been to throw more and more resources at a project. Throwing more resources ad-hoc at project tasks without proper planning, training, and team building is the surest way to hurt a team and guarantee a missed deadline. As indicated in Figure 1.6, productivity crashes with the more people there are on a project. A limit in the number of communication channels can happen through more than one (>1) smaller sub-teams, as long as:

- it makes sense for the embedded systems product being designed, i.e.,
  - not dozens of developers and several line/project managers for a few MB of code
  - not when few have embedded systems experience and/or experience building the product
  - not for corporate empire-building! – which results in costly project problems and delays = bad for business!

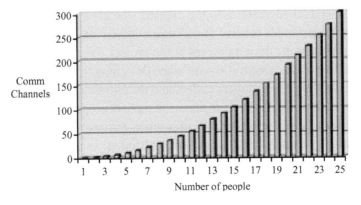

Figure 1.6: Too Many People[4]

- in a healthy team environment
- no secretiveness
- no hackers
- best practices and processes not ignored
- team members have sense of professional responsibility, alignment, and trust with each other, leadership and the organization.

While more related to this discussion will be covered in the last chapter of this book, ultimately the most powerful way to meet project schedules and successfully take an embedded system middleware-based solution to production is:

- by shipping a very high-quality product on time
- have a strong technical foundation
- sacrificing less essential features in the first release
- start with skeleton, then hang code off skeleton
- Do not overcomplicate the design!
- Systems integration, testing and verification from Day 1.

The rest of this chapter and most of this book are dedicated to supplying the reader with a strong, pragmatic technical foundation relative to embedded systems middleware. The last section of this book will pull it all together to link in what was introduced in this section.

1.3 Why is a Strong Technical Foundation Important in Middleware Design?

One of the biggest myths propagated by inexperienced team members and mistakes made in the industry is assuming that the embedded systems programmers of a middleware layer can afford to think as abstractly as PC developers and/or the application developers using that middleware layer. There are too many examples of stressed-out engineers, millions of dollars in project overruns, and failed ventures in the industry that are a result of team members not understanding the fundamentals relative to utilizing middleware within an embedded system at the start and throughout the design process of the project. When it comes to understanding the underlying hardware and system software when designing middleware software, it is critical that, at the very least, developers understand the entire design at a systems level. In fact, one of the most common mistakes made on an embedded project that makes it much tougher to successfully build a complex design is when engineers and programmers on the team do not investigate or understand the type of embedded system they are trying to build, the components that can make up the device, and/or the impact individual components have on each other.

Thus, this book is a springboard from *'Embedded Systems Architecture: A Practical Guide for Engineers and Programmers'*. This book takes a more detailed and practical approach of discussing all layers relative to the Embedded Systems Model, shown in Figure 1.1a, when introducing principles and major elements of embedded systems middleware. This is because it is critical to the success of any project team that introduces middleware into the architecture that all team members understand all layers of an embedded system because *all* layers of an embedded system are *impacted* by middleware and vice versa.

Introducing middleware software to an embedded system introduces an **additional overhead** that will impact everything from memory requirements to performance, reliability, as well as scalability, for instance. The goal of this book is not just about introducing some of the most common types of embedded systems middleware, but **more importantly** *to show the reader the pattern behind different types of embedded middleware designs and to help teach the reader an approach to understanding and applying this knowledge to any embedded system's middleware component encountered in the future.*

The Embedded Systems Model represents the layers in which all components existing within an embedded system design can reside. This model is a powerful tool utilized within the scope of this book because it not only provides a clear visual representation of the various middleware elements of an embedded system, their interrelationships, and functionality – this model also provides a basis for modular architectural representations that commonly are used to successfully structure an embedded systems project. At the highest level, there are three layers:

- **hardware**, which contains all the physical components located on an embedded systems board
- **system software**, which is the device's application-independent software
- **application software**, which is the device's application-specific software.

As shown in Figure 1.7, a middleware component – whether it is a file system, database, or networking protocol – that resides in an embedded system's middleware software layer typically resides on top of 'some' combination of other middleware, an operating system, device drivers, and hardware. This means middleware implemented in the system software layer exists either as:

- middleware that sits on top of the operating system layer, or device driver layer for systems with no operating system
- middleware that sits on top of other middleware components, for example a Java-based database or file system that resides over a Java Virtual Machine (JVM)
- middleware that has been tightly integrated and provided with a particular operating system distribution.

In some embedded systems, there may even be more than one different middleware component, as well as more than one of the same type of middleware in the embedded device (see Figure 1.8). In short, whatever the combination of middleware – in co-operation with

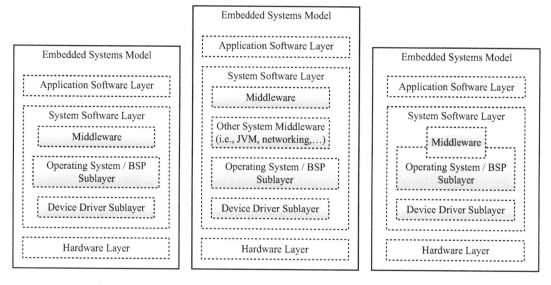

Figure 1.7: System Components and the Embedded Systems Model

the underlying embedded software and hardware – these components act as an abstraction layer that provides various data management functions to the other system software layer components, application software layer in the system, and even other computer systems that have remote access to the device.

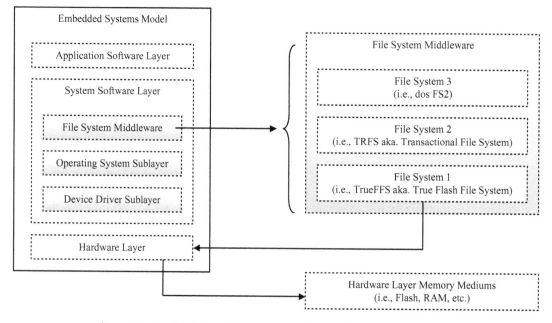

Figure 1.8: Multiple File Systems in an Embedded System Example

1.4 Summary

Middleware is increasingly becoming a required component in embedded systems designs due to the increase in the types of complex, distributed embedded systems, the number of applications found on embedded systems, and the desire for customizable embedded software applications for embedded devices. In this chapter, middleware was defined relative to the Embedded Systems Model, and the types of middleware introduced in this book were also discussed. Finally, some initial guidelines of whether using middleware within an embedded systems design should even be entertained as an option are discussed.

Chapters 4–7 cover core middleware components, specifically file systems, networking, and databases. Chapters 3, 8 and 9 go on to discuss middleware that builds on the core components, as well as pulls all the concepts together in discussing overall design implementations, approaches, and risk mitigation for utilizing middleware in real-world embedded designs.

The next chapter of this book introduces core components that underlie middleware commonly found in embedded systems. Chapter 2, specifically, introduces the hardware and underlying system software required by core middleware.

1.5 End Notes

1 Systems Architecture, Noergaard, 2005. Elsevier.
2 The six stages of creating an architecture outlined and applied to embedded systems in this book are inspired by the Architecture Business Cycle developed by SEI. For more on this brainchild of SEI, read 'Software Architecture in Practice,' by Bass, Clements, and Kazman.
3 Based on the chapter 'Legal Consequences of Defective Software' by Cem Kaner. Testing Computer Software. 1999.
4 'Better Firmware, Faster'. Jack Ganssle. 2007.

CHAPTER 2

The Foundation

Chapter Points

- Defines what components are required and underlie middleware
- Introduces fundamental hardware concepts and terminology
- Identifies the major elements of most underlying system software designs

Regardless of what middleware is in an embedded system, one of the most powerful approaches is to take the systems approach. This means having a solid technical foundation via defining and understanding all required components that underlie the particular middleware software. Meaning:

1. Understanding the hardware. If the reader comprehends the hardware, it is easier to understand why a particular middleware component implements functionality in a certain way relative to the storage medium, as well as the hardware requirements of a particular middleware implementation.
2. Defining and understanding the specific underlying system software components, such as the available device drivers supporting the storage medium(s) and the operating system API. Underlying system software will be discussed later in this chapter.

Why start with understanding the hardware? Because some of the most common mistakes programmers designing complex embedded systems make that lead to costly delays and problems include:

- being intimidated by the embedded hardware and tools
- treating all embedded hardware like it is a PC-Windows Desktop
- waiting for the hardware
- using PCs in place of 'available' embedded systems target hardware to do development and testing
- NOT using embedded hardware similar to production hardware, mainly similar I/O, processing power, and memory.

Demystifying Embedded Systems Middleware. DOI: 10.1016/B978-0-7506-8455-2.00002-9

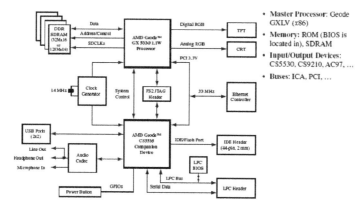

* **Master Processor:** Net+ARM ARM7
* **Memory:** Flash, RAM
* **Input/Output Devices:** 10Base-T transceiver, Thinnet transceiver, 100Base-T transceiver, RS-232 transceiver, 16646 transceiver , ….
* **Buses:** System Bus, MII, …

Figure 2.1a: Net Silicon ARM7 Reference Board[1]

* **Master Processor:** Geode GXLV (x86)
* **Memory:** ROM (BIOS is located in), SDRAM
* **Input/Output Devices:** CS5530, CS9210, AC97, …
* **Buses:** ICA, PCI, …

Figure 2.1b: AMD Geode Reference Board[2]

Developing software for embedded hardware is *not* the same as developing software for a PC or a larger computer system – especially when it comes to including the additional layer of complexity when introducing a middleware component. The embedded systems boards shown in Figures 2.1a–d demonstrate this point of how drastically embedded boards can vary in design.

This means each of the boards shown widely varies in terms of the software that can be supported because the major hardware components are different, from the type of master processor to the available memory to the I/O (input/output) devices. Target system hardware requirements depend on the software, especially complex systems that contain an operating system, middleware components, in addition to the overlying application software. So, middleware developers must learn to read the hardware schematics and datasheets to

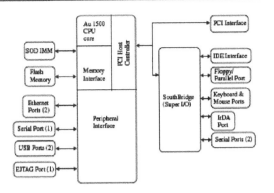

Figure 2.1c: Ampro MIPS Reference Board[3]

- Master Processor: Encore M3 (Au1500-based) processor
- Memory: Flash, SODIMM
- Input/Output Devices: Super I/O,...
- Buses: PCI, ...

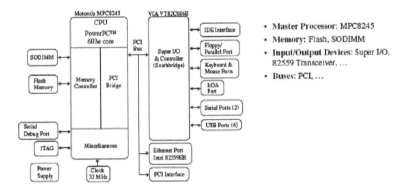

Figure 2.1d: Ampro PowerPC Reference Board[4]

- Master Processor: MPC8245
- Memory: Flash, SODIMM
- Input/Output Devices: Super I/O, 82559 Transceiver, ...
- Buses: PCI, ...

understand and verify all the major components found on an embedded board. This is to insure that the processor design is powerful enough to support the requirements of the software stack, the embedded hardware contains the required I/O, and the hardware has enough of the *right* type of memory.

2.1 A Middleware Programmer's Viewpoint – Why Care about Processor Design and I/O?

From the middleware programmer's point of view, it is critical to care about the processor design and I/O on the target hardware. In the case of processors (whether they are master and/or slave I/O CPUs), there are literally thousands of embedded processors that are differentiated according to their ISAs (instruction set architectures). A processor's ISA defines everything from the available operations to the operands to addressing modes to interrupt handling, for example. Most embedded processors fall under one of three ISA models:

- *Application-specific*, such as controller, datapath, finite state machine w/datapath (FSMD), and Java virtual machine (JVM)
- *General purpose*, such as complex instruction set computing (CISC) and reduced instruction set computing (RISC)
- *Instruction-level parallelism*, such as single instruction multiple data (SIMD), superscalar machine, very long instruction word computing (VLIW).

It is important for programmers to understand the processors and the ISA design they are based upon. This is because the ability to support a complex middleware solution, and the time it takes to design and develop it, will be impacted by the ISA in terms of available functionality, the cost of the chip, and most importantly the performance of the processor. For example, a programmer's ability to understand processor performance, and what to look for in a processor's design according to what needs to be accomplished via software. Processor performance is most commonly defined as some combination of the following:

- *Responsiveness*, length of elapsed time the processor takes to respond to some event, a.k.a latency
- *Availability*, the amount of time the processor runs normally without failure
- *Reliability*, the average time between failures, a.k.a. the MTBF (mean time between failures)
- *Recoverability*, the average time the processor takes to recover from failure, a.k.a. the MTTR (mean time to recover)
- *Throughput*, the amount of work the processor completes in a given period of time, a.k.a. the average execution rate (Figure 2.2).

CPU throughput = 1 / CPU execution time
CPU execution time   = (total number of instructions) * (CPI) * (clock period) = ((instruction count) * (CPI) / (clock rate))

i.e., Performance (Processor "A") / Performance (Processor "B")
= Execution Time (Processor "B") / Execution Time (Processor "A")
= "X", therefore, Processor "A" is "X" times faster than Processor "B"

CPU throughput = bytes/sec or MB/sec
CPU execution time = seconds per total # bytes
CPI = number of cycle cycles/instruction
clock period = seconds per cycle
clock rate = MHz

Figure 2.2: Processor Performance and Throughput

So, for example, given processor performance relative to throughput and managing instruction processing – specifically, the number of clock cycles per second (clock rate), as well as the number of cycle per instruction (CPI). Any internal processor design feature that allows for either an *increase* in the clock rate or *decrease* in the CPI will increase the overall performance of a processor. This could include anything from pipelining within the processor's ALU to selecting a processor based on the instruction-level parallelism ISA model.

In the case of IO subsystems consisting of some combination of transmission medium, ports and interfaces, IO controllers, buses, and the master processor integrated I/O – I/O subsystem performance in terms of throughput, execution time, and response time is key. Programmers need to pay attention not only to the speed of the master processor, but the data rates of the I/O devices, how to synchronize the speed of the master processor to the speeds of I/O, and how I/O and the master processor communicate. Programmers even need to pay attention to buses, meaning from a developer's viewpoint bus arbitration, handshaking, signal lines, and timing. Bus performance is typically measured via bandwidth where both physical design and associated protocols matter. For example:

- the simpler the bus handshaking scheme, the higher the bandwidth
- the shorter the bus, the fewer connected devices, and the more data lines typically means the faster the bus and the higher its bandwidth.
- more bus lines means the more data that can be physically transmitted at any one time, in parallel
- the bigger the bus width means the fewer the delays, and the greater the bandwidth.

Finally, *benchmarks*, such as EEMBC (Embedded Microprocessor Benchmark Consortium), Whetstone, and Dhrystone programs, are commonly used in the embedded space to provide some measure of processor performance such as determining latency and efficiency of individual features. Benchmarks typically report MIPS (Millions of Instructions per Second) = Instruction Count/(CPU execution time $\times 10^6$) = Clock Rate/(CPI $\times 10^6$).

The key for middleware programmers to remember is the importance of understanding what the benchmarks being executed are, and to use these benchmarks wisely.

This means that benchmarks give the *illusion* that faster CPUs have higher MIPS, because the MIPS formula is inversely proportional to execution time. MIPS cannot compare different ISAs, because instruction complexity and functionality are not considered in the formula. MIPS will also vary on the same CPU with different programs made up of different instructions. So, in short, ask the right questions and interpret benchmarks accurately to understand exactly what is being run and measured. Benchmarks are suitable in some cases as a starting point, but at the end of the day it is better for middleware programmers to use *real* embedded programs to measure a processor's performance in this regard.

2.2 The Memory Map, Storage Mediums, and Middleware

It is critical for middleware programmers to define and understand the board's memory map, specifically:

- Amount of memory matters (i.e., is there enough for run time needs?)
- Location of memory and how to reserve it
- Performance matters (gap between processor and memory speeds)
- Internal design of memory matters
- Type of memory matters (i.e., Flash versus RAM).

Why should a middleware programmer care? Take memory and performance, for example. Memory impacts board performance when memory has lower bandwidth than master CPU, thus it is important for programmers to understand memory timing parameters (performance indicators) such as memory access times and refresh cycle times. Memory performance can be better based on the internal design, such as:

- utilizing independent instruction and data memory buffers and ports
- integrating bus signals into one line to decrease the time it takes to arbitrate the memory bus to access memory
- having more memory interface connections (pins), increasing transfer bandwidth
- having a higher signaling rate on memory interface connections (pins)
- implementing a memory hierarchy, with multiple levels of cache.

Another example is that while middleware that utilizes different hardware storage devices is transparent to middleware users and higher layers of software, the underlying hardware of the different storage mediums available today is often quite different in terms of how they work, their performance, and how they physically store the data. Thus, it is important for embedded developers to understand the differences in the hardware in order to understand the implementation of a middleware component on these various underlying technologies. In other words, hardware features, quirks and/or limitations will dictate the type of file system(s) required and/or what modifications must be implemented in a particular middleware design to support this hardware.

If a programmer learns the features of the various hardware storage mediums available, then it will be much simpler for the programmer to understand a particular middleware implementation, how to modify a particular middleware design in support of a storage medium, as well as determine which middleware is the best 'fit' for the device. In short, it is important for the reader to understand the middleware relevant features of a storage medium(s) – and use this understanding when analyzing the middleware implementation that needs to support the particular storage medium.

In terms of hardware storage mediums used by middleware in the embedded systems arena, essentially if data can be stored on a hardware component, middleware can be designed and

USB Flash Memory Stick[2] CD[3] Smart Cards[4]

f0020 **Figure 2.3: Examples of Embedded System Hardware Storage Mediums Used To Store Data**

configured to use that storage medium. Examples of hardware storage mediums used by embedded middleware, such as file systems and databases today, are shown in Figure 2.3. Examples of hardware supported include hard drives, RAM, Flash, tape, CD, and floppy to name just a few.

As shown in Figure 2.4, middleware, like file systems, typically view and refer to physical hardware storage mediums as *raw devices*, *drives*, and/or *disks*. At the highest level, a raw

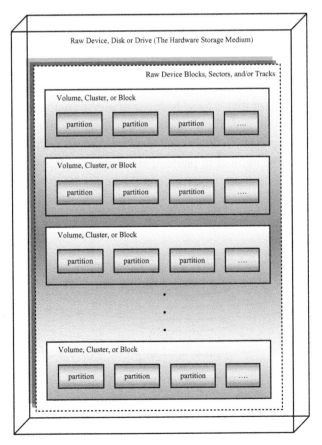

Figure 2.4: Hardware Storage Medium

device is then broken down into some combination of *blocks*, *tracks*, and/or *sectors*, terms used to represent addressable storage units on a raw device, disk, or drive. Middleware logical units, such as file system volumes or clusters, then reside within these storage units.

The next few hardware examples demonstrate some relevant differences between storage mediums that can be found in embedded system designs today. The reader can use these examples to understand the importance of learning about different hardware storage mediums, the differences between middleware software supporting various storage mediums, what is required to port a type of middleware to these various hardware storage elements, and/or to understand features of a storage medium that are relevant to middleware software. The reader can then apply this process of thinking to working with different hardware storage components and middleware software in the future.

2.2.1 Example of Hard Disk Hardware

While there are several different types of hard disk technologies on the market today, such as SCSI (Small Computer Systems Interface) and ATA (Advanced Technology Attachment) types of hard disk drives to name a few, in general many internals of traditional hard disks deployed today are similar. As shown in Figure 2.5a, most hard drives on the market are made up of *platters*, circular disks made from metal and covered with a magnetic material. This film of magnetic material is one of the main components that allows data to be recorded on a hard disk's platter. A hard disk's *head* is a type of electromagnet to process the data located on the associated platter. An arm supports each head, and the arm(s) is (are) attached to an actuator which is responsible for arm and head movement to the desired location on a platter to process data. The number of platters, associated heads, and arms in a hard drive is dependent on the size of the hard disk, meaning the larger the drive the more platters, associated heads, and arms exist.

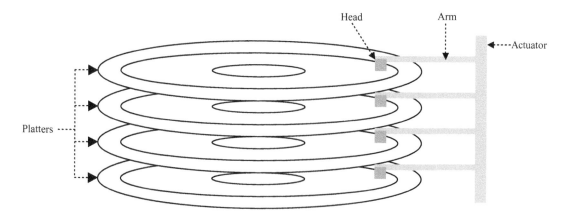

Figure 2.5a: Internals of a Hard Disk Drive[5]

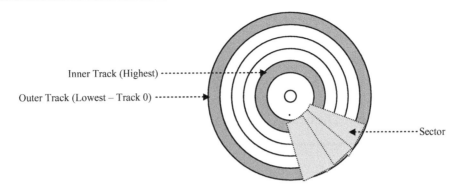

Figure 2.5b: Hard Disk Drive Platter[5]

A *low-level format* (LLF) creates *tracks, cylinders*, and *sectors* on each platter (see Figure 2.5b). An LLF is performed on most modern hard disks by the manufacturer before the hard disks are deployed into the field. Some hard drive manufacturers also provide tools to do an LLF in cases where everything needs to be removed from a hard disk without damage to the boot sector, such as when installing a new operating system or removing virus infection.

Tracks are concentric rings located on each platter that subdivide a platter for data recording. As shown in Figure 2.5c, a cylinder is a logical cross-section of tracks across all the hard disk's platters. Tracks are further broken down into sectors, which are data blocks on a platter that allow for simultaneous access to multiple tracks for data processing.

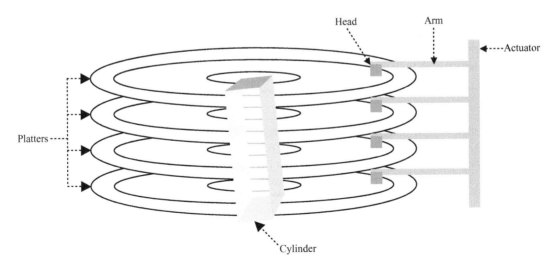

Figure 2.5c: Hard Disk Drive Cylinder

Accessing a data block on a hard disk is done via specifying the *CHS*, cylinder, head, and sector numbers.

Refer to a hard disk manufacturer's datasheet to determine detailed information of a particular hard disk's specifications. The real-world hard disk datasheets shown in Figures 2.6a and 2.6b are examples of how to find some of the hardware specification information that is useful for developers to know regarding hard disks (see highlighted portions of datasheets).

Helpful Hint

A datasheet is always a good starting point for understanding any hardware's general functions and features, but keep in mind this type of document is typically used for sales and marketing of the device as well. So it is always a good idea to review any available highly technical and in-depth users' guides and specifications for the particular storage medium to review specifics.

2.2.2 Example of USB Flash Memory

USB flash memory is simply a data storage device that contains non-volatile flash memory and an integrated USB interface. Relative to middleware, some of the key features of interest regarding USB Flash memory include:

- **Capacity**. The size of the USB flash memory.
- **Operating System (Device Driver) Support**. What operating system distributions include device drivers for the USB Flash memory. If the embedded system's operating system is not on that list, then a device driver will need to be created/ported and integrated.
- **Formatted**. Does the USB Flash memory come pre-formatted, in support of a particular file system, for example. The USB Flash memory may need to be erased and reprogrammed, as necessary, in support of a particular middleware.
- **Sector Size**. The smallest block of Flash that can be erased and/or programmed. The reader should also note whether there are any restrictions when reading the Flash.

Author Note

USB Flash memory can also be referred to by other names in the field, such as USB Flash Memory *Keys*, USB Flash Memory *Drives*, USB Flash Memory *Sticks*, and USB Flash Memory *Pen Drives* to name a few. If it is Flash memory that is hot-swappable into a USB port, then it falls under this category of USB Flash memory hardware.

As shown in Figure 2.7a, USB Flash memory is a small PCB (printed circuit board) that is enclosed in a durable chassis, and is powered via the connection to the embedded system's USB port. A standard USB interface that adheres to the industry standard USB specification,

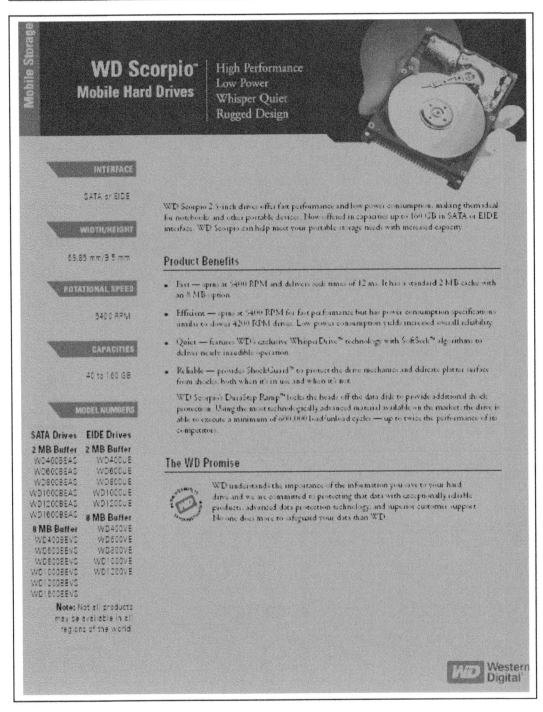

Figure 2.6a: Western Digital Hard Disk Datasheet Example[6]

WD Scorpio
Mobile Hard Drives

Drive Specifications

	40 GB	60 GB	80 GB	100 GB	120 GB	160 GB
Formatted capacity[1]	40,020 MB	60,022 MB	80,026 MB	100,030 MB	120,034 MB	160,041 MB
User sectors per drive	78,140,160	117,210,240	156,301,488	195,371,568	234,441,648	312,581,808
WD model number						
SATA						
2 MB cache buffer	WD400BEAS	WD600BEAS	WD800BEAS	WD1000BEAS	WD1200BEAS	WD1600BEAS
8 MB cache buffer	WD400BEVS	WD600BEVS	WD800BEVS	WD1000BEVS	WD1200BEVS	WD1600BEVS
EIDE						
2 MB cache buffer	WD400UE	WD600UE	WD800UE	WD1000UE	WD1200UE	
8 MB cache buffer	WD400VE	WD600VE	WD800VE	WD1000VE	WD1200VE	

SATA Hard Drives

Bytes per sector (GTO)	512
Form factor	2.5-inch
Interface	SATA 150 MB/s
RoHS compliant[2]	Yes

EIDE Hard Drives

Bytes per sector (GTO)	512
Form factor	2.5-inch
Interface	44-pin EIDE
RoHS compliant[2]	Yes

Performance Specifications

SATA Hard Drives

Average latency	4.2 ms
Cache	
Read	Adaptive
Write	Yes
Data transfer rate	
Buffer to disk	600 Mbits/s maximum[3]
Buffer to host	150 MB/s (maximum)
Drive ready time	4.0 sec (average)
Error rate (non-recoverable)	<1 in 10^14 bits read
Rotational speed	5400 RPM
Seek time	
Full stroke read	21 ms (average)
Read	12.0 ms (average)
Track-to-track	2.0 ms (average)
Load/unload cycles[5]	600,000

EIDE Hard Drives

Average latency	5.5 ms
Cache	
Read	Adaptive
Write	Yes
Data transfer rate (buffer to host)	
Mode 5 Ultra ATA[4]	100 MB/s
Mode 4 Ultra ATA[4]	66.6 MB/s
Mode 2 Ultra ATA[4]	33.3 MB/s
Mode 2 Ultra DMA[4]	16.6 MB/s
Mode 4 Ultra PIO[4]	16.6 MB/s
Buffer to disk	500 Mbit/s (maximum)
Drive ready time	4.0 sec (average)
Error rate (non-recoverable)	<1 in 10^14 bits read
Rotational speed	5400 RPM
Seek time	
Full stroke read	21 ms (average)
Read	12.0 ms (average)
Track-to-track	2.0 ms (average)
Load/unload cycles[5]	600,000

For service and literature:

support.wdc.com
www.westerndigital.com

800 ASK 4WDC	North America
949 672.7199	Spanish
+800.6008.6008	Asia Pacific
+31.20.4467651	EMEA

Western Digital, WD, and the WD logo are registered trademarks; and WD Scorpio, DuraStep Ramp, FIT Lab, ShockGuard, SoftSeek, and WhisperDrive are trademarks of Western Digital Technologies, Inc. Other marks may be mentioned herein that belong to other companies. Product specifications subject to change without notice.

© 2006 Western Digital Technologies, Inc. All rights reserved.

Western Digital
20511 Lake Forest Drive
Lake Forest, California 92630
U.S.A.

2879-001121 A10   Dec 2006

[1] One gigabyte (GB) = 1,000,000,000 bytes. Total accessible capacity varies depending on operating environment.
[2] WD complies with the Restriction of Hazardous Substances (RoHS) Directive 2002/95/EC of the European Parliament, which is effective in the EU beginning July 1, 2006. RoHS aims to protect human health and the environment by restricting the use of certain hazardous substances in new equipment, and consists of restrictions on lead, mercury, cadmium, and other substances.
[3] Buffer to disk transfer rate for WDxxxBEAS/BEVS ML80 drives. ML60 WDxxxBEAS/BEVS and WDxxxUE/VE drives have a buffer to disk transfer rate of 500 MB/s max. ML40 WDxxxUE/VE drives have a buffer to disk transfer rate of 421 MB/s max.
[4] Maximum burst rate running the specified PIO, DMA, or Ultra ATA transfer mode.
[5] Controlled unload at ambient condition.

Figure 2.6a continued: Western Digital Hard Disk Datasheet Example

Barracuda 7200.8

High-capacity Ultra ATA and Serial ATA desktop drives

Reliable, Top-Performing Storage for High-End PCs and Advanced Applications, Including Entry-Level RAID Servers and Gaming Systems

400, 300, 250 and 200 Gbytes • 7200 RPM • SATA/150 NCQ and Ultra ATA/100

Key Advantages

- Highest available capacity—up to 400 Gbytes, enabling OEMs and system builders to qualify one family of drives for a wide range of personal storage, entry-level server and fixed-content storage applications
- Native SATA interface with Native Command Queuing (NCQ) enables fast data transfer rates for high-performing, low-cost servers and hot-rod gaming systems
- A proven design for shorter qualification times and top reliability
- Inaudible acoustics with sound barrier technology (SBT), including the Seagate®-exclusive SoftSonic™ motor
- RoHS-compliant—significantly reduced levels of lead to comply with all environmental legislation

Best Fit Applications

- Mainstream and high-performance PCs
- Small workgroup servers in businesses, education and government
- Video, digital photo and e-mail storage on PC-based home media servers
- Storage of fixed-content information that is actively referenced but changed infrequently (medical records, active data archives, financial statements, videos, digital photos)
- PCs optimized for gaming

Figure 2.6b: Seagate Hard Disk Datasheet Example[7]

Seagate

We turn on ideas

Barracuda 7200.8

High-capacity, high-speed ATA drives for PCs and advanced storage applications

Seagate Makes the Best Even Better With a 5-Year Warranty

Seagate offers the industry's leading warranty to demonstrate our commitment to product reliability and our customers' success. Every Seagate internal hard drive for PCs, notebook computers and entry-level servers is covered under our unprecedented five-year warranty.

Seagate 5-Year Warranty

Highest Available Capacity Enables New Opportunities

- Industry's highest areal density lowers cost per Gbyte of storage
- High capacity demand continues to increase as storage needs proliferate
- Increased demand coming from users requiring more storage capacity on their PCs and small businesses implementing RAID servers
- Various-sized enterprises are replacing tape with high-capacity HDDs in low-cost backup or fixed content applications

Native Serial ATA Interface With Native Command Queuing (NCQ)

- Native SATA interface with NCQ and support for latest PC technology, including Intel Hyper-Threading
- Point-to-point interface eliminates need for jumper settings
- Easy-to-use connectors for simple installation
- 100 percent software compatible with existing PCs
- Thinner, longer cable provides improved system airflow

World-Class Technical Support

- Certified, experienced support staff
- Rated "Above Average to Excellent"
- Support lines with shortest wait times
- Individually archived case histories
- Web-based Q&A forum and autoreply e-mail
- Seagate Design Service Centers (DSC) to help companies transform innovative ideas into viable products

www.seagate.com
1-800-732-4283 (1-800-SEAGATE)

Specifications	400GB[1]	300GB[1]	250GB[1]	200GB[1]
Model Number	ST3400832AS ST3400832A	ST3300831AS ST3300831A	ST3250823AS ST3250823A	ST3200826AS
Interface/External Transfer Rate (Mbytes/sec)	SATA/150 NCQ Ultra ATA/100	SATA/150 NCQ Ultra ATA/100	SATA/150 NCQ Ultra ATA/100	SATA/150 NCQ
Performance				
Transfer Rate Maximum Internal (Mbytes/sec)	95	95	95	95
Maximum External (Mbytes/sec)	150/100	150/100	150/100	150
Sustained Transfer Rate OD (Mbytes/sec)	65	65	65	65
Cache, Multisegmented (Mbytes)	8	8	8	8
Average Seek (msec)	8	8	8	8
Average Latency (msec)	4.16	4.16	4.16	4.16
Spindle Speed (RPM)	7200	7200	7200	7200
Configuration/Organization				
Available Sectors	781,422,768	586,072,368	488,397,168	390,721,968
Bytes per Sector	512	512	512	512
Logical CHS	16,383/16/63	16,383/16/63	16,383/16/63	16,383/16/63
Recording Method	EPRML (16/17)	EPRML (16/17)	EPRML (16/17)	EPRML (16/17)
Reliability/Data Integrity				
Contact Start-Stops	50,000	50,000	50,000	50,000
Nonrecoverable Read Errors per Bits Read	1 per 10^{14}	1 per 10^{14}	1 per 10^{14}	1 per 10^{14}
Limited Warranty (years)	5	5	5	5
Power Management				
+12 VDC ±10% (amps peak)	2.8	2.8	2.8	2.8
Power Management (watts) Seek Avg	12.4	12.4	12.4	12.4
Operating Avg	12.8	12.8	12.8	12.8
Idle Avg	7.2	7.2	7.2	7.2
Standby Avg (SATA/PATA)	1.4/0.8	1.4/0.8	1.4/0.8	1.4/0.8
Environmental				
Temperature, Operating (°C)	0°C to 60°C	0°C to 60°C	0°C to 60°C	0°C to 60°C
Temperature, Nonoperating (°C)	–40 to 70	–40 to 70	–40 to 70	–40 to 70
Shock, Operating: 2 msec (Gs)	63	63	63	63
Shock, Nonoperating: 2 msec (Gs)	300	300	≥300	≥300
Acoustics Idle (bels—sound power)	2.8	2.8	≤2.8	≤2.8
Quiet Seek (bels—sound power)	3.2	3.2	≤3.2	≤3.2
Performance Seek (bels—sound power)	3.7	3.7	≤3.7	≤3.7
Physical				
Height (in/mm)	1/26.11	1/26.11	1/26.11	1/26.11
Width (in/mm)	4/101.85	4/101.85	4/101.85	4/101.85
Depth (in/mm)	5.78/146.99	5.78/146.99	5.78/146.99	5.78/146.99
Weight (lb/g)	1.4/635	1.4/635	1.4/635	1.4/635

[1] Capacity calculated as 1 Gbyte = 10^9 bytes

AMERICAS Seagate Technology LLC 920 Disc Drive, Scotts Valley, California 95066, United States, 831-438-6550
ASIA/PACIFIC Seagate Technology International Ltd. 7000 Ang Mo Kio Avenue 5, Singapore 569877, 65-6485-3888
EUROPE, MIDDLE EAST AND AFRICA Seagate Technology SA 130-136, rue de Silly, 92773 Boulogne-Billancourt Cedex, France, 33 1-41 86 10 00

Figure 2.6b continued: Seagate Hard Disk Datasheet Example

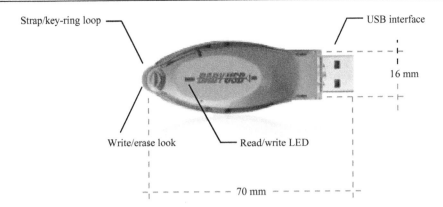

Figure 2.7a: BabyUSB USB Flash Memory Stick[8]

such as USB 1.1 or USB 2.0, extends from this small chassis that allows the stick to be plugged into a board's USB drive port as shown in Figure 2.7b. This device is typically smaller than other portable storage mediums, and is hot-swappable into a board's USB port that has device driver support for the particular type of USB Flash memory.

The real-world USB Flash memory datasheets shown in Figures 2.8a and 2.8b show some additional flash specification information that is useful for programmers to know regarding support of Flash types of storage mediums (see highlighted portions of datasheets).

2.3 Device Drivers and Middleware

Software that directly interfaces with the hardware in an embedded system is commonly referred to as a *device driver*. With some embedded operating systems that provide device drivers with their distributions, particular storage-medium-specific drivers can be referred to by other names, such as some Flash driver codes can be commonly referred to as MTDs (*memory technology drivers*). In the case of Flash, for example, MTDs are device drivers responsible for low-level mapping, reading, writing, and erasing of Flash. In short, as shown in Figure 2.9a, device drivers – including MTDs or **whatever** the particular device driver libraries are called in a distribution – manage the hardware and act as the interface to the hardware for higher layers of software.

For any embedded system that requires software, including higher-level software access to the hardware, these devices *all* have some type of device driver library. What is very important to remember as a programmer when trying to understand middleware support for a particular storage medium and its associated device driver library is that:

1. Different types of storage mediums will have different device driver requirements that need to be met

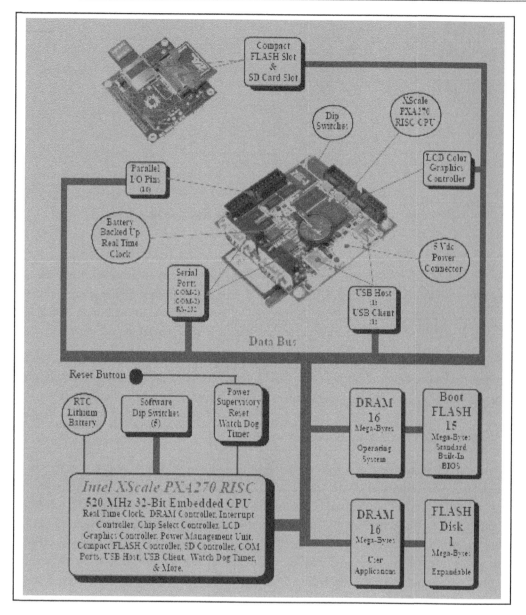

Figure 2.7b: USB Flash Memory Stick and Embedded Board Example[9]

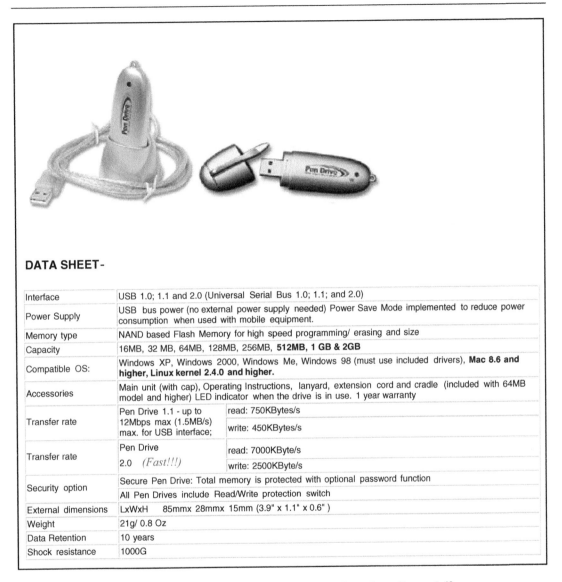

DATA SHEET-

Interface	USB 1.0; 1.1 and 2.0 (Universal Serial Bus 1.0; 1.1; and 2.0)	
Power Supply	USB bus power (no external power supply needed) Power Save Mode implemented to reduce power consumption when used with mobile equipment.	
Memory type	NAND based Flash Memory for high speed programming/ erasing and size	
Capacity	16MB, 32 MB, 64MB, 128MB, 256MB, **512MB, 1 GB & 2GB**	
Compatible OS:	Windows XP, Windows 2000, Windows Me, Windows 98 (must use included drivers), **Mac 8.6 and higher, Linux kernel 2.4.0 and higher.**	
Accessories	Main unit (with cap), Operating Instructions, lanyard, extension cord and cradle (included with 64MB model and higher) LED indicator when the drive is in use. 1 year warranty	
Transfer rate	Pen Drive 1.1 - up to 12Mbps max (1.5MB/s) max. for USB interface;	read: 750KBytes/s
		write: 450KBytes/s
Transfer rate	Pen Drive 2.0 *(Fast!!!)*	read: 7000KByte/s
		write: 2500KByte/s
Security option	Secure Pen Drive: Total memory is protected with optional password function	
	All Pen Drives include Read/Write protection switch	
External dimensions	LxWxH 85mmx 28mmx 15mm (3.9" x 1.1" x 0.6")	
Weight	21g/ 0.8 Oz	
Data Retention	10 years	
Shock resistance	1000G	

Figure 2.8a: PSI USB Flash Memory Pen Datasheet Example[10]

2. Even the same type of storage medium, such as USB Flash memory, that is created by different manufacturers can require different device drivers in support.

The reader must always check the details about the particular hardware if the part is not 100% identical to what is currently supported by the device, and not assume existing device drivers in the embedded system will be compatible for a particular storage medium part – even if the hardware is the same type of storage medium that the embedded device currently supports!

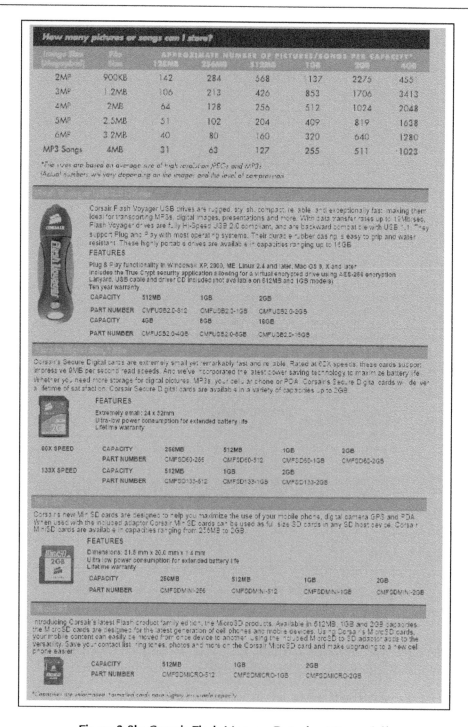

Figure 2.8b: Corsair Flash Memory Datasheet Example[11]

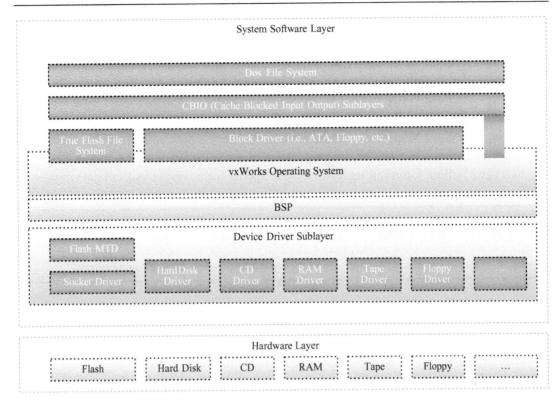

System Software Layer

Dos File System

CBIO (Cache Blocked Input Output) Sublayers

True Flash File System

Block Driver (i.e., ATA, Floppy, etc.)

vxWorks Operating System

BSP

Device Driver Sublayer

Flash MTD

Socket Driver

HardDisk Driver

CD Driver

RAM Driver

Tape Driver

Floppy Driver

Hardware Layer

Flash | Hard Disk | CD | RAM | Tape | Floppy | ...

Figure 2.9a: Device Drivers and vxWorks Example[12]

At a systems level, what specific middleware components exist and how they interface to the hardware will vary depending on the underlying device driver API for the particular storage medium(s). While, of course, libraries will vary between systems, in general hardware storage medium drivers will include some combination of:

- **Storage Medium Installation**, code that creates support of a storage medium in the embedded system
- **Storage Medium Uninstall**, code for removing the support of a storage medium in the embedded system
- **Storage Medium Startup**, initialization code for the storage medium upon reset and/or power-on
- **Storage Medium Shutdown**, termination code for the storage medium for entering into a power-off state
- **Storage Medium Enable**, code for enabling of the storage medium
- **Storage Medium Disable**, code for disabling the storage medium
- **Storage Medium Acquire**, code that provides other system software access to the storage medium

- **Storage Medium Release**, code that provides other system software the ability to free the storage medium
- **Storage Medium Read**, code that provides other system software the ability to read data from the storage medium
- **Storage Medium Write**, code that provides other system software the ability to write data to the storage medium
- **Storage Medium Mapping**, code for address mapping to and from the storage medium when reading, writing, and/or deleting data
- **Storage Medium Unmapping**, code for unmapping (removing) blocks of data in the storage medium.

Reminder

Different device driver libraries may have additional functions, but most device drivers in support of storage mediums will include some combination of the above functionality.

Figures 2.9b, 2.9c and 2.9d are real-world examples of device driver APIs for Flash and ATA storage mediums that demonstrate the type of functionality introduced above and found in device driver libraries for these particular storage mediums. Later sections of this chapter will demonstrate examples of how these device drivers are utilized for implementing a middleware in an embedded device.

Note: please refer to the CD that accompanies this text or the Elsevier website link for this book (if no CD has been included) to see all open-source files for Linux Flash examples referenced in Figures 2.9b and 2.9c. Also, remember that the JFS implementation is just an open-source reference, and that to support a particular hardware platform requires updating and/or replacing the reference JFS device driver-specific calls with the required device driver-specific calls of a particular platform throughout the JFS source.

2.4 The Role of an Embedded System's Operating System and Middleware-specific Code

The purpose of an embedded operating system is:

- to insure the embedded system operates in an efficient and reliable manner by managing hardware and software resources
- to provide an abstraction layer to simplify the process of developing higher layers of software
- to act as a partitioning tool.

The embedded OS (operating system) achieves these functions via a ***kernel*** that includes, at a minimum: process management, memory management, and I/O system management components (Figure 2.10).

```
/*
 * $Id: pcmciamtd.c,v 1.59 2006/03/29 08:31:11 dwmw2 Exp $
 *
 * pcmciamtd.c - MTD driver for PCMCIA flash memory cards
 *
 * Author: Simon Evans <spse@secret.org.uk>
 *
 * Copyright (C) 2002 Simon Evans
 *
 * Licence: GPL
 *
 * This program is free software; you can redistribute it and/or modify
 *
 * it under the terms of the GNU General Public License version 2 as
 *
 * published by the Free Software Foundation.
 */
```

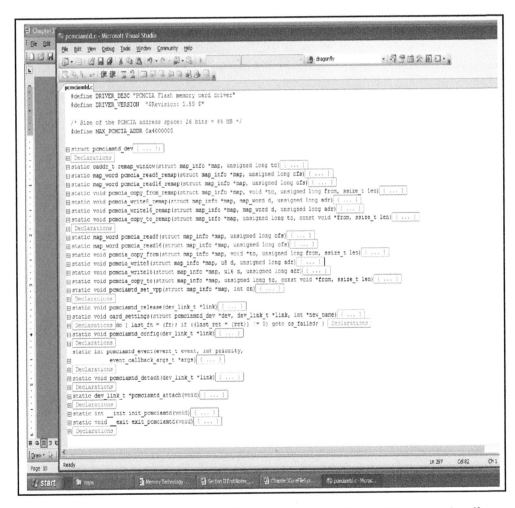

Figure 2.9b: Example of PCMCIA Flash Memory Card Device Driver Functions[13]

```
/*
 * MTD map driver for AMD compatible flash chips (non-CFI)
 *
 * Author: Jonas Holmberg <jonas.holmberg@axis.com>
 *
 * $Id: amd_flash.c,v 1.29 2006/03/29 08:31:10 dwmw2 Exp $
 *
 * Copyright (c) 2001 Axis Communications AB
 *
 * This file is under GPL.
 *
 */
```

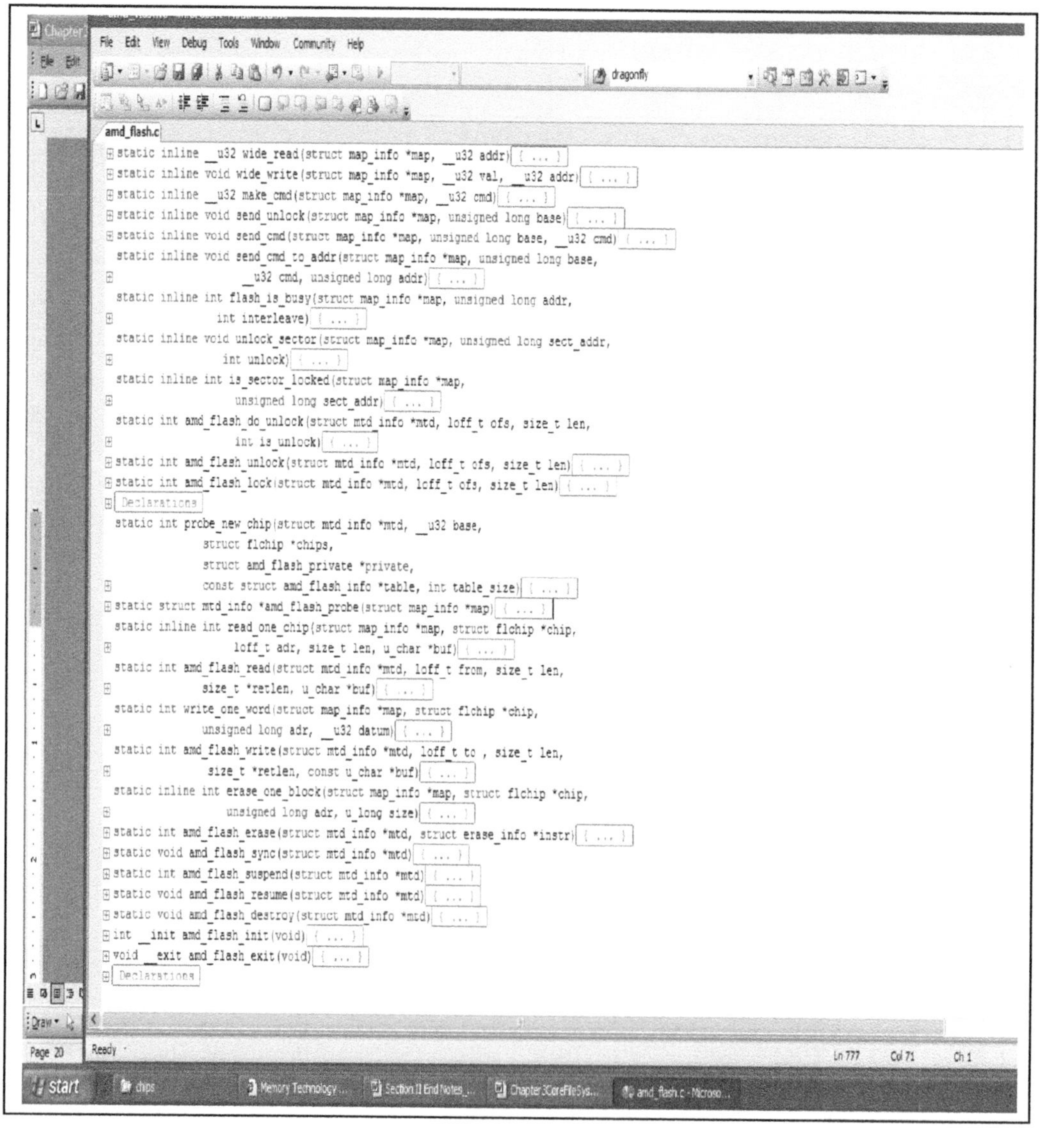

Figure 2.9c: Example of AMD Flash Device Driver Code[13]

VxWorks API Reference : Driver Libraries

2 ataDrv

2.1.1.1 NAME

ataDrv - ATA/IDE and ATAPI CDROM (LOCAL and PCMCIA) disk device driver

ROUTINES

ataDriveInit( ) - initialize ATA drive
ataDrv( ) - initialize the ATA driver
ataDevCreate( ) - create a device for a ATA/IDE disk
ataRawio( ) - do raw I/O access

DESCRIPTION

This is a driver for ATA/IDE and ATAPI CDROM devices on PCMCIA, ISA, and other buses. The driver can be customized via various macros to run on a variety of boards and both big-endian, and little endian CPUs.

ataDriveInit( )

NAME

ataDriveInit( ) - initialize ATA drive

SYNOPSIS

```
STATUS ataDriveInit
    (
    int ctrl,
    int drive
    )
```

DESCRIPTION

This routine checks the drive presents, identifies its type, initializes the drive controller and driver control structures.

RETURNS

OK if drive was initialized successfully, or ERROR.

Figure 2.9d: Example of ATA Device Driver Public APIs under vxWorks[12]

2   ataDrv()

4.1.1.1   NAME

ataDrv( ) - initialize the ATA driver

SYNOPSIS
```
STATUS ataDrv
    (
    int ctrl,              /* controller no. */
    int drives,            /* number of drives */
    int vector,            /* interrupt vector */
    int level,             /* interrupt level */
    int configType,        /* configuration type */
    int semTimeout,        /* timeout seconds for sync semaphore */
    int wdgTimeout         /* timeout seconds for watch dog */
    )
```

DESCRIPTION

This routine initializes the ATA/IDE/ATAPI CDROM driver, sets up interrupt vectors, and performs hardware initialization of the ATA/IDE chip. This routine must be called exactly once, before any reads, writes, or calls to **ataDevCreate( )**. Normally, it is called by **usrRoot( )** in **usrConfig.c**.

RETURNS

OK, or ERROR if initialization fails.

ataDevCreate( )

NAME

ataDevCreate( ) - create a device for a ATA/IDE disk

SYNOPSIS
```
BLK_DEV *ataDevCreate
    (
    int ctrl,              /* ATA controller number, 0 is the primary controller */
    int drive,             /* ATA drive number, 0 is the master drive */
    int nBlocks,           /* number of blocks on device, 0 = use entire disc */
    int blkOffset          /* offset BLK_DEV nBlocks from the start of the drive */
    )
```

DESCRIPTION

This routine creates a device for a specified ATA/IDE or ATAPI CDROM disk. *ctrl* is a controller number for the ATA controller; the primary controller is 0. The maximum is specified via **ATA_MAX_CTRLS**. *drive* is the drive number for the ATA hard drive; the master drive is 0. The maximum is specified via **ATA_MAX_DRIVES**. The *nBlocks* parameter specifies the size of the device in blocks. If *nBlocks* is zero, the whole disk is used. The *blkOffset* parameter specifies an offset, in blocks, from the start of the device to be used when writing or reading the hard disk. This offset is added to the block numbers passed by the file system during disk accesses. (VxWorks file systems always use block numbers beginning at zero for the start of a device.)

RETURNS

A pointer to a block device structure (**BLK_DEV**) or NULL if memory cannot be allocated for the device structure.

Figure 2.9d continued: Example of ATA Device Driver Public APIs under vxWorks

```
6      ataDevCreate( )


6.1.1.1     NAME

ataDevCreate( ) - create a device for a ATA/IDE disk

SYNOPSIS
BLK_DEV *ataDevCreate
    (
    int ctrl,               /* ATA controller number, 0 is the primary */
                            /* controller */
    int drive,              /* ATA drive number, 0 is the master drive */
    int nBlocks,            /* number of blocks on device, 0 = use */
                            /* entire disc */
    int blkOffset           /* offset BLK_DEV nBlocks from the start of */
                            /* the drive */

    )
```

DESCRIPTION

This routine creates a device for a specified ATA/IDE or ATAPI CDROM disk. *ctrl* is a controller number for the ATA controller; the primary controller is 0. The maximum is specified via **ATA_MAX_CTRLS**. *drive* is the drive number for the ATA hard drive; the master drive is 0. The maximum is specified via **ATA_MAX_DRIVES**. The *nBlocks* parameter specifies the size of the device in blocks. If *nBlocks* is zero, the whole disk is used. The *blkOffset* parameter specifies an offset, in blocks, from the start of the device to be used when writing or reading the hard disk. This offset is added to the block numbers passed by the file system during disk accesses. (VxWorks file systems always use block numbers beginning at zero for the start of a device.)

RETURNS

A pointer to a block device structure (**BLK_DEV**) or NULL if memory cannot be allocated for the device structure.

ataRawio( )

Figure 2.9d continued: Example of ATA Device Driver Public APIs under vxWorks

A kernel's process management mechanisms are what provide the functionality that secures the illusion of simultaneous multitasking over a single processor. Kernel functionality that is relevant to middleware development ranges from task implementation to scheduling to synchronization to intertask communication. Middleware programmers need to note that embedded operating systems, and even different versions of the same embedded operating system, will vary widely in their process management schemes. For example, the types and number of operating system tasks:

- WindRiver's vxWorks 6.4 (1)
  - one type of task that implements one 'thread of execution' (task's Program Counter)
- WindRiver's vxWorks 653 (1)
  - core OS vThreads based on vxWorks 5.5 multithreading, like vxWorks 6.4 one type
- Timesys Linux (2)
  - Linux fork
  - Periodic task

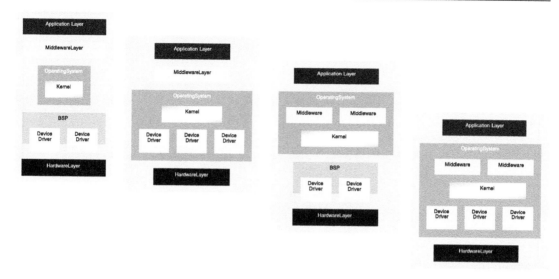

Figure 2.10: Embedded Operating Systems

- Esmertec's Jbed (6)
  - OneshotTimer Task, task that is run only once
  - PeriodicTimer Task, task that is run after a particular set time interval
  - HarmonicEvent Task, task that runs alongside a periodic timer task
  - JoinEvent Task, task that is set to run when an associated task completes
  - InterruptEvent Task, task that is run when a hardware interrupt occurs
  - UserEvent Task, task that is explicitly triggered by another task.

It comes down to balancing between utilizing the system's resources (i.e., keeping the CPU, I/O, etc. as busy as possible) – with task throughput to process as many tasks as possible in a given amount of time – with fairness and ensuring that task starvation does not occur when trying to achieve a maximum task throughput. The key for developers to note relative to embedded operating systems is what impacts effectiveness and performance, and not to underestimate the impact of an embedded OS's internal design. The key differentiators between embedded operating systems in this regard are:

1. *Memory Management Scheme*, i.e., virtual memory swapping scheme and page faults
2. *Scheduling Scheme*, i.e., throughput, execution time, and wait time
3. *Performance*, i.e.,
   - Response time, to make the context switch to a ready task and waiting time of task in ready queue
   - Turnaround time, how long a process takes to complete running
   - Overhead, the time and data needed to determine which tasks will run next
   - Fairness, what are the determining factors as to which processes get to run.

The key questions middleware developers need to ask of embedded OS support include: What hardware can this support? Are there any performance limitations? How about memory footprint? Middleware that resides on an OS needs an embedded OS that has been *stably* ported and is *supporting* the hardware.

How about what features you need given cost, schedule, requirements, etc.? Do you just need a kernel or more? How scalable should the embedded OS be? This is because in addition to a kernel, embedded OS distributions may also provide additional integrated components, such as networking, file system, and database support. These components allow the overlying middleware layers to be ported to the OS kernel design, as well as the underlying system software and hardware (see examples in Figures 2.11a and 2.11b).

For example, a file system *interface* is some subset of OS functionality that can be utilized by the ported file system. When porting a file system to a different OS, it is important to understand what (if any) interfaces are available to the file system since the OS APIs available to a file system will vary from one OS to another, and what APIs a file system requires will differ from one file system implementation to another. For example, in Figure 2.11c, the JFS open-source file system provided on this textbook's CD utilizes several different Linux-specific files (see source code on CD for complete overview of all required Linux APIs for JFS). To port JFS to an unsupported OS requires replacing the current OS-specific calls, such as the Linux-specific code shown in Figure 2.11c, with the new OS-specific file system interface calls throughout the JFS source.

2.5 Operating Systems and Device Driver Access for Middleware

While middleware can access device drivers directly, as introduced in the previous section, an embedded OS can also include an abstraction layer API that allows for device driver access. When providing device access, or any type of I/O access to middleware, most OS APIs categorize their associated device drivers as some combination of:

- **Character**, a driver that allows hardware access via a (character) byte stream
- **Block**, a driver that allows hardware access via some smallest addressable set of bytes at any given time
- **Network**, a driver that allows hardware access via data in the form of networking packets
- **Virtual**, a driver that allows I/O access to virtual (software) devices
- **Miscellaneous Monitor and Control**, a driver that allows I/O access to hardware that is not accessible via the other categories above.

For an example of an OS block device interface, vxWorks provides an I/O interface, called CBIO (cache blocked input output), that allows different file systems, such as JFS, dosFS, etc., to be ported to one standard vxWorks interface regardless of the underlying hardware storage medium (see Figures 2.11d and 2.11e). As stated in the previous section, to port

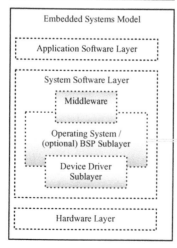

Figure 2.11a: Example OS Permutations

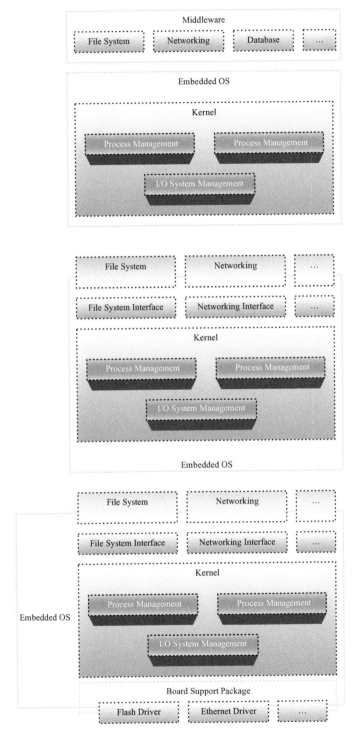

Figure 2.11b: Example OS Components

```
/*      super.c
 *
 *      Copyright (C) International Business Machines Corp., 2000-2003
 *      Portions Copyright (C) Christoph Hellwig, 2001-2002
 *
 *      This program is free software;  you can redistribute it and/or modify
 *      it under the terms of the GNU General Public License as published by
 *      the Free Software Foundation; either version 2 of the License, or
 *      (at your option) any later version.
 *
 *      This program is distributed in the hope that it will be useful,
 *      but WITHOUT ANY WARRANTY;  without even the implied warranty of
 *      MERCHANTABILITY or FITNESS FOR A PARTICULAR PURPOSE.  See
 *      the GNU General Public License for more details.
 *
 *      You should have received a copy of the GNU General Public License
 *      along with this program;  if not, write to the Free Software
 *      Foundation, Inc., 59 Temple Place, Suite 330, Boston, MA 02111-1307 USA
 */
#include <linux/config.h>
#include <linux/fs.h>
#include <linux/module.h>
#include <linux/blkdev.h>
#include <linux/completion.h>
#include <asm/uaccess.h>
#include "jfs_incore.h"
#include "jfs_filsys.h"
#include "jfs_metapage.h"
#include "jfs_superblock.h"
#include "jfs_dmap.h"
#include "jfs_imap.h"
#include "jfs_debug.h"

....
```

Figure 2.11c: Example of JFS Usage of Linux File System Interface

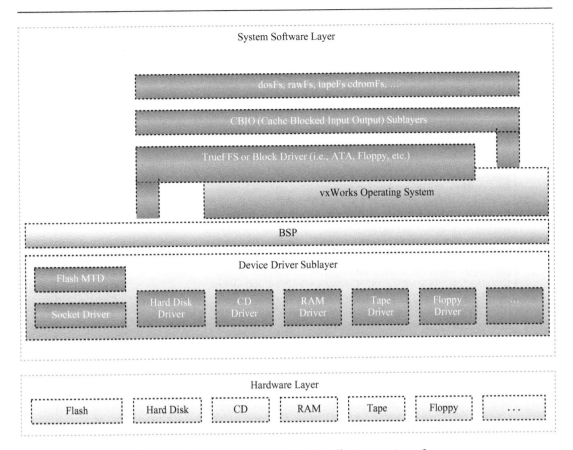

Figure 2.11d: Example of vxWorks File System Interface

cbioLib

NAME

cbioLib - Cached Block I/O library

ROUTINES

cbioLibInit( ) - Initialize CBIO Library
cbioBlkRW( ) - transfer blocks to or from memory
cbioBytesRW( ) - transfer bytes to or from memory
cbioBlkCopy( ) - block to block (sector to sector) transfer routine
cbioIoctl( ) - perform ioctl operation on device
cbioModeGet( ) - return the mode setting for CBIO device
cbioModeSet( ) - set mode for CBIO device
cbioRdyChgdGet( ) - determine ready status of CBIO device
cbioRdyChgdSet( ) - force a change in ready status of CBIO device
cbioLock( ) - obtain CBIO device semaphore.
cbioUnlock( ) - release CBIO device semaphore.
cbioParamsGet( ) - fill in **CBIO_PARAMS** structure with CBIO device parameters
cbioShow( ) - print information about a CBIO device
cbioDevVerify( ) - verify **CBIO_DEV_ID**
cbioWrapBlkDev( ) - create CBIO wrapper atop a **BLK_DEV** device
cbioDevCreate( ) - Initialize a CBIO device (Generic)

DESCRIPTION

This library provides the Cached Block Input Output Application Programmers Interface (CBIO API). Libraries such as **dosFsLib**, **rawFsLib**, and **usrFdiskPartLib** use the CBIO API for I/O operations to underlying devices.

This library also provides generic services for CBIO modules. The libraries dpartCbio, dcacheCbio, and ramDiskCbio are examples of CBIO modules that make use of these generic services.

This library also provides a CBIO module that converts blkIo driver **BLK_DEV** (**blkIo.h**) interface into CBIO API compliant interface using minimal memory overhead. This lean module is known as the basic **BLK_DEV** to CBIO wrapper module.

CBIO MODULES AND DEVICES

A CBIO module contains code for supporting CBIO devices. The libraries **cbioLib**, dcacheCbio, dpartCbio, and ramDiskCbio are examples of CBIO modules.

A CBIO device is a software layer that provide its master control of I/O to it subordinate. CBIO device layers typically reside logically below a file system and above a storage device. CBIO devices conform to the CBIO API on their master (upper) interface.

CBIO modules provide a CBIO device creation routine used to instantiate a CBIO device. The CBIO modules device creation routine returns a **CBIO_DEV_ID** handle. The **CBIO_DEV_ID** handle is used to uniquely identify the CBIO device layer instance. The user of the CBIO device passes this handle to the CBIO API routines when accessing the device.

The libraries **dosFsLib**, **rawFsLib**, and **usrFdiskPartLib** are considered users of CBIO devices because they use the CBIO API on their subordinate (lower) interface. They do not conform to the CBIO API on their master interface, therefore they are not CBIO modules. They are users of CBIO devices and always reside above CBIO devices in the logical stack.

Figure 2.11e: vxWorks CBIO Library[13]

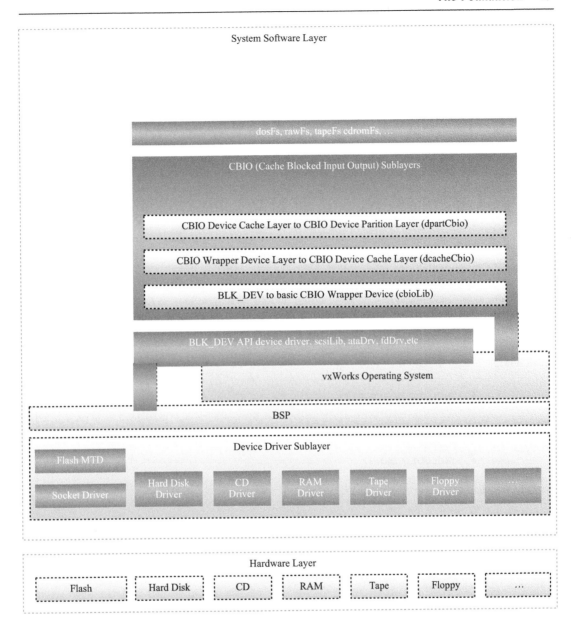

Figure 2.11f: Logical Layers of CBIO-based vxWorks System[13]

JFS to an unsupported OS, such as vxWorks in this case, requires replacing the current OS-specific calls, such as the Linux-specific code shown in Figure 2.11c, with the vxWorks specific code and utilizing the CBIO library throughout the JFS source.

In vxWorks, calling some of the CBIO APIs is part of the process of setting up a file system, such as dosFS on a hard disk, floppy drive or any other storage medium accessed as a block device under vxWorks.

As shown in Figure 2.11f, when utilizing the CBIO APIs in vxWorks an example process is as follows:

Step 1. Configure vxWorks to support the:

* Block Device
* CBIO Library
* File System, i.e., dosFS.

< From vxWorks Programmer's Guide 5.5>

5.2.2 Configuring Your System

To include dosFs in your VxWorks-based system, configure the kernel with the appropriate components for the dosFs file system.

Required Components

The following components are required:

* **INCLUDE_DOSFS_MAIN.**

dosFsLib (2)

* **INCLUDE_DOSFS_FAT.**

dosFs FAT12/16/32 FAT handler

* **INCLUDE_CBIO.**

CBIO API module

And, either one or both of the following components are required:

* **INCLUDE_DOSFS_DIR_VFAT.**

Microsoft VFAT direct handler

* **INCLUDE_DOSFS_DIR_FIXED.**

Strict 8.3 & VxLongNames directory handler

In addition, you need to include the appropriate component for your block device; for example, **INCLUDE_SCSI** or **INCLUDE_ATA**. Finally, add any related components that are required for your particular system.

Figure 2.11g: Example of Configuring vxWorks[12]

Optional dosFs Components

Optional dosFs components are:

- ∞ **INCLUDE_DOSFS.** usrDosFsOld.c wrapper layer

- ∞ **INCLUDE_DOSFS_FMT.** dosFs2 file system formatting module

- ∞ **INCLUDE_DOSFS_CHKDSK.** file system integrity checking

- ∞ **INCLUDE_DISK_UTIL.** standard file system operations, such as **ls**, **cd**, **mkdir**, **xcopy**, and so on

- ∞ **INCLUDE_TAR.** the **tar** utility

Optional CBIO Components

Optional CBIO components are:

- ∞ **INCLUDE_DISK_CACHE.** CBIO API disk caching layer

- ∞ **INCLUDE_DISK_PART.** disk partition handling code

- ∞ **INCLUDE_RAM_DISK.** CBIO API RAM disk driver

Figure 2.11g continued: Example of Configuring vxWorks

Step 2. Create the Block Device.

< From vxWorks Programmer's Guide 5.5>

5.2.4 Creating Block Device

Next, create one or more block devices. To create the device, call the routine appropriate for that device driver. The format for this routine is *xxx***DevCreate( )** where *xxx* represents the device driver type; for example, **scsiBlkDevCreate( )** or **ataDevCreate( )**.

The driver routine returns a pointer to a block device descriptor structure, **BLK_DEV**. This structure describes the physical attributes of the device and specifies the routines that the device driver provides to a file system.

Figure 2.11h: Example of Creating Block Device in vxWorks[12]

Step 3. Create the CBIO Block Driver Wrapper. The CBIO block driver wrapper layer wraps the block driver with a CBIO API compatible layer using the cbioWrapBlkDev() function.

---

cbioWrapBlkDev( )

NAME

cbioWrapBlkDev( ) - create CBIO wrapper atop a **BLK_DEV** device

SYNOPSIS
```
CBIO_DEV_ID cbioWrapBlkDev
    (
    BLK_DEV * pDevice          /* BLK_DEV * device pointer */
    )
```

DESCRIPTION

The purpose of this function is to make a blkIo (**BLK_DEV**) device comply with the CBIO interface via a wrapper.

The device handle provided to this function, *device* is verified to be a blkIo device. A lean CBIO to **BLK_DEV** wrapper is then created for a valid blkIo device. The returned **CBIO_DEV_ID** device handle may be used with **dosFsDevCreate( )**, **dcacheDevCreate( )**, and any other routine expecting a valid **CBIO_DEV_ID** handle.

To verify a blkIo pointer we see that all mandatory functions are not NULL.

Note that if a valid **CBIO_DEV_ID** is passed to this function, it will simply be returned without modification.

The **dosFsLib**, dcacheCbio, and dpartCbio CBIO modules use this function internally, and therefore this function need not be otherwise invoked when using those CBIO modules.

RETURNS

a CBIO device pointer, or NULL if not a blkIo device

Figure 2.11i: CBIO Block Device Wrapper in vxWorks[1]

Step 4. Create the CBIO Cache Layer.

dcacheCbio
NAME

dcacheCbio - Disk Cache Driver

ROUTINES

dcacheDevCreate( ) - Create a disk cache
dcacheDevDisable( ) - Disable the disk cache for this device
dcacheDevEnable( ) - Reenable the disk cache
dcacheDevTune( ) - modify tunable disk cache parameters
dcacheDevMemResize( ) - set a new size to a disk cache device
dcacheShow( ) - print information about disk cache
dcacheHashTest( ) - test hash table integrity

DESCRIPTION

This module implements a disk cache mechanism via the CBIO API. This is intended for use by the VxWorks DOS file system, to store frequently used disk blocks in memory. The disk cache is unaware of the particular file system format on the disk, and handles the disk as a collection of blocks of a fixed size, typically the sector size of 512 bytes.

The disk cache may be used with SCSI, IDE, ATA, Floppy or any other type of disk controllers. The underlying device driver may be either comply with the CBIO API or with the older block device API.

This library interfaces to device drivers implementing the block device API via the basic CBIO **BLK_DEV** wrapper provided by **cbioLib**.

Because the disk cache complies with the CBIO programming interface on both its upper and lower layers, it is both an optional and a stackable module. It can be used or omitted depending on resources available and performance required.

The disk cache module implements the CBIO API, which is used by the file system module to access the disk blocks, or to access bytes within a particular disk block. This allows the file system to use the disk cache to store file data as well as Directory and File Allocation Table blocks, on a Most Recently Used basis, thus keeping a controllable subset of these disk structures in memory. This results in minimized memory requirements for the file system, while avoiding any significant performance degradation.

The size of the disk cache, and thus the memory consumption of the disk subsystem, is configured at the time of initialization (see **dcacheDevCreate**( )), allowing the user to trade-off memory consumption versus performance. Additional performance tuning capabilities are available through **dcacheDevTune**( ).

Briefly, here are the main techniques deployed by the disk cache:

- Least Recently Used block re-use policy
- Read-ahead
- Write-behind with sorting and grouping
- Hidden writes
- Disk cache bypass for large requests
- Background disk updating (flushing changes to disk) with an adjustable update period (ioctl flushes occur without delay.)

Figure 2.11j: CBIO Cache Layer Using vxWorks CBIO Library

Step 5. Implement CBIO Partition Manager.

---

dpartCbio

NAME

dpartCbio - generic disk partition manager

ROUTINES

dpartDevCreate( ) - Initialize a partitioned disk
dpartPartGet( ) - retrieve handle for a partition

DESCRIPTION

This module implements a generic partition manager using the CBIO API (see **cbioLib**) It supports creating a separate file system device for each of its partitions.

This partition manager depends upon an external library to decode a particular disk partition table format, and report the resulting partition layout information back to this module. This module is responsible for maintaining the partition logic during operation.

When using this module with the dcacheCbio module, it is recommended this module be the master CBIO device. This module should be above the cache CBIO module layer. This is because the cache layer is optimized to function efficiently atop a single physical disk drive. One should call dcacheDevCreate before dpartDevCreate.

An implementation of the de-facto standard partition table format which is created by the MSDOS FDISK program is provided with the **usrFdiskPartLib** module, which should be used to handle PC-style partitioned hard or removable drives.

EXAMPLE

The following code will initialize a disk which is expected to have up to 4 partitions:

```
usrPartDiskFsInit( BLK_DEV * blkDevId )
  {
  const char * devNames[] = { "/sd0a", "/sd0b", "/sd0c", "/sd0d" };
  CBIO_DEV_ID cbioCache;
  CBIO_DEV_ID cbioParts;

  /* create a disk cache atop the entire BLK_DEV */

  cbioCache = dcacheDevCreate ( blkDevId, NULL, 0, "/sd0" );

  if (NULL == cbioCache)
      {
      return (ERROR);
      }

  /* create a partition manager with a FDISK style decoder */

  cbioParts = dpartDevCreate( cbioCache, 4, usrFdiskPartRead );

  if (NULL == cbioParts)
      {
      return (ERROR);
      }
```

Figure 2.11k: CBIO Partition Layer Using vxWorks CBIO Library[12]

```
    /* create file systems atop each partition */

    dosFsDevCreate( devNames[0], dpartPartGet(cbioParts,0), 0x10, NONE);
    dosFsDevCreate( devNames[1], dpartPartGet(cbioParts,1), 0x10, NONE);
    dosFsDevCreate( devNames[2], dpartPartGet(cbioParts,2), 0x10, NONE);
    dosFsDevCreate( devNames[3], dpartPartGet(cbioParts,3), 0x10, NONE);
    }
```
Because this module complies with the CBIO programming interface on both its upper and lower layers, it is both an optional and a stackable module.

SEE ALSO

dcacheLib, <u>dosFsLib</u>, <u>usrFdiskPartLib</u>

Figure 2.11k continued: CBIO Partition Layer Using vxWorks CBIO Library

An example of source code using the CBIO APIs in vxWorks is shown in Figure 2.11l.

```
STATUS sampleAtaHardDriveConfig
    (
    int     address,     // primary or secondary address (0 or 1)
    int     driveNumber, // hard drive number (0 or 1)
    char    *driveName   // partition (s) mount point (s)
    )
    {
    BLK_DEV           *pointerBlockDrive;
    CBIO_DEV_ID       cbioDevice;
    CBIO_DEV_ID       masterCbioDevice;
    int               ataCacheSize = ATA_CACHE_SIZE;
    …

    // Step 2. block device creation
    pointerBlockDrive = ataDevCreate (address, driveNumber, 0, 0);
    if (pointerBlockDrive == (BLK_DEV *)null)
    {
        printf ("sampleAtaHardDriveConfig : Error Creating Block Device \n");
        return (ERROR);
    }

    // Step 3. block device in a CBIO wrapper
    cbioDevice = cbioWrapBlkDev (pointerBlockDrive);
    if (cbioDevice == null)
    {
        printf ("sampleAtaHardDriveConfig : Error Creating CBIO Wrapper \n");
        return (ERROR);
    }

    // Step 4. disk cache creation
    cbioDevice = dcacheDevCreate ((CBIO_DEV_ID) cbioDevice,
                                        NULL,
                                        ataCacheSize,
                                        driveName);
    if (cbioDevice == null)
    {
        printf ("sampleAtaHardDriveConfig : Error Creating CBIO Disk Cache \n");
        return (ERROR);
    }
```

Figure 2.11l: vxWorks CBIO APIs Source Code Example

```
// Step 5. partition manager creation
masterCbioDevice = dpartDevCreate (cbioDevice,4, usrFdiskPartRead);

if (masterCbioDevice == null)
{
    printf ("sampleAtaHardDriveConfig : Error Creating CBIO Partition Manager \n");
    return (ERROR);
}

...

return (OK);
}
```

Figure 2.11l continued: vxWorks CBIO APIs Source Code Example

2.6 A Brief Comment on Multiple Middleware Components

There is middleware that requires other middleware components in the embedded device in order to function. In the case of a network file system, for example, since it is a file system scheme that allows for access to files, a.k.a. file sharing, across networked computer systems it requires compatible, underlying networking protocols in support of file management and transmission (see Figure 2.12a).

Another example, shown in Figures 2.12b and 2.12c, is in the instance in which some type of virtual machine is integrated in the system software in support middleware, such as a database or file system, written in a non-native language such as C# or Java. Refer to the chapter discussing the particular middleware components in these examples for more information.

2.7 Summary

In order to understand a particular middleware design, to determine which middleware design is the right choice for an embedded device, as well as understand the impact of middleware software on a particular device, it is important to first understand the foundation that underlies the middleware. This foundation includes some combination of the hardware, as well as device drivers, operating systems, and other required middleware components. The reader can then apply these fundamentals to analyzing what would be required to get a particular middleware component running in an embedded system, to determine which middleware design is the right one for a particular system, as well as the impact of the file system on the embedded device.

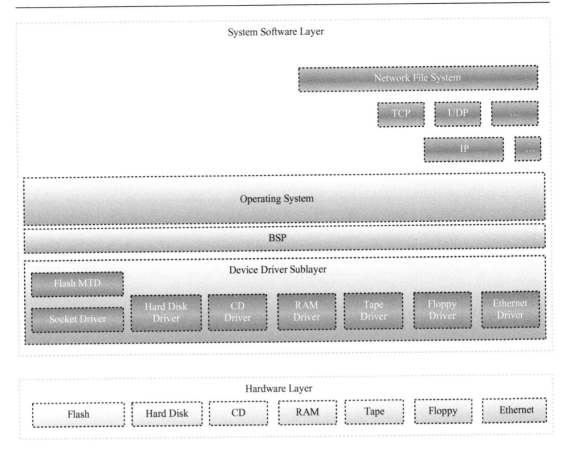

Figure 2.12a: Example of Underlying Networking Middleware for a Network File System

Chapter 3 introduces middleware standards and the importance of these standards within the context of any design.

2.8 Problems

1. Name three underlying components that could act as a foundation to an embedded system with middleware. Draw an example.
2. Middleware can reside directly over device driver software (True/False).
3. Why is it important for middleware programmers to understand the hardware of an embedded system?
4. One or more middleware component can be implemented in an embedded system (True/False).
5. How does middleware view the hardware storage medium? Draw an example.

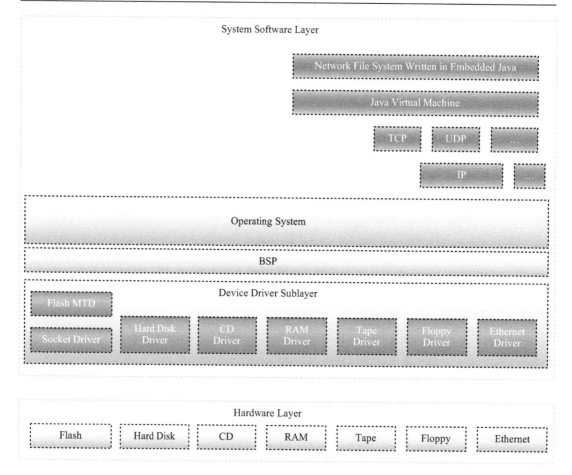

Figure 2.12b: Example of Underlying JVM Middleware for a Java-based File System

6. Middleware can manage data on the following hardware:
   A. RAM
   B. CD
   C. Smart card
   D. Only B and C
   E. All of the above.
7. List and describe six types of device driver API functionality typically found in hardware storage medium device drivers.
8. What is the difference between an operating system character device and a block device?
9. Middleware never requires other underlying middleware components (True/False).
10. Draw a high-level diagram of a type of middleware that requires a Java Virtual Machine (JVM).

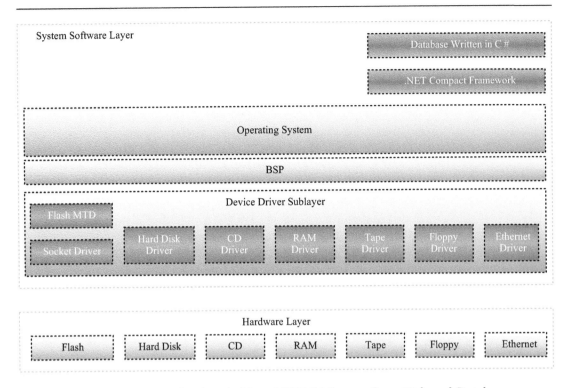

Figure 2.12c: Example of Underlying. NET Middleware for a C#-based Database

2.9 End Notes

1 Microsoft Extensible Firmware Initiative FAT32 File System Specification. *Version 1.03, December 6, 2000. Microsoft Corporation.*

2 http://redhat.brandfuelstores.com/.

3 www.microsoft.com.

4 http://shop.cxtreme.de

5 'Embedded Systems Architecture: A Comprehensive Guide for Engineers and Programmers'. T. Noergaard. Elsevier 2005, p. 245.

6 http://www.westerndigital.com/en/products/Products.asp?DriveID=104

7 http://www.seagate.com/cda/products/discsales/marketing/detail/0,1081,771,00.html

8 http://www.babyusb.com/flashspecs2.htm

9 'XScale Lite Datasheet' RLC Enterprises, Inc.

10 http://www.psism.com/pendrive.htm

11 'Corsair USB Flash Memory Datasheet'. Corsair.

12 http://www.linux-mtd.infradead.org/archive/

13 'vxWorks API Reference Guide : Device Drivers'. Version 5.5.

14 National Semiconductor, 'Geode User Manual,' Rev. 1, p. 13.

15 Net Silicon, 'Net+ARM40 Hardware Reference Guide,' pp. 1–5.

16 'EnCore M3 Embedded Processor Reference Manual,' Revision A, p. 8.

17 'EnCore PP1 Embedded Processor Reference Manual,' Revision A, p. 9.

Middleware and Standards in Embedded Systems

Chapter Points

- Defining what middleware standards are
- Listing examples of different types of middleware standards
- Providing examples of middleware standards that derive embedded components

3.1 What are Standards for Middleware Software?

One of the first steps to understanding an embedded middleware solution is to, first, *know your standards*! *Standards* are documented methodologies that can define some of the most important, as well as required, components within an embedded system. Embedded systems that share similar end-user and/or technical characteristics are typically grouped into *market-specific* categories within the embedded systems industry. Thus, there exists middleware that is utilized for a particular market category of embedded devices.

In short, middleware standards can either exist for a particular market category of embedded devices, whereas other standards are utilized across all market segments. The most common types of middleware standards in the embedded systems arena can typically fall under one or some combination of the following categories:

- **Emergency Services, Police, and Defense,** middleware standards which are implemented within embedded systems used by the police or military, such as within 'smart' weapons, police patrol, ambulances, and radar systems to name a few.
- **Aerospace and Space,** middleware standards which are implemented within aircrafts, as well as embedded systems that must function in space, such as on a space station or within an orbiting satellite.

Demystifying Embedded Systems Middleware. DOI: 10.1016/B978-0-7506-8455-2.00003-0

- **Automotive,** middleware standards that are implemented within cars, trucks, vans, and so on. This can include anything from security and engineer controls to a DVD entertainment center.
- **Commercial and Home Office Automation,** middleware standards that are implemented in appliances used in professional corporate and home offices, such as: fax machines, scanners, and printers, for example.
- **Consumer Electronics,** middleware standards that are implemented in devices used by consumers in everyday personal activities, such as in kitchen appliances, washing machines, televisions, and set top boxes.
- **Energy and Oil,** middleware standards implemented within embedded systems used in the power and energy industries, such as control systems within power plant ecosystems for wind turbine generators and solar, for example.
- **Industrial Automation and Control,** middleware standards implemented within robotic devices typically used in the manufacturing industries to execute cyclic work processes on an assembly line.
- **Medical,** middleware standards implemented in devices used to aid in providing medical treatments, such as infusion pumps, prosthetics, dialysis machines, and drug-delivery devices to name a few.
- **Networking and Communications,** middleware standards implemented in audio/video communication devices, such as cell phones and pagers, middleware standards used within network-specific devices, such as in hubs and routers, as well as the standards used in any embedded device to implement network connectivity.
- **General Purpose,** middleware standards that are generically utilized within any type of embedded system, and are even implemented or have originated in non-embedded computer systems, such as standards for programming languages and virtual machines, for example.

Embedded system market segments and their associated devices are always changing as new devices emerge and other devices are phased out. The market definitions can also vary from company to company semantically as well as how the devices are grouped by market segment. Remember, this does not mean that any middleware that falls under a market-specific category can never be utilized in other types of devices, or cannot be adapted to another type of design that falls under a different market; only that there is a lot of middleware that has been designed and intended to target a particular type of device with certain types of requirements.

3.2 Real-world Middleware Standards Implemented in Embedded Systems

As shown in Figure 3.1, functionality defined in standards can be specific to a particular layer, reside across multiple layers, as well as indirectly derive what additional components are required to allow for successful integration.

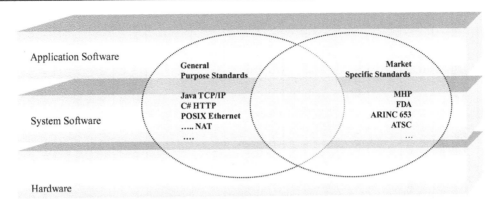

Figure 3.1: General Standards Diagram

Table 3.1 contains a list of some standards organizations, commonly utilized real-world standards in the embedded market space, as well as a general description of the purposes the standards and organizations serve. Keep in mind that Table 3.1 is a dynamic table meant as a guideline for the reader to start with and includes standards relevant to the different layers of an embedded system's architecture. It is important for the reader to think of the overall device when thinking of what standards are relevant, because, for example, other computer systems the embedded device needs to network successfully with, as well as standards explicitly required within the embedded system, itself will implicitly derive what middleware standards need to be adhered to within the design. Also, the embedded market is always changing, so the reader should take the time to research, in addition to starting with Table 3.1, in order to stay up-to-date on what those changes are relative to the required standards to be adhered to.

Note that some market-specific standards in Table 3.1 have been adopted, and may even have originated, within other market segments. Moreover, note that for the same type of device, different standards can exist depending on the country and even the region within a country. There are also industries in which multiple competing standards exist, each supported by competing business interests. So, it is recommended that the readers do their research to determine what standards are out there, who supports them and why, as well as how they differ.

At this time, there is not one single middleware software standards organization that defines and manages middleware standards within the embedded systems space. Thus, it is recommended that the reader research what middleware standards are out there via any means available, such as:

- using the Internet to google the various standards bodies and access their documentation
- looking up within published trade magazines, datasheets and manuals of the relevant industry and device
- by attending industry-specific tradeshows, seminars, and/or conferences. For example, the Embedded Systems Conference (ESC), Real-time Embedded Computing Conference, and Java One to name a few.

Table 3.1: Examples of Real-world Standards Organizations and Middleware Standards in Embedded Systems Market

Standard type	Standard	General description
Aerospace and Defense	Aerospace Industries, Association of America, Inc. (AIA/NAS)	Association representing the nation's major aerospace and defense manufacturers, helping to establish industry goals, strategies, and standards. Related to national and homeland security, civil aviation, and space (www.aia-aerospace.org)
	ARINC (Avionics Application Standard Software Interface)	ARINC standards specify air transport avionics equipment and systems used by commercial and military aircraft worldwide (www.arinc.com)
	DOD (Department of Defense) – JTA (Joint Technical Architecture)	DOD initiative that supports the smooth flow of information via standards, necessary to achieve military interoperability (www.disa.mil)
	Multiple Independent Levels of Security/Safety (MILS)	Middleware framework for creating security-related and safety critical embedded systems
	SAE (Society of Automotive Engineers)	Defining aerospace standards, reports, and recommended practices (www.sae.org)
Automotive	Federal Motor Vehicle Safety Standards (FMVSS)	The Code of Federal Regulations are regulations issued by various agencies within the US Federal government (http://www.nhtsa.dot.gov/cars/rules/standards/)
	Ford Standards	From the engineering material specifications and laboratory test methods volumes, the approved source list collection, global manufacturing standards, non-production material specifications, and the engineering material specs and lab test methods handbook (www.ihs.com/standards/index.html)
	GM Global	Used in the design, manufacturing, quality control, and assembly of General Motors automotives (www.ihs.com/standards/index.html)
	ISO/TS 16949 – The Harmonized Standard for the Automotive Supply Chain	Developed by the International Automotive Task Force (IATF), based on ISO9000, AVSQ (Italy), EAQF (France), QS-9000 (USA), and VDA6.1 (Germany), for example (www.iaob.org)
	Jaguar Procedures and Standards Collection	Contains Jaguar standards including Jaguar-Test Procedures Collection, Jaguar-Engine and Fastener Standards Collection, for example (www.ihs.com/standards/index.html)
Commercial and Home Office Automation	ANSI/AIM BC3-1995, Uniform Symbology Specification for Bar Codes	Specifies encoding general purpose all-numeric types of data, reference decode algorithm, and optional character calculation. This standard is intended to be identical to the CEN (commission for European normalization) specification (www.aimglobal.org/standards/)
	IEEE Std 1284.1-1997 IEEE Standard for Information Technology Transport Independent Printer/System Interface (TIP/SI)	Standard defining a protocol for printer manufacturers, software developers, and computer vendors that defines how data should be exchanged between printers and other devices (www.ieee.org)

Table 3.1: Examples of Real-world Standards Organizations and Middleware Standards in Embedded Systems Market *continued*

Standard type	Standard	General description
Commercial and Home Office Automation	Postscript	Major printer manufacturers make their printers to support postscript printing and imaging standard (www.adobe.com)
Consumer Electronics	ARIB-BML (Association of Radio Industries and Business of Japan)	Responsible for establishing standards in the telecommunications and broadcast arena in Japan[5] (http://www.arib.or.jp/english/)
	ATSC (Advanced Television Standards Committee) DASE (Digital TV Application Software Environment)	Defines middleware that allows programming content and applications to run on DTV receivers. This environment provides content creators the specifications necessary to ensure that their applications and data will run uniformly on all hardware platforms and operating systems for receivers[6] (www.atsc.org)
	ATVEF (Advanced Television Enhancement Forum) – SMPTE (Society of Motion Picture and Television Engineers) DDE-1	The Advanced Television Enhancement Forum (ATVEF) is a cross-industry group that created an enhanced content specification defining fundamentals necessary to enable creation of HTML-enhanced television content. The ATVEF specification for enhanced television programming delivers enhanced TV programming over both analog and digital video systems using terrestrial, cable, satellite and Internet networks[7] (http://www.atvef.com/)
	CEA (Consumer Electronics Association)	An association for the CE industry that develops essential industry standards and technical specifications to enable interoperability between new products and existing devices[8] –Audio and Video Systems Committee –Television Data Systems Subcommittee –DTV Interface Subcommittee –Antennas Committee –Mobile Electronics Committee –Home Network Committee –HCS1 Subcommittee –Cable Compatibility Committee –Automatic Data Capture Committee (www.ce.org)
	DTVIA (Digital Television Alliance of China)	An organization made up of broadcasting academics, research organizations, and TV manufacturers targeting technology and standards within the TV industry in China (http://www.dtvia.org.cn/)
	DVB (Digital Video Broadcasting) – MHP (Multimedia Home Platform)	The collective name for a compatible set of Java-based open middleware specifications developed by the DVB Project, designed to work across all DVB transmission technologies (see www.mhp.org)
	GEM (Globally Executable MHP)	A core of MHP APIs, where the DVB-transmission-specific elements were removed. This allows other content delivery platforms that use other transmission systems to adopt MHP middleware (see www.mhp.org)

(continued)

Table 3.1: Examples of Real-world Standards Organizations and Middleware Standards in Embedded Systems Market *continued*

Standard type	Standard	General description
Consumer Electronics	HAVi (Home Audio Video Initiative)	Digital AV home networking software specification for seamless interoperability among home entertainment products. HAVi has been designed to meet the particular demands of digital audio and video by defining an operating-system-neutral middleware that manages multidirectional AV streams, event schedules, and registries, while providing APIs for the creation of a new generation of software applications[3] (www.havi.org)
	ISO/IEC 16500 DAVIC (Digital Audio Visual Council)	Open interfaces and protocols that maximize interoperability, not only across geographical boundaries but also across diverse of interactive digital audio-visual applications and services (www.davic.org)
	JavaTV	Java-based API for developing interactive TV applications within digital television receivers. Functionality provided via the JavaTV API includes audio/video streaming, conditional access, access to in-band/out-of-band data channels, access to service information, tuner control for channel changing, on-screen graphics control, media synchronization, and control of the application life-cycle, for example[2] (see java.sun.com)
	MicrosoftTV	Interactive TV systems software layer that contains middleware that provides a standard which combines analog TV, digital TV, and internet functionality (http://www.microsoft.com/tv/default.mspx)
	OCAP (OpenCable Application Forum)	System software, middleware layer that provides a standard that allows for application portability over different platforms. OCAP is built on the DVB-MHP Java-based standard, with some modifications and enhancements to MHP (www.opencable.com)
	OpenTV	DVB compliant system software, middleware standard and software for interactive digital television receivers. Based on the DVB-MHP specification with additional available enhancements (www.opentv.com)
	OSGi (Open Services Gateway Initiative)	OSGi provides *Universal Middleware* for service-oriented, component-based environments across a range of markets (www.osgi.org)

Table 3.1: Examples of Real-world Standards Organizations and Middleware Standards in Embedded Systems Market *continued*

Standard type	Standard	General description
Energy and Oil	AWEA (American Wind Energy Association)	Organization that develops standards for the USA wind turbine market (www.awea.org)
	International Electrotechnical Commission (IEC)	One of the world's leading organizations that prepares and publishes international standards for all electrical, electronic and related technologies – such as in the wind turbine generator arena (www.iec.ch)
	International Standards Organization (ISO)	One of the world's leading organizations that prepares and publishes international standards for energy and oil systems – such as in the nuclear energy arena (www.iso.org)
Industrial Automation and Control	International Electrotechnical Commission (IEC)	One of the world's leading organizations that prepares and publishes international standards for all electrical, electronic and related technologies – including in industrial machinery and robotics (www.iec.ch)
	International Standards Organization (ISO)	One of the world's leading organizations that prepares and publishes international standards for energy and oil systems – including in industrial machinery and robotics (www.iso.org)
	Object Management Group (OMG)	An international, open membership consortium developing middleware standards and profiles that are based on the Common Object Request Broker Architecture (CORBA®) and support a wide variety of industries, including for the field of robotics via the OMG Robotics Domain Special Interest Group (DSIG) (www.omg.org)
Medical	Department of Commerce, USA – Office of Microelectronics, Medical Equipment and Instrumentation	Website that lists the medical device regulatory requirements for various countries (www.ita.doc.gov/td/mdequip/regulations.html)
	Digital Imaging and Communications in Medicine (DICOM)	Standard for transferring images and data between devices used in the medical industry (medical.nema.org)
	Food and Drug Administration (FDA) USA	Among other standards, includes US government standards for medical devices, including class I non-life sustaining, class II more complex non-life sustaining, and class III life sustaining and life support devices (www.fda.gov)
	IEEE1073 Medical Device Communications	Standard for medical device communication for plug-and-play interoperability for point-of-care/acute care environments (www.ieee1073.org)
	Medical Devices Directive (EU)	Standards for medical devices for EU states for various classes of devices (europa.eu.int)

(continued)

Table 3.1: Examples of Real-world Standards Organizations and Middleware Standards in Embedded Systems Market *continued*

Standard type	Standard	General description
Networking and Communication	Cellular	Networking standards implemented for cellular phones (www.cdg.org and www.tiaonline.org)
	IP (Internet Protocol)	OSI Network layer protocol implemented within various network devices based on RFC 791 (www.faqs.org/rfcs)
	TCP (Transport Control Protocol)	OSI Transport layer protocol implemented within various network devices based on RFC 793 (www.faqs.org/rfcs)
	Bluetooth	Standards developed by the Bluetooth Special Interest Group (SIG) which allows for developing applications and services that are interactive via interoperable radio modules and data communication protocols (www.bluetooth.org)
	UDP (User Datagram Protocol)	OSI Transport layer protocol implemented within various network devices based on RFC 768 (www.faqs.org/rfcs)
	HTTP (Hypertext Transfer Protocol)	A WWW (world wide web) standard defined via a number of RFC (request for comments), such as RFC2616, 2016, 2069 to name a few (www.w3c.org/Protocols/Specs.html)
	DCE (Distributed Computing Environment)	Defined by the Open Group, the Distributed Computing Environment is a framework that includes RPC (remote procedure call), various services (naming, time, authentication), and a file system to name a few (http://www.opengroup.org/dce/)
	SOAP (Simple Object Access Protocol)	WWW Consortium specification that defines an XML-based networking protocol for exchange of information in a decentralized, distributed environment (http://www.w3.org/TR/soap/)

Table 3.1: Examples of Real-world Standards Organizations and Middleware Standards in Embedded Systems Market *continued*

Standard type	Standard	General description
General Purpose	Networking and Communication Standards	TCP, Bluetooth, IP, etc.
	C# and .NET Compact Framework	Microsoft-based standard and middleware system for portable application development. Evolution of COM (www.microsoft.com)
	HTML (Hyper Text Markup Language)	A WWW (world wide web) standard for a scripting language processed by an interpreter on the device (www.w3c.org)
	Java and the Java Virtual Machine	Various standards and middleware systems from Sun Microsystems targeted for application development in different types of embedded devices (java.sun.com) Personal Java (pJava) Embedded Java, Java 2 Micro Edition (J2ME) The Real Time Specification for Java From J Consortium Real Time Core Specification
	SSL (Secure Socket Layer) 128-bit encryption	Security standard providing data encryption, server authentication, and message integrity, for example for a TCP/IP-based device (wp.netscape.com)
	Filesystem Hierarchy Standard	Standard that defines a file system directory structure hierarchy (http://www.linuxfoundation.org/)
	COM (Component Object Model)	Originally from Microsoft, a standard that allows for interprocess communication and dynamic object creation independent of underlying hardware and system software
	DCOM (Distributed COM)	Based on DCE-RPC and COM, that allows for interprocess communication and dynamic object creation across networked devices

3.3 The Contribution of Standards to an Embedded System

This section illustrates that to begin the process of demystifying the software within an embedded device, it is useful to simply derive from the standards what the system requirements would be and then determine where in the architecture of the embedded device these components belong. To demonstrate how middleware standards can define some of the most critical components of an embedded system software design, examples of:

- an operating system standard
- programming language standards
- industry-specific standards

are introduced in the next sections of this chapter.

3.3.1 Why have a POSIX Middleware Layer?

Middleware developers who want the flexibility of porting and utilizing their stack on more than one embedded operating system commonly take the approach of creating a middleware layer that abstracts out the operating system APIs commonly used by overlying libraries. These APIs include process management (i.e., creating and deleting tasks), memory management, and I/O management functionality. This middleware layer is implemented by wrapping an embedded OS's functions in a common API that overlying software uses instead of the functions provided by an embedded OS directly. Many off-the-shelf embedded OSs today support such an abstraction layer called the portable operating system interface (POSIX), summarized in Table 3.2 and in the real-world implementation of POSIX in Figure 3.2.

Additional custom POSIX wrappers can also be useful to extend and to abstract out device driver libraries for overlying software layers that need access to managing the hardware (Figure 3.3). For example, if higher-level middleware and/or application software requires access to low-level driver Flash routines to read/write data to Flash directly, then POSIX wrappers can be added to abstract out device driver APIs when porting from one target to another with vastly different BSPs (and internal functions).

It is also useful when designing to use an embedded operating system that implements a partitioning protection scheme for mission critical-type devices (such as vxWorks653 shown in Figure 3.4). These types of OSs *require* that there be some type of middleware abstraction layer for 'protected' partitions that contain software that can access to lower level drivers directly.

3.3.2 When the Programming Language Impacts the Middleware Layer

Relative to programming languages, standards, and middleware there is not one programming language that is a perfect fit for all embedded systems designs, and this reality

Table 3.2: Example of POSIX Functionality[13]

OS Subsystem	Function	Definition
Process Management	Threads	Functionality to support multiple flows of control within a process. These flows of control are called threads and they share their address space and most of the resources and attributes defined in the operating system for the owner process. The specific functional areas included in threads support are: • Thread Management: the creation, control, and termination of multiple flows of control that share a common address space. • Synchronization primitives optimized for tightly coupled operation of multiple control flows in a common, shared address space.
	Semaphores	A minimum synchronization primitive to serve as a basis for more complex synchronization mechanisms to be defined by the application program.
	Priority scheduling	A performance and determinism improvement facility to allow applications to determine the order in which threads that are ready to run are granted access to processor resources.
	Real-time signal extension	A determinism improvement facility to enable asynchronous signal notifications to an application to be queued without impacting compatibility with the existing signal functions.
	Timers	A mechanism that can notify a thread when the time as measured by a particular clock has reached or passed a specific value, or when a specified amount of time has passed.
	IPC	A functionality enhancement to add a high-performance, deterministic interprocess communication facility for local communication.
Memory Management	Process memory locking	A performance improvement facility to bind application programs into the high-performance random access memory of a computer system. This avoids potential latencies introduced by the operating system in storing parts of a program that were not recently referenced on secondary memory devices.
	Memory mapped files	A facility to allow applications to access files as part of the address space.
	Shared memory objects	An object that represents memory that can be mapped concurrently into the address space of more than one process.
I/O Management	Synchronionized I/O	A determinism and robustness improvement mechanism to enhance the data input and output mechanisms, so that an application can ensure that the data being manipulated is physically presented on secondary mass storage devices.
	Asynchronous I/O	A functionality enhancement to allow an application process to queue data input and output commands with asynchronous notification of completion.

1   posixScLib

1.1.1.1   NAME

posixScLib - POSIX message queue and semaphore system call documentation

1.1.1.2   ROUTINES

pxOpen( ) - open a POSIX semaphore or message queue (syscall)
pxClose( ) - close a reference to a POSIX semaphore or message queue (syscall)
pxUnlink( ) - unlink the name of a POSIX semaphore or message queue (syscall)
pxMqReceive( ) - receive a message from a POSIX message queue (syscall)
pxMqSend( ) - send a message to a POSIX message queue (syscall)
pxSemPost( ) - post a POSIX semaphore (syscall)
pxSemWait( ) - wait for a POSIX semaphore (syscall)
pxCtl( ) - control operations on POSIX semaphores and message queues (syscall)

1.1.1.3   DESCRIPTION

This module contains system call documentation for POSIX message queue and semaphore system calls.

2   pxOpen( )

2.1.1.1   NAME

pxOpen( ) - open a POSIX semaphore or message queue (syscall)

2.1.1.2   SYNOPSIS

```
OBJ_HANDLE pxOpen
 (
 PX_OBJ_TYPE type,
 const char * name,
 int mode,
 void * attr
 )
```

2.1.1.3   DESCRIPTION

This routine opens a POSIX object

3   pxClose( )

3.1.1.1   NAME

pxClose( ) - close a reference to a POSIX semaphore or message queue (syscall)

3.1.1.2   SYNOPSIS

```
int pxClose (
 OBJ_HANDLE handle)
```

3.1.1.3   DESCRIPTION

This routine closes the specified *handle* to the underlying POSIX object. If the handle refers to an unnamed semaphore, then the object is deleted, provided no task is blocked on it. If tasks are blocked on the semaphore then this function returns **ERROR** with **EBUSY** errno.

......

Figure 3.2: POSIX Functionality and vxWorks[14]

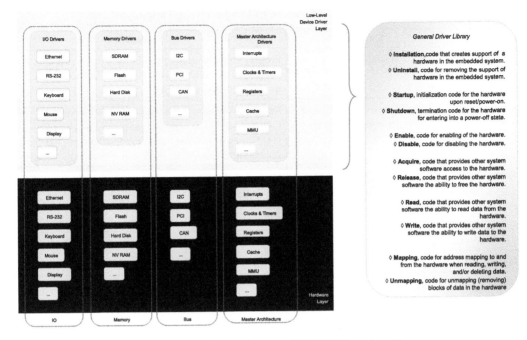

Figure 3.3: Device Drivers and POSIX Functionality

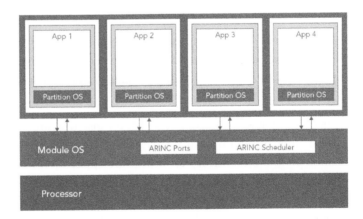

Figure 3.4: vxWorks653 Protected Application within Partitions[15]

is reflected by the fact that different languages are used in designing various embedded systems today. In many real-world embedded devices, more than one programming language has been utilized.

Typically, it is a fourth-generation or higher type of programming language standard (see Table 3.3) that can introduce this additional middleware element within an embedded system's architecture design. Of course, languages like C, a third-generation language, can

Table 3.3: General Evolution of Programming Languages[4]

	Language	Details
5th Generation	Natural languages	Programming languages similar to conversational languages typically used for AI (artificial intelligence) programming and design
4th Generation	Very high level (VHLL) and non-procedural languages	Very high level languages that are object-oriented, like C++, C#, and Java, scripting languages, such as Perl and HTML – as well as database query languages, like SQL, for example
3rd Generation	High-order (HOL) and procedural languages, such as C and Pascal for example	High-level programming languages with more English-corresponding phrases. More portable than 2nd and 1st generation languages
2nd Generation	Assembly language	Hardware dependent, representing machine code
1st Generation	Machine code	Hardware dependent, binary zeros (0s) and ones (1s)

be based on standards such as ANSI C or Kernighan and Ritchie C, for example – but these types of standards usually do not introduce an additional middleware component when using a language based on them in an embedded system design.

To support a fourth-generation language like Java within an embedded system, for example, requires that a JVM (Java virtual machine) reside within the deployed device. As shown in Figure 3.5a, real-world embedded systems currently contain JVMs in their hardware layer, as *middleware* within their system software layer, or within their application layer.

So, where standards make a difference relative to a JVM, for instance, are with the JVM classes. These classes are compiled libraries of Java byte code, commonly referred to as *Java APIs* (application program interfaces). Java APIs are application-independent libraries provided by the JVM to, among other things, allow programmers to execute system functions and reuse code. Java applications require the Java API classes, in addition to their own code, to successfully execute. The size, functionality, and constraints provided by these APIs differ according to the Java specification they adhere to, but can include memory management features, graphics support, networking support, and so forth. Different standards with their corresponding APIs are intended for different families of embedded devices (see Figure 3.5b).

In the embedded market, recognized embedded Java standards include J Consortium's *Real-Time Core Specification*, and *Personal Java* (pJava), *Embedded Java, Java 2 Micro Edition* (J2ME) and *The Real-Time Specification for Java* from Sun Microsystems. Figure 3.5c shows the differences between the APIs of two different embedded Java standards.

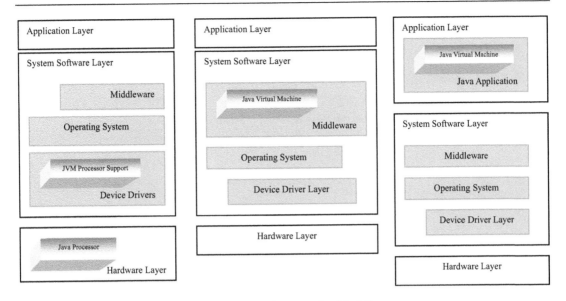

Figure 3.5a: JVMs in an Embedded System

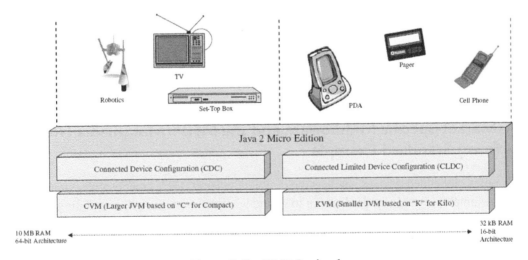

Figure 3.5b: J2ME Devices[1]

For another fourth-generation language, C#, regarding supporting of its usage on an embedded WinCE device – Microsoft, for example, supplies a .NET Compact Framework (see Figure 3.6) to be included in the middleware layer of an embedded system similar to the manner in which a JVM can be integrated into an embedded device's system software layer.

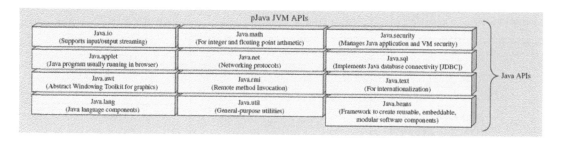

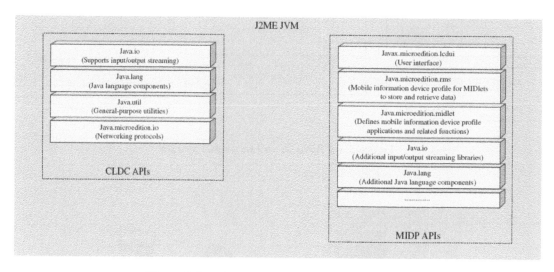

Figure 3.5c: pJava versus J2ME Sample APIs[3-1]

3.4 Market-specific Middleware and the MHP (Multimedia Home Platform) Standard Example

In complex embedded devices, such as the digital television (DTV) receiver shown in Figure 3.7 for example, several standards serve to define what components will be residing within the middleware software stack. While there are several types of DTV receivers on the market today, from enhanced broadcast receivers that provide traditional broadcast television to interactive broadcast receivers providing services including video-on-demand, web browsing, and email, a DTV receiver serves as a good example of an embedded system that can require some subset of multiple general-purpose and market-specific standards (see Table 3.4) and how these standards can be used to derive what components are required within the device.

Analog TVs process incoming analog signals of traditional TV video and audio content, whereas digital TVs (DTVs) process both incoming analog and digital signals of TV video/ audio content, as well as application data content that is embedded within the entire digital data stream (a process called data broadcasting or data casting). This application data can

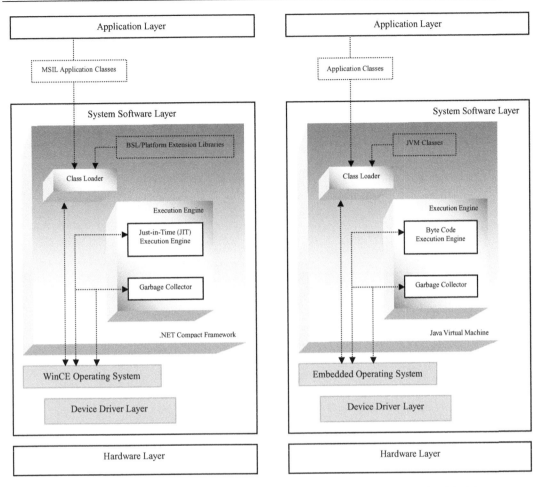

Figure 3.6: NET Compact Framework vs. Java Virtual Machine in an Embedded System

either be unrelated to the video/audio TV content (non-coupled), related to video/audio TV content in terms of content but not in time (loosely coupled), or entirely synchronized with TV audio/video (tightly coupled).

The type of application data embedded is dependent on the capabilities of the DTV receiver itself. While there are a wide variety of DTV receivers, most fall under one of three categories:

- enhanced broadcast receivers, which provide traditional broadcast TV enhanced with graphics controlled by the broadcast programming
- interactive broadcast receivers, capable of providing e-commerce, video-on-demand, email, and so on through a return channel on top of 'enhanced' broadcasting
- multinetwork receivers that include internet and local telephony functionality on top of interactive broadcast functionality.

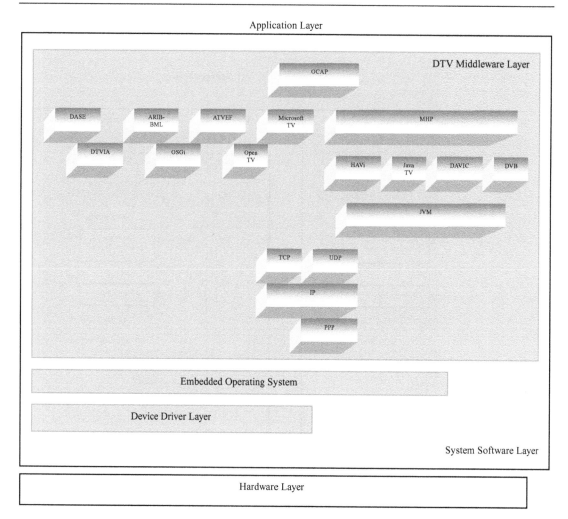

Figure 3.7: DTV Receiver Example of Several Middleware Standards

Depending on the type of receiver, DTVs can implement general-purpose, market-specific, and/or application-specific standards all into one DTV/set-top box (STB) system architecture design (shown in Table 3.4). These standards then can define several of the major components that are implemented in all layers of the DTV Embedded Systems Model, as shown in Figure 3.7. The Digital Video Broadcasting (DVB) – Multimedia Home Platform (MHP) platform is one example of real-world market-specific middleware software that is targeted for the DTV embedded systems market, and used as the real-world example in this chapter.

MHP is a Java-based middleware solution based upon the *Digital Video Broadcasting (DVB) – Multimedia Home Platform (MHP) Specification*. MHP implementations in digital television are a powerful example to learn from when designing or using just about any

Table 3.4: Examples of Digital Television (DTV) Receiver Standards

Standard Type	Standards
Market Specific	ATVEF (Advanced Television Enhancement Forum)
	ATSC (Advanced Television Standards Committee)/DASE (Digital TV Applications Software Environment)
	ARIB-BML (Association of Radio Industries and Business of Japan)
	DAVIC (Digital Audio Video Council)
	DTVIA (Digital Television Industrial Alliance of China)
	DVB (Digital Video Broadcasting)/MHP (Multimedia Home Platform)
	HAVi (Home Audio Video Interoperability)
	JavaTV
	MicrosoftTV
	OCAP (OpenLabs Opencable Application Platform)
	OSGi (Open Services Gateway Initiative)
	OpenTV
General Purpose	Java
	Networking (TCP/IP over terrestrial, cable, and satellite, for example)

market-specific middleware solution, because it incorporates many complex concepts and challenges that must be addressed in its approach.

3.4.1 Initial Steps: Understanding Underlying MHP System Requirements

In general, as shown in Figure 3.8, hardware boards that support MHP include:

- Master processor
- Memory subsystem
- System buses
- I/O subsystem
  - tuner/demodulator
  - de-multiplexer
  - decoders/encoders
  - graphics processor
  - communication interface/modem
  - Conditional Access (CA) module
  - a remote control receiver module.

Of course, there can be additional components, and these components will differ in design from board to board, but these elements are generally what are found on most boards targeted

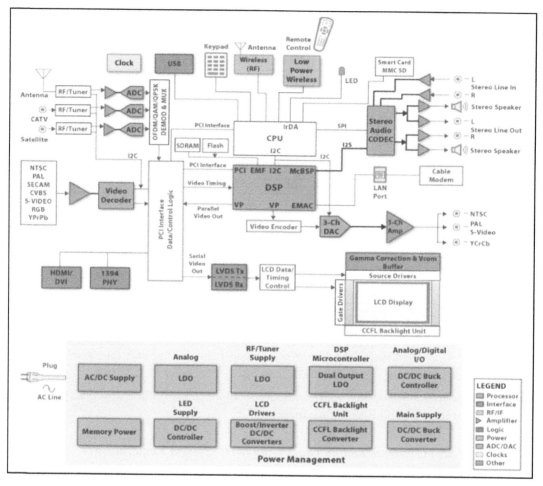

Figure 3.8: Texas Instruments DTV Block Diagram [3-9]

for this market. MHP and associated system software APIs typically require a minimum of 16 MB of RAM, 8–16 MB of Flash, and depending on how the JVM and OS are implemented and integrated can require a 150–250+ MHz CPU to run in a practical manner. Keep in mind that depending on the type of *applications* that will be run over the system software stack, memory and processing power requirements for these applications need to be taken into consideration, thus they may require a change to this 'minimum' baseline memory and processing power requirements for running MHP.

The flow of video data originates with some type of input source. As shown in Figure 3.9, in the case of an analog video input source, for example, each is routed to the analog video decoder. The decoder then selects one of three active inputs and quantizes the video signal, which is then sent to some type of MPEG-2 subsystem. An MPEG-2 decoder is responsible

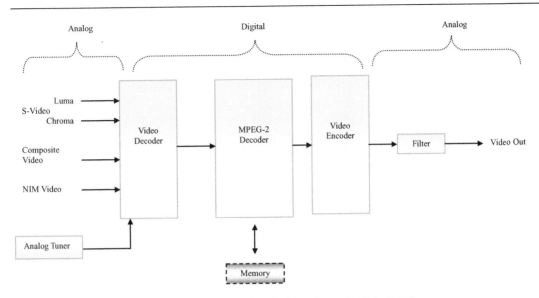

Figure 3.9: Example of Video Data Path in DTV

for processing the video data received to allow for either standard-definition or high-definition output. In the case of standard-definition video output, it is encoded as either S-video or composite video using an external video encoder. No further encoding or decoding is typically done to the high-definition output coming directly from the MPEG-2 subsystem.

The flow of transport data originating from some type of input source is passed to the MPEG-2 decoder subsystem (see Figure 3.10). The output information from this can be processed and displayed.

In the case of audio data flow it originates at some type of analog source such as the analog audio input sources shown in Figure 3.11. The MPEG-2 subsystem receives analog data from the A/D converters that translated the incoming analog sources. Audio data can be merged with other data, or transmitted as-is to D/A converters to be then routed to some type of audio output ports.

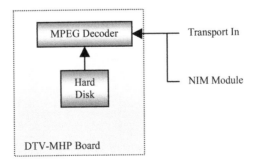

Figure 3.10: Example of Transport Data Path in DTV

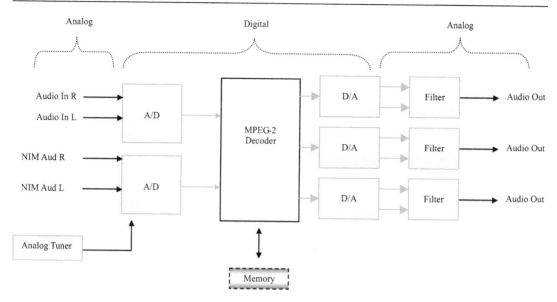

Figure 3.11: Example of Audio Data Path in DTV

An MHP hardware subsystem will then require some combination of device driver libraries to be developed, tested, and verified within the context of the overlying MHP compliant software platform. Like the hardware, these low-level device drivers generally will fall under general master processor-related device drivers (see Figure 3.12), memory and bus device drivers (see Figure 3.13), and I/O subsystem drivers.

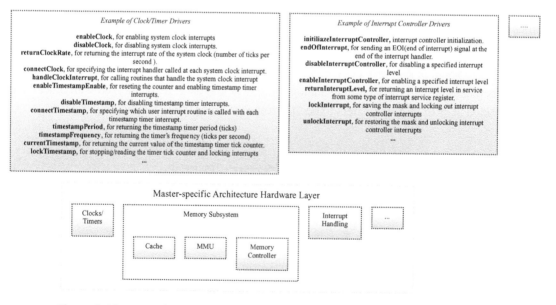

Figure 3.12: Example of General Architecture Device Drivers on MHP Platform

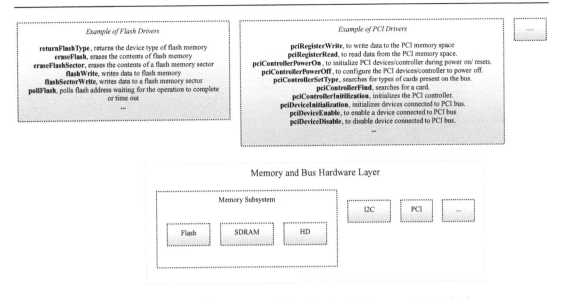

Figure 3.13: Example of Memory and Bus Device Drivers on MHP Platform

The I/O subsystem drivers include Ethernet, keyboard/mouse, video subsystem, and audio subsystem drivers to name a few. Figures 3.14a–c show a few examples of MHP I/O subsystem device drivers.

Because MHP is Java-based, as the previous section of this chapter indicated and shown in Figure 3.15, a Java Virtual Machine (JVM) and ported operating system must then reside on the embedded system that implements an MHP stack and underlie this MHP stack. This JVM must meet the Java API specification required by the particular MHP implementation, meaning the underlying Java functions that the MHP implementation calls down for must reside in some form in the JVM that the platform supports.

Example of Ethernet Drivers

loadEthernetCard, initializes the Ethernet driver and the device.
unloadEthernetCard, removes the Ethernet driver and the device.
initializeEthernetMemory, initializes required memory.
startEthernetCard, starts the device as running and enables related interrupts.
stopEthernetCard, shuts down device and disables interrupts.
restartEthernetCard, restarts a stopped device after cleaning up the receive/transmit queues.
parseEthernetCardInput, parses the input string to retrieve data.
transmitEthernetData, transmits data over Ethernet
receiveEthernetReceive, processes incoming data over Ethernet.
...

Example of Keyboard/Mouse Drivers

getKMEvent, reads keyboard or mouse events from queue.
putKMEvent, puts keyboard or mouse events onto queue.
processMouse, takes mouse input and puts event in queue
findMouseMove, finds a mouse move event in the queue
...

....

General IO Hardware Layer

| Ethernet | Serial | Keyboard | Mouse | ... |

Figure 3.14a: Example of MHP General I/O Device Drivers

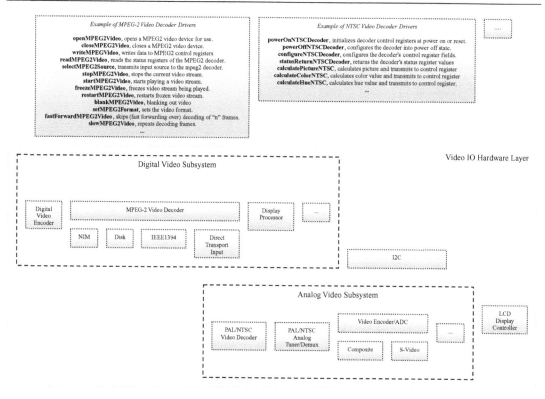

Figure 3.14b: Example of MHP Video I/O Device Drivers

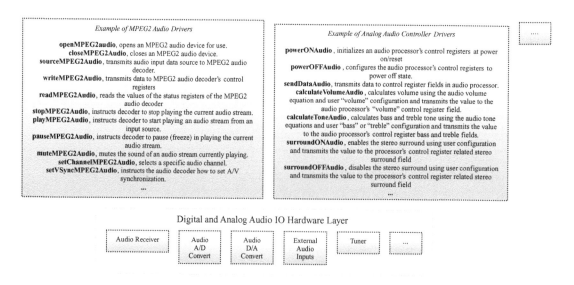

Figure 3.14c: Example of MHP Audio I/O Device Drivers

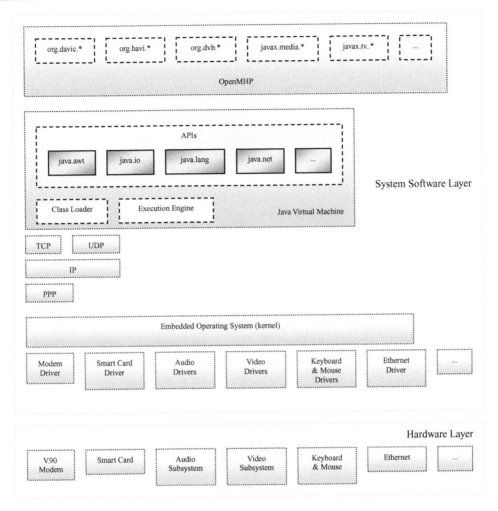

Figure 3.15: MHP-based System Architecture

The open source example, openMHP, shows how some JVM APIs in its implementation, such as the org.havi.ui library translate, into source code in this particular package (see Figure 3.16).

3.4.2 Understanding MHP Components, MHP Services, and Building Applications

As shown in Figure 3.17, the MHP standard is made up of a number of different sub-standards which contribute to the APIs, including:

```
package org.dvb.ui;

/* Copyright 2000-2003 by HAVi, Inc. Java is a trademark of Sun Microsystems, Inc. All rights reserved.  This program is free software; you can redistribute it and/or modify
 * it under the terms of the GNU General Public License as published by the Free Software Foundation; either version 2 of the License, or (at  your option) any later version.
 * This program is distributed in the hope that it will be useful, but WITHOUT ANY WARRANTY;  without even the implied warranty of MERCHANTABILITY or FITNESS
 * FOR A PARTICULAR PURPOSE.  See the GNU General Public License for more details.
 * You should have received a copy of the GNU General Public License along with this program ; if not, write to the Free Software Foundation, Inc., 59 Temple Place, Suite 330,
 * Boston, MA 02111-1307 USA */

import java.awt.Graphics2D;
import java.awt.Graphics;
import java.awt.Dimension;
import javax.media.Clock;
import javax.media.Time;
import javax.media.IncompatibleTimeBaseException ;

/** A <code>BufferedAnimation</code> is an AWT component that maintains a queue of one or more image buffers.  This permits efficient flicker-free animation by allowing a
 * caller to draw to an off-screen buffer, which the system then copies to the framebuffer in coordination with the video output subsystem. This class also allows an application
 * to request a series of buffers, so that it can get a small number of frames ahead in an animation.  This allows an application to be robust in the presence of short delays, e.g. from
 * garbage collection. A relatively small number of buffers is recommended, perhaps three or  four. A BufferedAnimation with one buffer provides little or no protection
 * from pauses, but does provide double-buffered animation. .... **/
....

public class BufferedAnimation extends java.awt.Component {

/**
 * Constant representing a common video framerate, approximately 23.98 frames per second, and equal to <code>24000f/1001f</code>.
 *
 * @see #getFramerate()
 * @see #setFramerate(float)
 **/
static public float FRAME_23_98 = 24000f/1001f;

/**
 * Constant representing a common video framerate, equal to <code>24f</code>.
 *
 * @see #getFramerate()
 * @see #setFramerate(float)
 **/
static public float FRAME_24 = 24f;

/**
 * Constant representing a common video framerate, equal to <code>25f</code>.
 *
 * @see #getFramerate()
 * @see #setFramerate(float)
 **/
static public float FRAME_25 = 25f;

/**
 * Constant representing a common video framerate, approximately 29.97 frames per second, and equal to<code>30000f/1001f</code>.
 *
 * @see #getFramerate()
 * @see #setFramerate(float)
 **/
static public float FRAME_29_97 = 30000f/1001f;

/**
 * Constant representing a common video framerate, equal to <code>50f</code>.
 *
 * @see #getFramerate()
 * @see #setFramerate(float)
 **/
static public float FRAME_50 = 50f;

/**
 * Constant representing a common video framerate, approximately 59.94 frames per second, and equal to<code>60000f/1001f</cod e>.
 *
 * @see #getFramerate()
 * @see #setFramerate(float)
 **/
static public float FRAME_59_94 = 59.94f;

....
}
```

Figure 3.16: openMHP org.havi.ui Source Example[10]

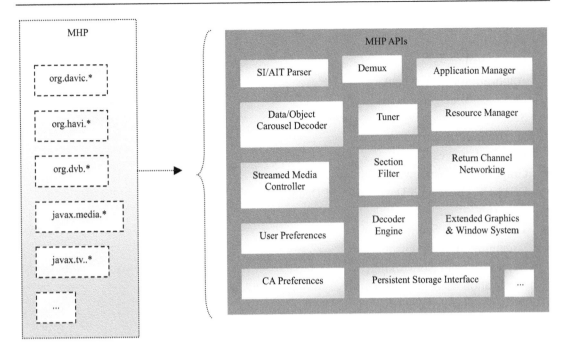

Figure 3.17: MHP APIs

- Core MHP (varies between implementations)
  - DSMCC
  - BIOP
  - Security
- HAVi UI
  - HAVi Level 2 User Interface (org.havi.ui)
  - HAVi Level 2 User Interface Event (org.havi.ui.event)
- DVB
  - Application Listing and Launching (org.dvb.application)
  - Broadcast Transport Protocol Access (org.dvb.dsmcc)
  - DVB-J Events (org.dvb.event)
  - Inter-application Communication (org.dvb.io.ixc)
  - DVB-J Persistent Storage (org.dvb.io.persistent)
  - DVB-J Fundamental (org.dvb.lang)
  - Streamed Media API Extensions (org.dvb.media)
  - Datagram Socket Buffer Control (org.dvb.net)
  - Permissions (org.dvb.net.ca and org.dvb.net.tuning)
  - DVB-J Return Connection Channel Management (org.dvb.net.rc)
  - Service Information Access (org.dvb.si)
  - Test Support (org.dvb.test)
  - Extended Graphics (org.dvb.ui)
  - User Settings and Preferences (org.dvb.user)

- JavaTV
- DAVIC
- Return Path
- Application Management
- Resource Management
- Security
- Persistent Storage
- User Preferences
- Graphics and Windowing System
- DSM-CC Object and Data Carousel Decoder
- SI Parser
- Tuning, MPEG Section Filter
- Streaming Media Control
- Return Channel Networking
- Application Manager and Resource Manager Implementation
- Persistent Storage Control
- Conditional Access support and Security Policy Management
- User Preference Implementations.

Within the MHP world, content of the end-user of the system it interacts with is grouped and managed as *services*. Content that makes up a service can fall under several different types, such as applications, service information, and data/audio/video streams to name a few. In addition to platform-specific requirements and end-user preferences, the different types of content in services are used to manage data. For example, when a digital TV allows support for more than one type of different video stream, service information can be used to determine which stream actually gets displayed.

MHP applications can range from browsers to email to games to EPGs (electronic program guides) to advertisements, to name a few. At the general level, all these different types of MHP applications will typically fall under one of three general types of profile:

- *Enhanced broadcasting*, where the digital broadcast contains a combination of audio services, video services, and executable applications to allow end-users to interact with the system locally
- *Interactive broadcasting*, where the digital broadcast contains a combination of audio services, video services, executable applications, as well as interactive services and channels that allow end-users to interact with residing applications remotely to their digital TV device
- *Internet access*, where the system implements functionality that allows access to the internet.

An important note is that while MHP is Java-based, the MHP DVB-J type of applications are not regular Java applications, but are executed within the context of a Java servlet (Xlet) similar to the concept behind the Java applet. MHP applications communicate and interact with their external environment via the Xlet context. For example, Figures 3.18a and 3.18b

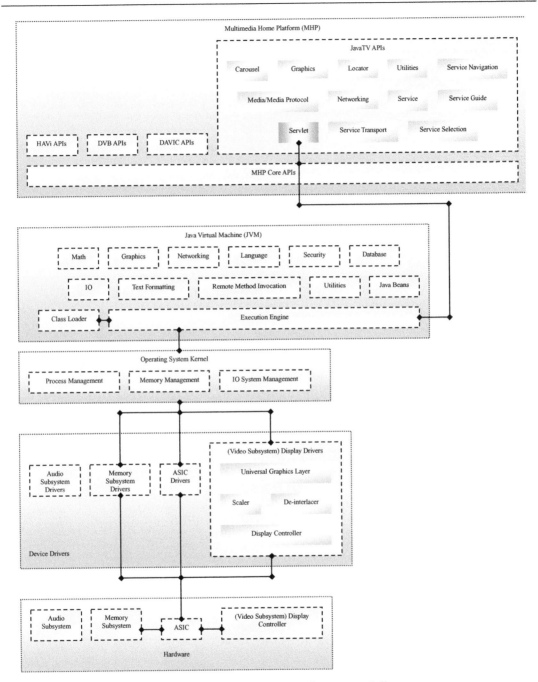

Figure 3.18a: Simple Xlet Flow Example[11]

```
import javax.tv.xlet.*;

// The main class of every MHP Xlet must implement this interface
public class XletExample implements javax.tv.xlet.Xlet
{

// Every Xlet has an Xlet context, created by the MHP middleware and passed in to the Xlet as a parameter to the initXlet() method.
private javax.tv.xlet.XletContext context;

// A private field to hold the current state. This is needed because the startXlet() method is called both start the Xlet for the first time and also to make the Xlet resume from //
the paused state. This filed lets us keep track of whether we're starting for the first time.
private boolean hasBeenStarted;

// Every Xlet should have a default constructor that takes no arguments. The constructor should contain nothing. Any initialisation should be done in the initXlet() method, //
or in the startXlet method if it's time- or resource-intensive. That way, the MHP middleware can control when the initialisation happens in a much more predictable way
// Initializing the Xlet.
public XletExample()

// store a reference to the Xlet context that the Xlet is executing in this .context = context;
 public void initXlet(javax.tv.xlet.XletContext context) throws javax.tv.xlet.XletStateChangeException

// The Xlet has not yet been started for the first time, so set this variable to false.
hasBeenStarted = false;

// Start the Xlet. At this point the Xlet can display itself on the screen and start interacting with the user, or do any resource-intensive tasks.
public void startXlet() throws javax.tv.xlet.XletStateChangeException
{
if (hasBeenStarted)
{
System.out.println ("The startXlet() method has been called to resume the Xlet after it's been paused. Hello again, world!");
}
else
{
System.out.println("The startXlet() method has been called to start the Xlet for the first time. Hello, world!");

// set the variable that tells us we have actually been started
hasBeenStarted = true;
}
}

// Pause the Xlet and free any scarce resources that it's using, stop any unnecessary threads and remove itself from the screen.
public void pauseXlet()
{
System.out.println("The pauseXlet() method has been called. to pause the Xlet...");
}

// Stop the Xlet.
public void destroyXlet(boolean unconditional) throws javax.tv.xl et.XletStateChangeException
{
if (unconditional)
{
System.out.println("The destroyXlet() method has been called to unconditionally destroy the Xlet." ");
}
else
{
System.out.println("The destroyXlet() method has been called requesting that the Xlet stop, but giving it the choice. Not Stopping.");
throw new XletStateChangeException("Xlet Not Stopped");
}
}

}
```

**application example based upon MHP open source by Steven Morris available for download at www.interactivetvweb.org

Figure 3.18b: Simple Xlet Source Example[12]

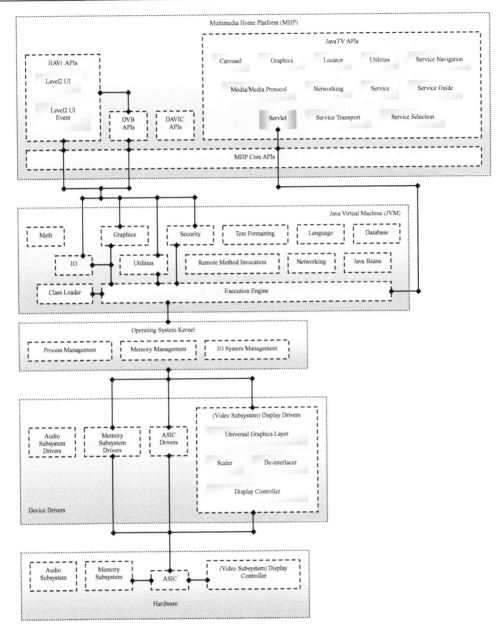

Figure 3.19a: Simple MHP HAVi Xlet Flow Example[11]

show an application example where a simple Xlet is created, initialized, and can be paused or destroyed via an MHP Java TV API package 'javax.tv.xlet'.

The next example shown in Figures 3.19a and 3.19b is a sample application which uses the

- JVM packages java.io, java.awt, and java.awt.event
- MHP Java TV API package 'javax.tv.xlet'

```
// Import required MHP JavaTV package
import javax.tv.xlet.*;

//import required MHP HAVi packages
import org.havi.ui.*;
import org.havi.ui.event.*;

//import required MHP DVB package
import org.dvb.ui.*;

// import required non-MHP pJava packages
import java.io.*;
import java.awt.*;
import java.awt.event.*;

// This Xlet will be visible on-screen, so we extend org.havi.ui.Hcomponent and it also implements java.awt.KeyListener to receive

// Something went wrong reading the message file.
System.out.println("I/O exception reading message.txt"); } }

// Start the Xlet.
public void startXlet() throws javax.tv.xlet.XletStateChangeException
{

// startXlet() should not block for too long
myWorkerThread = new Thread(this);
myWorkerThread.start();
}

// Pause the Xlet.
public void pauseXlet()
{

// do what we need to in order to pause the Xlet.
doPause();
}

// Destroy the Xlet.
public void destroyXlet(boolean unconditional) throws javax.tv.xlet.XletStateChangeException
{
if (unconditional)
{
// Unconditional Termination
doDestroy();
}
else
{

// Conditional Termination
throw new XletStateChangeException("Termination Not Required");
}
}

// Before we can draw on the screen, we need an HScene to draw into. This variable will hold a reference to our Hscene
private HScene scene;

// The image that we will show
private Image image;

// The message that will get printed. This is read from a file in initXlet()
private String message;

// this holds the alpha (transparency) level that we will be using
private int alpha = 0;

// this object is responsible for displaying the background I-frame
private HaviBackgroundController backgroundManager;

// The main method for the worker thread.
public void run()
{
// We need quite a few resources before we can start doing anything.
getResources();

// This component should listen for AWT key events so that it can respond to them.
addKeyListener(this);
// This adds the background image to the display. The background image is displayed in the background plane.
displayBackgroundImage();

// The bitmap image is shown in the graphics plane.
displayForegroundBitmap();
}
```

Figure 3.19b: Simple MHP HAVi Xlet Source Example[12]

- MHP HAVi packages org.havi.ui and org.havi.ui.event
- MHP DVB package org.dvb.ui.

Finally, an *application manager* within an MHP system manages all MHP applications residing on the device both from information input from the end-user, as well as via the AIT (application information table) data within the MHP broadcast stream transmitted to the system. AIT data simply instructs the application manager as to what applications are actually available to the end-user of the device, and the technical details of controlling the running of the application.

3.5 Summary

Chapter 3 demonstrated the importance of understanding middleware standards relative to an embedded systems design. The different types and examples of middleware standards were defined according to industries, as well as general purpose standards that are utilized in a wide variety of embedded systems. General examples relative to programming languages and a digital television receiver were used to demonstrate that middleware standards can define important components within an embedded system's software stack. Only general examples were used in this chapter since a later chapter of this book continues with a more detailed discussion of programming languages that introduce middleware elements within an embedded system design.

The next section of this book, Section II, begins the detailed discussion of core middleware commonly found in embedded systems as well as being the foundation of more complex middleware software.

3.6 Problems

1. Which standard is not a standard typically implemented within an embedded system?
   A. MHP – Multimedia Home Platform
   B. HTTP – Hypertext Transfer Protocol
   C. J2EE – Java 2 Enterprise Edition
   D. FTP – File Transfer Protocol
   E. None of the above.
2. Give three examples of middleware standards implemented in embedded systems today.
3. How can middleware standards be classified?
4. Name and define four types of general purpose middleware standards implemented within embedded systems today.
5. Give three examples of standards that fall under the following markets:
   A. Consumer Electronics
   B. Networking and Communications.
6. Name two examples of standards which introduce middleware component(s) within an embedded system, and list what those middleware components are.

7. HTTP is an application layer standard that does not implicitly require any particular underlying middleware (True/False).

8. Give an example of an embedded device which adheres to standards that introduce several middleware components into the design. Draw the high-level diagram of an example of such a device.

9. Which middleware standards below are Java-based:
   A. HTML – Hypertext Markup Language
   B. CLDC – Connected Limited Device Configuration
   C. MHP – Multimedia Home Platform
   D. A and B only
   E. B and C only
   F. All of the above.

3.7 End Notes

1 Embedded Systems Architecture, Noergaard, 2005. Elsevier.
2 http://java.sun.com/products/javatv/overview.html
3 http://www.havi.org/
4 System Analysis and Design. Harris, David. page 17.
5 http://www.arib.or.jp/english/
6 www.atsc.org
7 http://www.atvef.com/
8 http://www.ce.org/
9 http://focus.ti.com/docs/solution/folders/print/327.html
10 openMHP API Documentation and Source Code.
11 Digital Video Broadcasting (DVB); Multimedia Home Platform (MHP) Specification 1.1.2. European Broadcasting Union.
12 Application examples based upon MHP open source by Steven Morris available for download at www.interactivetvweb.org
13 http://www.pasc.org/
14 WindRiver vxWorks API Reference Guide.
15 WindRiver vxWorks653 Datasheet.

CHAPTER 4

The Fundamentals in Understanding Networking Middleware

Chapter Points

- Introduce fundamental networking concepts
- Discuss the OSI model relevance to networking middleware
- Show examples of real-world networking middleware protocols

By definition, two or more devices that are connected in some fashion to allow for the transmission and/or reception of data are a ***network***. To successfully communicate, each system within a network must implement some set of compatible networking elements (Figure 4.1). Some of these mechanisms are implemented in the middleware layer of an embedded system, and many are based upon industry standards – typically referred to as networking ***protocols***. In fact, one of the most commonly included types of middleware in an embedded system is networking protocols, even if this code in the embedded device is only executed when connecting to a host at development time for developing and debugging the software on the device.

The first steps to learning about networking middleware within an embedded systems design include:

Step 1. Reviewing and using standard industry networking models, such as the *Open Systems Interconnection* (OSI) networking model, as tools to define and understand what internal networking components would be required by an embedded system to successfully function within a particular network.

Step 2. Having a clear understanding of the overall network an embedded device will be required to function properly within, specifically:
- The distance between the devices connected on the network
- The physical medium that connects the embedded device to the network
- The overall architecture (structure) of the network.

Step 3. Understanding the underlying hardware and system software layers, specifically:

Demystifying Embedded Systems Middleware. DOI: 10.1016/B978-0-7506-8455-2.00004-2

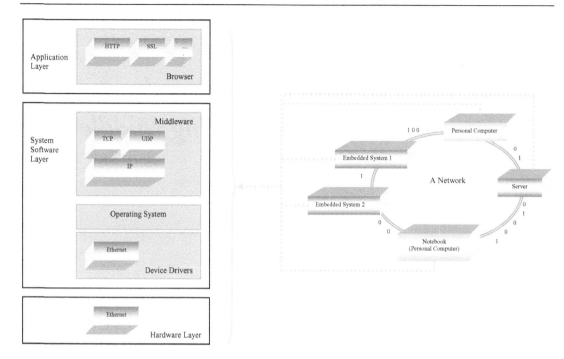

Figure 4.1: What is a Network?

- Know your networking-specific standards (introduced in Chapter 3).
- Understand the hardware (see Chapter 2). If the reader comprehends the hardware, it will be easier to understand the functionality of the overlying networking components.
- Define and understand the specific underlying system software components, such as the available device drivers supporting the networking hardware and the operating system API (Chapter 2).

Step 4. Using a networking model, such as OSI, define and understand what type of functionality and data exists at the middleware layer for a particular device and protocol stack.

Step 5. Define and understand different types of networking application requirements and corresponding protocols in order to ultimately be able to understand what middleware components are necessary within a particular system to support the overlying software layers.

4.1 Step 1 to Understanding Networking Middleware: Networking Models

The International Organization for Standardization's OSI (open systems interconnection) reference model from the early 1980s is a representation of what types of hardware and software networking components can be found in any computer system. Of the seven layers of the OSI model, protocols at the *upper data-link*, *network* and *transport* layers are typically implemented within some form of middleware software (see Figure 4.2).

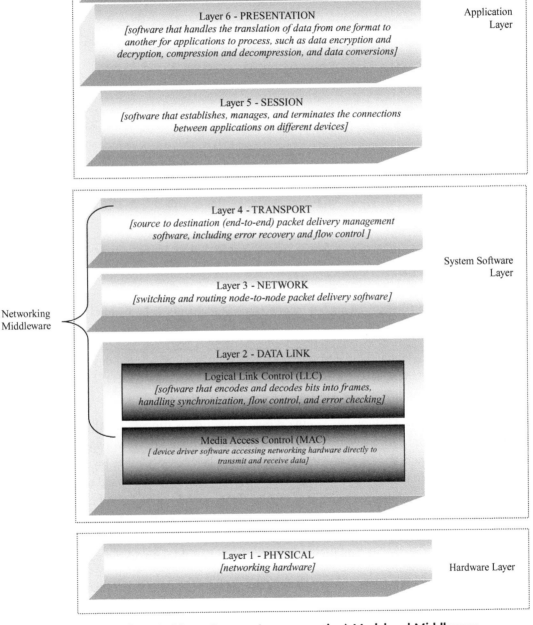

Figure 4.2: The OSI (Open Systems Interconnection) Model and Middleware

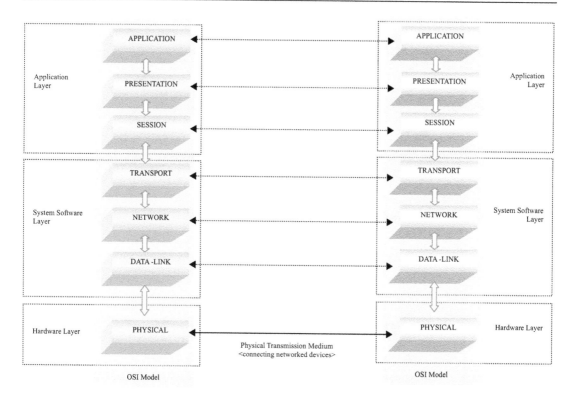

Figure 4.3: The OSI (Open Systems Interconnection) Model and the Embedded Systems Model

To fundamentally understand the purpose of each OSI layer in networked devices, it is important to understand that data are transmitted to be processed by peer OSI layers in other devices (see Figure 4.3).

Within the scope of the OSI model, a networking connection is triggered with data originating at the application layer of a device. These data, then, flow downward through all seven layers. Except for the physical and application layers every other layer appends additional information, called a *header*, to the data being transmitted down the stack. Via the transmission medium, data are transmitted over to the physical layer of another networked device, then up through the OSI layers of the receiving device. As the data flow upward, peer layers in receiving devices strip these headers, unwrapping the data for processing. Figure 4.4 provides a visual overview of data flowing up and down an OSI networking stack.

While the OSI model is a powerful tool that can be used by the reader to demystify networking fundamentals, keep in mind that it is not always the case that embedded devices contain 'exactly' seven 'distinct' networking layers. Meaning, in many real-world networking stacks, sometimes the functionality of more than one OSI layer is integrated into fewer layers, and/or the functionality of one OSI layer is split out to more than one layer. As an example,

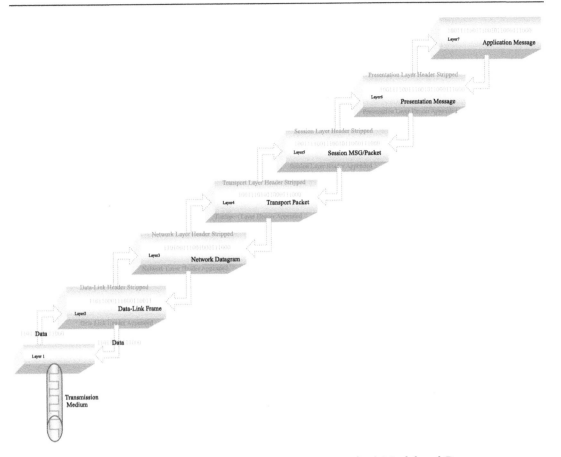

Figure 4.4: The OSI (Open Systems Interconnection) Model and Data

one of the most common real-world networking protocol stacks which deviates from the standard OSI model is the four-layer TCP/IP (Transmission Control Protocol/Internet Protocol) model shown in Figure 4.5. Under the TCP/IP model, OSI layers one and two are integrated into the TCP/IP network access layer, and OSI layers five, six, and seven are incorporated into the TCP/IP application layer.

In short, the important thing to note is that regardless of how a networking stack is implemented in the real world, once the reader can visualize and understand from the OSI model:

1. what is required to implement networking functionality within an embedded device
2. where these components can be located in the particular device
3. the purpose of networking protocols at various layers

the reader can then apply this fundamental understanding to any embedded system design – regardless of how many layers this functionality is implemented within a particular device or what these layers are called within a particular embedded design.

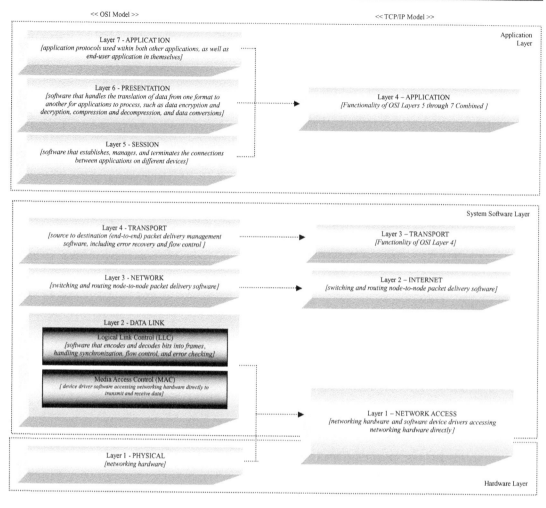

Figure 4.5: The OSI Model and TCP/IP

4.2 Step 2 to Understanding Networking Middleware: Understanding the Overall Network

In addition to software and/or hardware limitations dictated by the embedded device itself, the overall network the embedded device is a part of is what determines which middleware elements need to be implemented within the embedded system. Relative to this, as shown in Figure 4.6, there are at least three key features about the network that the reader needs to be familiar with at the start:

- The distance between the devices connected on the network
- The physical medium that connects the embedded device to the network
- The overall architecture (structure) of the network.

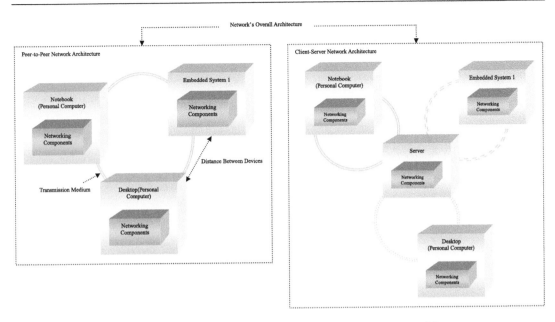

Figure 4.6: Features of an Embedded System's Network[5]

4.2.1 WAN versus LAN: The Distance Between Networked Systems

In terms of where devices are geographically located within a network, at the highest level networks can be divided into two types: *local area networks* (LANs) or *wide area networks* (WANs). LANs are networks with connected devices that are located within close proximity to each other, such as within the scope of the same building and/or the same room. WANs, on the other hand, are networks with connected devices that are geographically located outside the scope of the same building, such as across multiple buildings, across a city, and/or across the globe for example. Despite the endless acronyms used to refer to the different types of networks in the field, inherently all networks are either WANs, LANs, or some interconnected hybrid combination of both.

Within an embedded device, whether or not a device will be connected within an LAN and/or WAN will drive what networking technologies can be implemented within (see Figure 4.7). Given the compatible LAN or WAN physical layer hardware, overlying protocols in support of the physical layer are then implemented in the above software layers including any required middleware components.

4.2.2 Wired vs. Wireless: The Transmission Medium

In general, the transmission medium connecting devices in a network can be categorized as one of two possible types: *bound* (wired) and *unbound* (wireless). Bound transmission

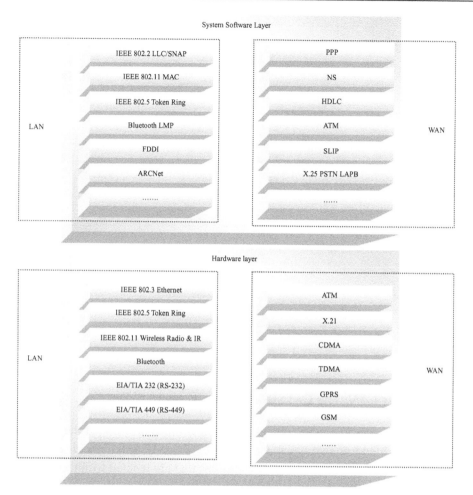

Figure 4.7: Examples of LAN versus WAN Networking Protocols

mediums interconnect devices via some type of *physical* cabling which guides electromagnetic waves along the physical path of the wires within the cable.

Unbound transmission mediums are mediums in which devices are not connected via any physical cable. Wireless transmission mediums utilize transmitted electromagnetic waves which are not guided by a physical path of wiring, but via mediums such as water, air, and/or a vacuum, to name a few.

Within an embedded device, whether or not a device will be connected via a wired versus wireless transmission medium will also drive what networking technologies can be implemented within (see Figure 4.8) as well as what performance can be expected. As stated within the previous section, networking software protocols that are implemented within a device need to be compatible with the underlying wired and/or wireless physical layer hardware.

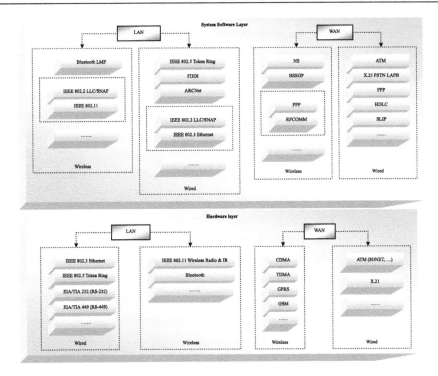

Figure 4.8: Examples of Wireless versus Wired Networking Protocols

4.2.3 Peer-to-Peer vs. Client–Server: The Network's Overall Architecture

A network's architecture essentially defines the relationship between devices on the network. To date, the most common types of structures are modeled after *client–server* architectures, *peer-to-peer* architectures, or some *hybrid* combination of both architectures.

A **client–server** architecture is a model in which one centralized device on the network has control in managing the network in terms of resources, security, and functions, for example. This centralized device is referred to as the *server* of the network. All other devices connected to the network are referred to as *clients*. Servers can manage clients' requests either iteratively, one at a time, or concurrently where more than one client request can be handled in parallel. A client contains fewer resources than the server, and it accesses the server to utilize additional resources and functionality.

On the flip-side, with a **peer-to-peer** architecture network implementation there is not one centralized device in control. Devices in a peer-to-peer network are more functionally independent and are responsible for managing themselves as equals.

Hybrid networks are networks that are structured on some combination of both peer-to-peer and client–server models. LANs and WANs can be based on either client–server or hybrid

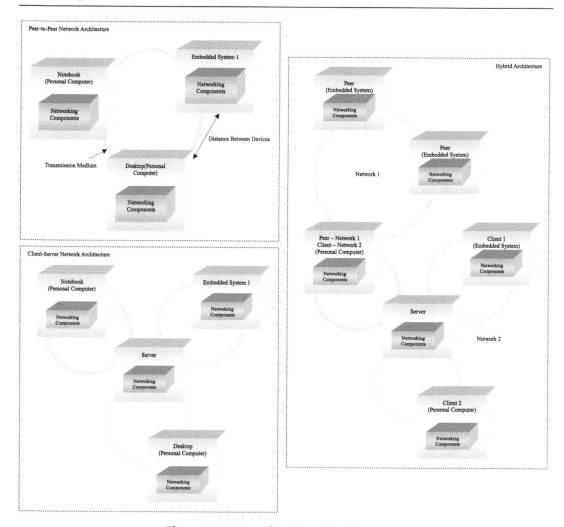

Figure 4.9: Network's Overall Architecture

architectures. Peer-to-peer networks, on the other hand, typically pose additional security and performance challenges that make them more likely to be implemented in LANs rather than WANs.

4.3 Step 3 to Understanding Networking Middleware: Understanding the Underlying Hardware and System Software Layers

Networking protocols implemented in an embedded system's middleware software layer typically reside on top of some combination of other middleware, an operating system, device drivers, and hardware (see Figure 4.10). Specifically, a networking protocol implemented as middleware in the system software layer exists either as:

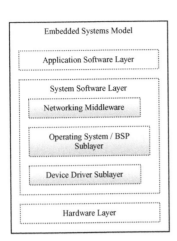
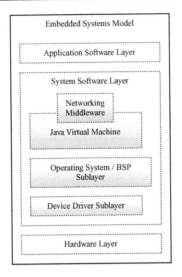
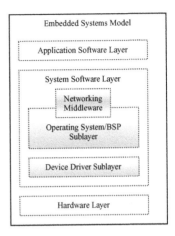

Figure 4.10: Networking System Components the Embedded Systems Model

- Independent middleware components that sit on top of the operating system layer, or directly over device drivers in a system with no operating system.
- Middleware that sits on top of and/or is integrated with other middleware components. For example, a networking stack integrated with an embedded Java Virtual Machine (JVM) distribution from a vendor.
- Middleware that has been tightly integrated and provided with a particular operating system distribution from a vendor.

As shown in Figure 4.11, in some embedded systems the system software can be a little more complex because of more than one implemented networking protocol stack in the embedded device, such as in support of different physical layers, for example.

4.3.1 About the Networking (Physical Layer) Hardware

Why Understand Networking Hardware?

Networking protocols residing at the higher layers of the OSI model view lower software layers that execute over different physical layer hardware as transparent. However, the underlying networking hardware available today is often quite different in terms of how it works. Thus, it is important for embedded developers to understand the differences in the hardware, in order to understand the implementation of a networking stack on which these various technologies reside. In other words, hardware features, quirks, and/or limitations will ultimately impact the type of networking library required and/or what modifications must be implemented in a particular networking stack to support this hardware.

Continued

In other words, when a programmer learns about the networking hardware of a device, then it will be much simpler for the programmer to understand a particular networking protocol implementation, how to modify a particular protocol in support of underlying technologies, as well as determine which middleware networking protocol is the best 'fit' for the device. In short, it is important for the reader to understand the networking relevant features of the hardware – and to use this understanding when analyzing the networking stack implementation that needs to support the particular underlying technology.

Networking hardware on a board falls under a type of I/O (input/output) hardware, and is responsible for transmitting data into and out of the device. At the highest level, I/O networking hardware can be classified according to how the hardware manages the transmission and reception of data, specifically whether the physical layer manages data in *serial*, in *parallel*,

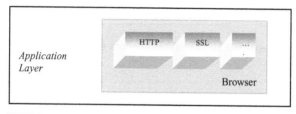
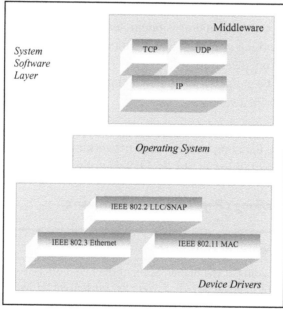
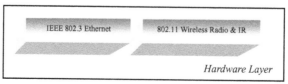

Figure 4.11: Example of Multiple Networking Protocol in an Embedded System

or some hybrid combination of *both*. Networking hardware that is classified as serial, such as EIA/RS-232, manages incoming and outgoing data one bit at a time. Hardware that can manage data in parallel is a physical layer which has the ability to manage multiple bits simultaneously. Hardware such as that based on IEEE 802.3 Ethernet has the capability of supporting both serial and parallel communication and can be configured to manage data either way.

Be it hardware that supports serial communication, parallel communication, or both – as shown in the example of real-world hardware in Figure 4.12 with RS-232 and Ethernet support – an I/O networking hardware subsystem on an embedded systems board is typically made up of some combination of the following six logical units:

- the *transmission medium*, as described in Section 4.2, wireless or wired medium(s) that connect the embedded system to a network
- the *communication (COM) port*, the component(s) on the embedded board in which a wired medium connects to or that receives the signal of a wireless transmission medium
- the *network controller*, a slave processor that manages the networking communication from the other logical units on the board
- the *master processor's integrated networking I/O*, master processor-specific networking components
- the *communication interface*, which manages data communication and the encoding/ decoding of data. It can be integrated into the master processor or another IC (integrated circuit) on the board
- the *I/O bus*, connects master processor to other networking I/O logical units on the board.

Given a serial networking subsystem, for example, that hardware would be made up of some combination of the above logical units, including a 'serial' interface and 'serial' port. A parallel networking subsystem would, instead, have a 'parallel' interface and a 'parallel' port.

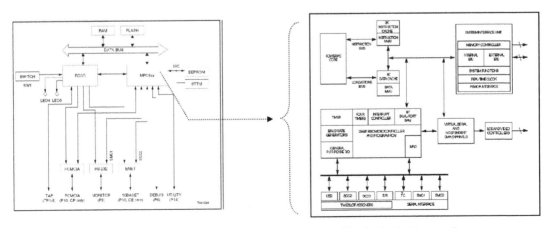

Figure 4.12: Embedded Planet PPC823 Simplified Block Diagram[3]

4.3.2 More on Serial versus Parallel Networking I/O

Whether or not a serial interface (shown in Figure 4.13) is integrated within the master processor or residing as a separate component on the target board, it is this interface that ultimately determines the serial handshaking involved in the transmission and reception of bits between connected devices. Serial handshaking is typically based upon one of three schemes:

- *Simplex*, where bits can only be transmitted and received in one direction, such as shown in Figure 4.13
- *Half Duplex*, where bits can be transmitted and received in either direction, but only specifically in one direction at any given time (see Figure 4.14)
- *Duplex*, where bits can be transmitted and received in either direction at any given time (see Figure 4.15).

Within the serial data stream itself, bits can be transmitted either *asynchronously* or *synchronously* depending on the hardware. With asynchronous data transmission, bits are transmitted at irregular intervals, randomly and intermittently. With synchronous data transmission, data transmission is regulated by a CPU clock resulting in a continuous and steady data stream transmission at regular intervals.

Asynchronous transmission requires that the data being transmitted be divided into groups, referred to as *packets*, of 4–8 bits per character or 5–9 bits per character, for example. These packets are encapsulated into frames that append START bit to indicate the start of the packet and one, one and a half, or two STOP bit(s) to indicate the end of the packet. An optional parity bit can also be appended to the packet for basic error checking, with values of either:

- NONE, meaning no parity bit appended
- ODD, meaning excluding the START and STOP bits, for transmission to be considered successful – the total number of bits set to one must be an odd number

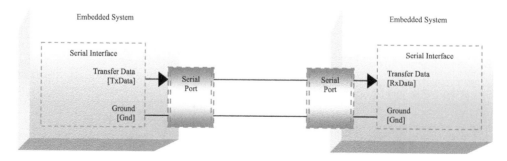

Figure 4.13: Example of Simplex Serial Networking I/O Block Diagram[4]

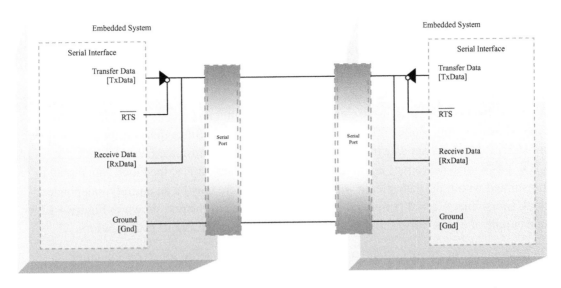

Figure 4.14: Example of Half-Duplex Serial Networking I/O Block Diagram[4]

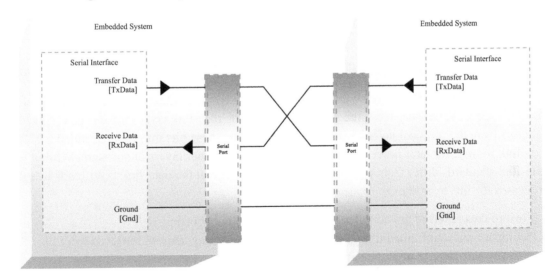

Figure 4.15: Example of Duplex Serial Networking I/O Block Diagram[4]

- EVEN, meaning excluding the START and STOP bits, for transmission to be considered successful – the total number of bits set to one must be an even number.

The key to successful asynchronous serial communication is that the *bit rate* of the transmitter and receiver must be synchronized, where

$$\text{Bit Rate (bandwidth)} = \text{Baud Rate}*(\text{\# of actual data bits per frame/total \# of bits per frame})$$
$$\text{and}$$
$$\text{Baud Rate} = \text{total \# of bits per unit time (i.e., kbits/s, Mbits/s, etc.)}$$

The serial interfaces within the transmitter and receiver then synchronize their transmissions to their own independent bit-rate clocks. When there is no transmission of data, the communication channel is in an idle state. The UART (universal asynchronous receiver-transmitter) is an example of a real-world serial interface that, as its name implies, supports asynchronous serial transmission.

With **synchronous** serial transmission, the data transmitter and receiver also must be in sync – however, this is done off one common clock for both. Since this common clock does not start or stop between data transmissions, data are not encapsulated with START and STOP bits with synchronous communication. In some subsystems, the clock signal may be transmitted within the data stream, whereas in others there may be an entirely independent clock signal line. A serial peripheral interface (SPI), such as the one shown in Figure 4.12, is an example of a real-world serial interface that supports synchronous transmission.

On a final note regarding *parallel* networking I/O – as with serial schemes – parallel communication schemes include simplex, half-duplex, duplex, as well as synchronous and asynchronous data transmission. It is because multiple bits can be transmitted and received simultaneously over parallel networking I/O which allows this hardware to have a greater bandwidth transmission capacity over serial hardware.

4.3.3 Device Drivers and Networking

As shown in Figure 4.16, I/O networking device drivers reside in the lower data-link layer of the OSI model. At the very least, the responsibility of the data-link layer includes receiving data bits from the physical layer hardware and formatting these bits into groups, called data-link *frames,* for later processing and transmission to higher layers of software. While data-link standards differ from protocol to protocol, in general the data-link layer reads in and processes the bits as frames to process the header to:

- insure data received are complete, free of errors, and not corrupted
- compare relevant frame bit field to the physical networking address retrieved from the hardware to determine if the data are intended for that device
- determine who transmitted the frame.

If the data are indeed intended for the device, the data-link header is stripped from the frame. The remaining data bits, commonly referred to as a datagram, are transmitted up the stack. With a datagram coming down the stack to the data-link layer, a data-link header with the above information is appended to the datagram, creating the data-link frame. The relevant I/O networking device drivers then transmit this frame to the I/O networking hardware (physical layer) for transmission outside the device. Figure 4.17 shows a high-level block diagram of this flow.

A lot of I/O networking hardware integrated in the master processor, as well as networking controllers that can reside independently on the embedded systems board, require some set of

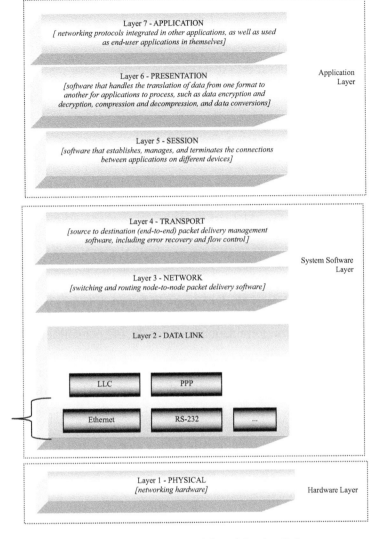

Figure 4.16: The OSI Model and Device Drivers

software functionality to function. Depending on the I/O networking subsystem, the device driver library will generally include some combination of:

- **I/O Networking Installation**, code that allows for on-the-fly support of I/O networking hardware in the embedded system
- **I/O Networking Uninstall**, code for removing the support of I/O networking hardware in the embedded system
- **I/O Networking Startup**, initialization code for the I/O networking hardware upon reset and/or power-on

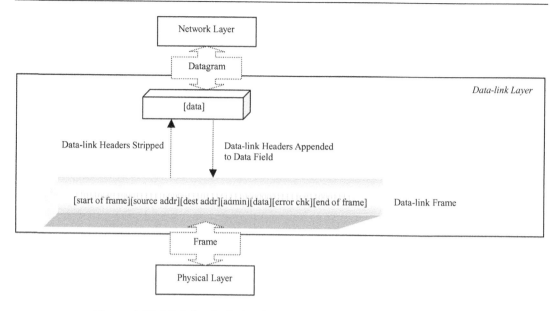

Figure 4.17: High-level Block Diagram of Data-link Layer Data Flow

- **I/O Networking Shutdown**, termination code for the I/O networking hardware for entering into a power-off state
- **I/O Networking Enable**, code for enabling of the I/O networking hardware
- **I/O Networking Disable**, code for disabling the I/O networking hardware
- **I/O Networking Acquire**, code that provides other system software access to the I/O networking hardware
- **I/O Networking Release**, code that provides other system software the ability to free the I/O networking hardware
- **I/O Networking Read**, code that provides other system software the ability to read data from the I/O networking hardware
- **I/O Networking Write**, code that provides other system software the ability to write data to the I/O networking hardware.

Reminder

Different device driver libraries may have additional functions, but most device drivers in support of I/O networking hardware will include some combination of the above functionality.

The device driver libraries are also the foundation on which the middleware functionality is built upon, so it is very important for the reader to insure the existence and stability of any networking device driver functionality the networking middleware requires. Figure 4.18 shows an example of a real-world, open-source Ethernet library and a snippet of some associated device driver function source code for reading and writing to the hardware layer. Overlying middleware layers then utilize functions, such as these types of function for reading, writing, etc. in addition to any other functions included in the device driver library for that particular hardware, to process and manage incoming and outgoing networking data.

4.4 An Embedded OS and Networking I/O APIs

A common method of providing an abstraction layer to simplify software development, managing an embedded device's hardware and software resources, as well as insuring efficient and reliable operation, is the utilization of an embedded operating system (OS) within a design. In addition to processes, memory, and I/O system management components within its *kernel*, an embedded OS may also provide additional I/O system management functionality for networking protocol libraries (see Figures 4.19a and 4.19b).

While networking middleware code can of course be written to access device driver functionality directly, an embedded OS can also include an abstraction layer API that allows for device driver access by middleware software. When providing device access, or any type of I/O access to overlying networking libraries, many OS APIs categorize and abstract their associated underlying device drivers as some combination of:

- **Character**, a driver that allows hardware access via a (character) byte stream
- **Block**, a driver that allows hardware access via some smallest addressable set of bytes at any given time
- **Network**, a driver that allows hardware access via data in the form of networking packets

```
/*
 * Copyright (C) 2008 by egnite GmbH. All rights
reserved.
 * Copyright (C) 2003-2005 by egnite Software
GmbH. All rights reserved.
 *
 * Redistribution and use in source and binary forms,
with or without
 * modification, are permitted provided that the
following conditions
 * are met:
 *
 * 1. Redistributions of source code must retain the
above copyright
 * notice, this list of conditions and the following
disclaimer.
 * 2. Redistributions in binary form must reproduce
the above copyright
 * notice, this list of conditions and the following
disclaimer in the
```

Figure 4.18: Open Source Ethernet Driver Library[6]

```
* OR TORT (INCLUDING NEGLIGENCE OR OTHERWISE) ARISING IN ANY WAY OUT
OF
* THE USE OF THIS SOFTWARE, EVEN IF ADVISED OF THE POSSIBILITY OF
* SUCH DAMAGE.
*
* For additional information see http://www.ethernut.de/
*
*/

....
.....

/*!
* \brief Read contents of PHY register.
*
* \param reg PHY register number.
*
* \return Contents of the specified register.
*/
static u_short phy_inw(u_char reg)
{
/* Select PHY register */
nic_outb(NIC_EPAR, 0x40 | reg);

/* PHY read command. */
nic_outb(NIC_EPCR, 0x0C);
NutDelay(1);
nic_outb(NIC_EPCR, 0x00);

/* Get data from PHY data register. */
return ((u_short) nic_inb(NIC_EPDRH) << 8) | (u_short) nic_inb(NIC_EPDRL);
}

/*!
* \brief Write value to PHY register.
*
* \note NIC interrupts must have been disabled before calling this routine.
*
* \param reg PHY register number.
* \param val Value to write.
*/
static void phy_outw(u_char reg, u_short val)
{
/* Select PHY register */
nic_outb(NIC_EPAR, 0x40 | reg);

/* Store value in PHY data register. */
nic_outb(NIC_EPDRL, (u_char) val);
nic_outb(NIC_EPDRH, (u_char) (val >> 8));

/* PHY write command. */
nic_outb(NIC_EPCR, 0x0A);
NutDelay(1);
nic_outb(NIC_EPCR, 0x00);
}

...

/*!
* \brief Reset the Ethernet controller.
*
* \return 0 on success, -1 otherwise.
*/
static int NicReset(void)
{
/* Hardware reset. */
#ifdef undef_NIC_RESET_BIT
sbi(NIC_RESET_DDR, NIC_RESET_BIT);
sbi(NIC_RESET_PORT, NIC_RESET_BIT);
NutDelay(WAIT100);
cbi(NIC_RESET_PORT, NIC_RESET_BIT);
NutDelay(WAIT250);
NutDelay(WAIT250);
#else
/* Software reset. */
nic_outb(NIC_NCR, NIC_NCR_RST | NIC_NCR_LBMAC);
NutDelay(1);
/* FIXME: Delay required. */
#endif

return NicPhyInit();
}
.....
....
```

Figure 4.18 continued: Open Source Ethernet Driver Library

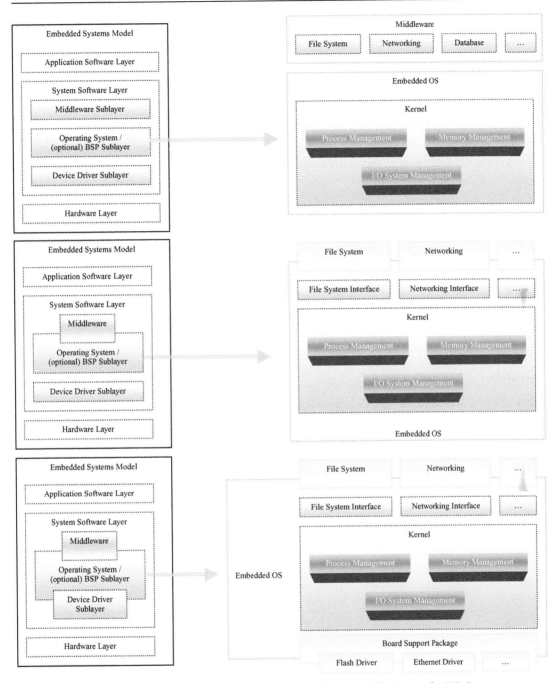

Figure 4.19a: Example OS Permutations

Figure 4.19b: Example OS Components

- **Virtual**, a driver that allows I/O access to virtual (software) devices
- **Miscellaneous Monitor and Control**, a driver that allows I/O access to hardware that is not accessible via the other categories above.

Figure 4.20 shows an example of a vxWorks network device interface library available to middleware for usage – this example is a subset of vxWorks available functionality for network interfacing, buffering, and monitoring. Overlying middleware software layers then have the option of utilizing functions, such as these types of functions provided by the OS layer, to process and manage incoming and outgoing networking data.

VxWorks API Reference : OS Libraries

netLib

NAME

netLib - network interface library

ROUTINES

netLibGeneralInit( ) - initialize the various network code
netLibInit( ) - initialize the network package
netTask( ) - network task entry point

DESCRIPTION

This library contains the network task that runs low-level network interface routines in a task context. The network task executes and removes routines that were added to the job queue. This facility is used by network interfaces in order to have interrupt-level processing at task level.

The routine netLibInit( ) initializes the network and spawns the network task netTask( ). This is done automatically when **INCLUDE_NET_LIB** is defined.

Figure 4.20: Example of Ethernet Device Driver Public Library under VxWorks[7]

The routine netHelp( ) in usrLib displays a summary of the network facilities available from the VxWorks shell.

INCLUDE FILES

netLib.h

SEE ALSO

routeLib, hostLib, netDrv, netHelp( ),

---

netLibGeneralInit( )

NAME

netLibGeneralInit( ) - initialize the various network code

SYNOPSIS
```
STATUS netLibGeneralInit (void)
```

DESCRIPTION

This code use to be in netLibInit. With virtual stacks, we need these specific routines to be executed on a per virtual stack bases.

RETURNS

OK/ERROR

SEE ALSO

netLib

---

netLibInit( )

NAME

netLibInit( ) - initialize the network package

SYNOPSIS
```
STATUS netLibInit (void)
```

DESCRIPTION

This creates the network task job queue, and spawns the network task netTask( ). It should be called once to initialize the network. This is done automatically when **INCLUDE_NET_LIB** is defined.

PROTECTION DOMAINS

This function can only be called from within the kernel protection domain.

RETURNS

OK, or ERROR if network support cannot be initialized.

Figure 4.20 continued: Example of Ethernet Device Driver Public Library under VxWorks

OS Libraries : Routines

netTask( )

NAME

netTask( ) - network task entry point

SYNOPSIS
```
void netTask (void)
```

DESCRIPTION

This routine is the VxWorks network support task. Most of the VxWorks network runs in this task's context.

NOTE

To prevent an application task from monopolizing the CPU if it is in an infinite loop or is never blocked, the priority of netTask( ) relative to an application may need to be adjusted. Network communication may be lost if netTask( ) is "starved" of CPU time. The default task priority of netTask( ) is 50. Use taskPrioritySet( ) to change the priority of a task.

This task is spawned by netLibInit( ).

PROTECTION DOMAINS

This function can only be called from within the kernel protection domain.

RETURNS

N/A

SEE ALSO

netLib, netLibInit( )

Figure 4.20 continued: Example of Ethernet Device Driver Public Library under VxWorks

4.5 Step 4: Networking Middleware

As shown in Figure 4.21, within the scope of this book, networking protocols that reside within the:

- upper data-link layer
- network layer
- transport layers

are defined as middleware software components.

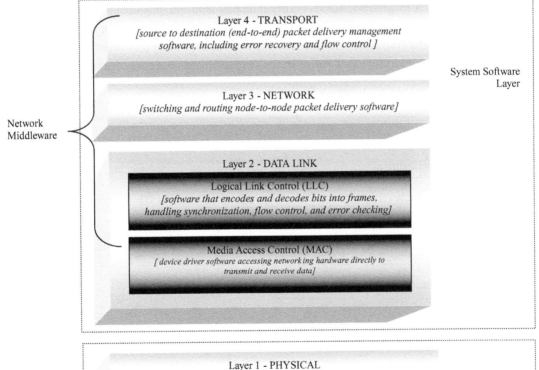

Figure 4.21: Middleware and the OSI Model

4.5.1 Upper Data-link Layer Middleware[5]

As shown in Figure 4.22, the *data-link* layer is the software closest to the hardware – the physical layer in OSI model terms. Thus, it includes, among other functions, any software needed to access, control, and manage the hardware. Bridging also occurs at this layer to allow networks interconnected with different physical layer protocols – for example, Ethernet LAN and an 802.11 LAN – to interconnect.

Like physical layer protocols, data-link layer protocols are classified as either LAN protocols, WAN protocols, or protocols that can be used for both LANs and WANs. Data-link layer protocols that are reliant on a specific physical layer may be limited to the transmission medium involved, but in some cases (for instance, PPP over RS-232 or PPP over Bluetooth's RFCOMM), data-link layer protocols can be ported to very different mediums if there is a layer that simulates the original medium the protocol was intended for, or if the protocol supports hardware-independent upper-data-link functionality.

The data-link layer is responsible for receiving data bits from the physical layer and formatting these bits into groups, called data-link frames. Different data-link standards have varying data-link frame formats and definitions, but in general this layer reads the bit fields of these frames to ensure that entire frames are received, that these frames are error

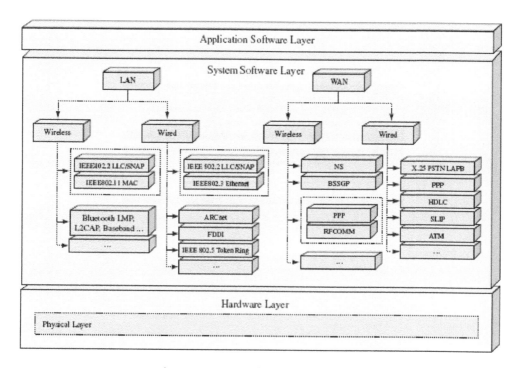

Figure 4.22: Data-link Layer Protocols

free, that the frame is meant for this device by using the physical address retrieved from the networking hardware on the device, and where this frame came from. If the data are meant for the device, then all data-link layer headers are stripped from the frame, and the remaining data field, called a ***datagram***, is passed up to the networking layer. These same header fields are appended to data coming down from upper layers by the data-link layer, and then the full data-link frame is passed to the physical layer for transmission (see Figure 4.23).

As shown in Figure 4.21, within the scope of the OSI model the data-link layer is logically split into two sublayers, a lower sublayer referred to as the media access control (MAC) and the upper sublayer called the logical link control (LLC). The upper data-link LLC sublayer is what is typically found at the middleware software layer, and can provide various functions depending on the protocol, including some combination of:

- multiplexing protocols overlaying the data-link layer
- managing the physical (MAC) addressing between systems and being passed to upper layers for translation to network addresses
- managing data flow and providing flow control of frames
- synchronization of data
- managing communication that is connectionless and/or connection-oriented (with acknowledgments of received frames)
- error recovery
- data-link addressing and control.

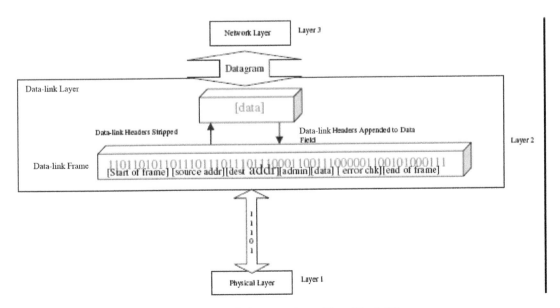

Figure 4.23: Data-link Layer Data Flow Block Diagram

4.5.2 Point-to-Point Protocol Example[5]

PPP (point-to-point protocol) is a common OSI data-link (or network access layer under the TCP/IP model) protocol that can encapsulate and transmit data to higher layer protocols, such as IP, over a physical serial transmission medium (see Figure 4.24). PPP provides support for both asynchronous (irregular interval) and synchronous (regular interval) serial communication.

PPP is responsible for processing data passing through it as frames. When receiving data from a lower layer protocol, for example, PPP reads the bit fields of these frames to insure that entire frames are received, that these frames are error free, that the frame is meant for this device (using the physical address retrieved from the networking hardware on the device), and to determine where this frame came from. If the data are meant for the device, then PPP strips all data-link layer headers from the frame, and the remaining data field, called a *datagram*, is passed up to a higher layer. These same header fields are appended to data coming down from upper layers by PPP for transmission outside the device.

In general, PPP software is defined via a combination of four submechanisms:

- The *PPP encapsulation mechanism* (in RFC1661) such as the high-level data-link control (HDLC) framing in RFC1662 or the link control protocol (LCP) framing defined in RFC1661 to process (i.e., demultiplex, create, verify checksum, etc.)
- *Data-link protocol handshaking*, such as the link control protocol (LCP) handshaking defined in RFC1661, responsible for establishing, configuring, and testing the data-link connection
- *Authentication protocols*, such as PAP (PPP authentication protocol) in RFC1334, used to manage security after the PPP link is established

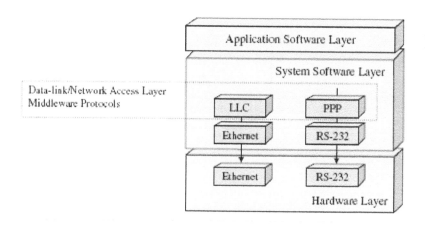

Figure 4.24: Data-link Middleware

Table 4.1: Phase Table[8]

Phase	Description
Link Dead	The link necessarily begins and ends with this phase. When an external event (such as carrier detection or network administrator configuration) indicates that the physical layer is ready to be used, PPP proceeds to the Link Establishment phase. During this phase, the LCP automaton (described later in this chapter) will be in the Initial or Starting states. The transition to the Link Establishment phase signals an Up event (discussed later in this chapter) to the LCP automaton.
Establish Link	The link control protocol (LCP) is used to establish the connection through an exchange of configuration packets. An Establish Link phase is entered once a Configure-Ack packet (described later in this chapter) has been both sent and received.
Authentication	Authentication is an optional PPP mechanism. If it does take place, it typically does so soon after the Establish Link phase.
Network Layer Protocol	Once PPP has completed the establish or authentication phases, each Network Layer Protocol (such as IP, IPX, or AppleTalk) MUST be separately configured by the appropriate Network Control Protocol (NCP).
Link Termination	PPP can terminate the link at any time, after which PPP should proceed to the Link Dead phase.

- *Network control protocols* (NCP), such as IPCP (Internet protocol control protocol) in RFC1332, that establish and configure upper-layer protocol (i.e., OP, IPX, etc.) settings.

These submechanisms work together in the following manner: a PPP communication link, connecting both devices, can be in one of five possible phases at any given time, as shown in Table 4.1. The current phase of the communication link determines which mechanism – encapsulation, handshaking, authentication, and so on – is executed.

How these phases interact to configure, maintain, and terminate a point-to-point link is shown in Figure 4.25.

As defined by PPP layer 1 (i.e., RFC1662), data are encapsulated within the PPP frame, an example of which is shown in Figure 4.26.

The *flag* bytes mark the beginning and end of a frame, and are each set to 0x7E. The *address* byte is a high-level data-link control (HDLC) broadcast address and is always set to 0xFF, since PPP does not assign individual device addresses. The *control* byte is an HDLC command for UI (unnumbered information) and is set to 0x03. The *protocol* field defines the protocol of the data within the information field (i.e., 0x0021 means the information field contains IP datagram, 0xC021 means the information field contains link control data, 0x8021 means the information field contains network control data – see Table 4.2). Finally, the *information* field contains the data for higher-level protocols, and the *FCS* (frame check sequence) field contains the frame's checksum value.

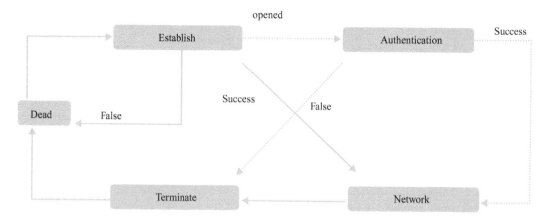

Figure 4.25: PPP Phases[8]

Flag	Address	Control	Protocol	Information	FCS	Flag
1 byte	1 byte	1 byte	2 bytes	Variable	2 bytes	1 byte

Figure 4.26: PPP HDLC-like Frame[8]

Table 4.2: Protocol Information[8]

Value (in hex)	Protocol Name
0001	Padding Protocol
0003 to 001 f	Reserved (transparency inefficient)
007d	Reserved (Control Escape)
00cf	Reserved (PPP NLPID)
00ff	Reserved (compression inefficient)
8001 to 801 f	Unused
807d	Unused
80cf	Unused
80ff	Unused
c021	Link Control Protocol
c023	Password Authentication Protocol
c025	Link Quality Report
c223	Challenge Handshake Authentication Protocol

Code	Identifier	Length	Data [variable in size]		
			Type	Length	Data
1 byte	1 byte	2 bytes			

Figure 4.27: LCP Frame[8]

The data-link protocol may also define a frame format. An LCP frame, for example, is as shown in Figure 4.27.

The **data** field contains the data intended for higher networking layers, and is made up of information (type, length, and data). The **length** field specifies the size of the entire LCP frame. The **identifier** is used to match client and server requests and responses. Finally, the **code** field specifies the type of LCP packet (indicating the kind of action being taken); the possible codes are summarized in Table 4.3. Frames with codes 1–4 are called link configuration frames, 5 and 6 are link termination frames, and the rest are link management packets.

The LCP code of an incoming LCP datagram determines how the datagram is processed, as shown in the pseudocode example below.

```
....
if (LCPCode)
{
        = CONFREQ:
                RCR(…); //see table 4-29
        end CONFREQ;

        = CONFACK:
                RCA(..); //see table 4-29
        end CONFACK;

        = CONFNAK or CONFREJ:
                RCN(…); //see table 4-29
        end CONFNAK or CONFREJ;

        = TERMREQ:
                event(RTR);
        end TERMREQ;

        = TERMACK:
                ....
}
.....
```

In order for two devices to be able to establish a PPP link, each must transmit a data-link protocol frame, such as LCP frames, to configure and test the data-link connection. As mentioned, LCP is one possible protocol that can be implemented for PPP, to handle PPP handshaking. After the LCP frames have been exchanged (and thereby a PPP link established), authentication can then occur. It is at this point where authentication protocols, such as PPP Authentication Protocol or PAP, can be used to manage security, through password authentication and so forth. Finally, Network Control Protocols (NCP) such as

Table 4.3: LCP Codes[8]

Code	Definition
I	Configure-Request
2	Configure-Ack
3	Configure-Nak
4	Configure-Reject
5	Terminate-Request
6	Terminate-Ack
7	Code-Reject
8	Protocol-Reject
9	Echo-Request
10	Echo-Reply
11	Discard-Request
12	Link Quality Report

IPCP (Internet Protocol Control Protocol) establish and configure upper-layer protocols in the network layer protocol settings, such as IP and IPX.

At any given time, a PPP connection on a device is in a particular *state*, as shown in Figure 4.28; the PPP states are outlined in Table 4.4.

Events (also shown in Figure 4.28) are what cause a PPP connection to transition from state to state. The LCP codes (from the RFC1661 spec) in Table 4.5 define the types of events that cause a PPP state transition.

As PPP connections transition from state to state, certain actions are taken stemming from these events, such as the transmission of packets and/or the starting or stopping of the Restart timer, as outlined in Table 4.6.

PPP states, actions, and events are usually created and configured by the platform-specific code at boot-time, some of which is shown in pseudocode form on the next several pages. A PPP connection is in an initial state upon creation; thus, among other things, the 'initial' state routine is executed. This code can be called later at runtime to create and configure PPP, as well as respond to PPP runtime events (i.e., as frames are coming in from lower layers for processing). For example, after PPP software demuxes a PPP frame coming in from a lower layer, and the checksum routine determines the frame is valid, the appropriate field of the frame can then be used to determine what state a PPP connection

Table 4.4: PPP States[8]

States	Definition
Initial	PPP link is in the Initial state, the lower layer is unavailable (Down), and no Open event has occurred. The Restart timer is not running in the Initial state.
Starting	The Starting state is the Open counterpart to the Initial state. An administrative Open has been initiated, but the lower layer is still unavailable (Down). The Restart timer is not running in the Starting state. When the lower layer becomes available (Up), a Configure-Request is sent.
Stopped	The Stopped state is the Open counterpart to the Closed state. It is entered when the automaton is waiting for a Down event after the This-Layer-Finished action, or after sending a Terminate-Ack. The Restart timer is not running in the Stopped state.
Closed	In the Closed state, the link is available (Up), but no Open has occurred. The Restart timer is not running in the Closed state. Upon reception of Configure-Request packets, a Terminate-Ack is sent. Terminate-Acks are silently discarded to avoid creating a loop.
Stopping	The Stopping state is the Open counterpart to the Closing state. A Terminate-Request has been sent and the Restart timer is running, but a Terminate-Ack has not yet been received.
Closing	In the Closing state, an attempt is made to terminate the connection. A Terminate-Request has been sent and the Restart timer is running, but a Terminate-Ack has not yet been received. Upon reception of a Terminate-Ack, the Closed state is entered. Upon the expiration of the Restart timer, a new Terminate-Request is transmitted, and the Restart timer is restarted. After the Restart timer has expired Max-Terminate times, the Closed state is entered.
Request-Sent	In the Request-Sent state an attempt is made to Configure the connection. A Configure-Request has been sent and the Restart timer is running, but a Configure-Ack has not yet been received nor has one been sent.
Ack-Sent	In the Ack-Received state, a Configure-Request has been sent and a Configure-Ack has been received. The Restart timer is still running, since a Configure-Ack has not yet been sent.
Opened	In the Opened state, a Configure-Ack has been both sent and received. The Restart timer is not running. When entering the Opened state, the implementation SHOULD signal the upper layers that it is now Up. Conversely, when leaving the Opened state, the implementation SHOULD signal the upper layers that it is now Down.

is in and thus what associated software state, event, and/or action function needs to be executed. If the frame is to be passed to a higher layer protocol, then some mechanism is used to indicate to the higher layer protocol that there are data to receive (*IPReceive* for *IP*, for example).

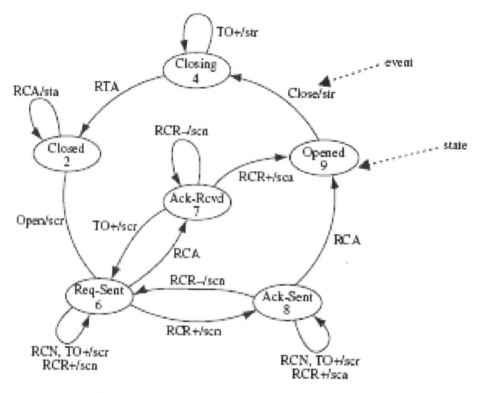

Figure 4.28: PPP Connection States and Events[8]

Table 4.5: PPP Events[8]

Event Label	Event	Description
Up	lower layer is Up	This event occurs when a lower layer indicates that it is ready to carry packets.
Down	lower layer is Down	This event occurs when a lower layer indicates that it is no longer ready to carry packets.
Open	administrative open	This event indicates that the link is administratively available for traffic; that is, the network administrator (human or program) has indicated that the link is allowed to be Opened. When this event occurs, and the link is not in the Opened state, the automaton attempts to send configuration packets to the peer.
Close	administrative close	This event indicates that the link is not available for traffic; that is, the network administrator (human or program) has indicated that the link is not allowed to be Opened. When this event occurs, and the link is not in the Closed state, the automaton attempts to terminate the connection. Further attempts to re-configure the link are denied until a new Open event occurs.

Table 4.5 continued: PPP Events

Event Label	Event	Description
TO+	timeout with counter > 0	This event indicates the expiration of the Restart timer.
TO−	timeout with counter expired	The Restart timer is used to time responses to Configure-Request and Terminate-Request packets. The TO+ event indicates that the Restart counter continues to be greater than zero, which triggers the corresponding Configure-Request or Terminate-Request packet to be retransmitted. The TO− event indicates that the Restart counter is not greater than zero, and no more packets need to be retransmitted.
RCR+	receive configure request good	An implementation wishing to open a connection MUST transmit a Configure-Request. The Options field is filled with any desired changes to the link defaults. Configuration Options SHOULD NOT be included with default values.
RCR−	receive configure request bad	
RCA	receive configure ack	This event occurs when a valid Configure-Ack packet is received from the peer. The Configure-Ack packet is a positive response to a Configure-Request packet. An out of sequence or otherwise invalid packet is silently discarded. If every Configuration Option received in a Configure-Request is recognizable and all values are acceptable, then the implementation MUST transmit a Configure-Ack. The acknowledged Configuration Options MUST NOT be reordered or modified in any way. On reception of a Configure-Ack, the Identifier field MUST match that of the last transmitted Configure-Request. Additionally, the Configuration Options in a Configure-Ack MUST exactly match those of the last transmitted Configure-Request. Invalid packets are silently discarded.
RCN	receive configure nak/rej	This event occurs when a valid Configure-Nak or Configure-Reject packet is received from the peer. The Configure-Nak and Configure-Reject packets are negative responses to a Configure-Request packet. An out of sequence or otherwise invalid packet is silently discarded.
RTR	receive terminate request	This event occurs when a Terminate-Request packet is received. The Terminate-Request packet indicates the desire of the peer to close the connection.
RTA	receive terminate ack	This event occurs when a Terminate-Ack packet is received from the peer. The Terminate-Ack packet is usually a response to a Terminate-Request packet. The Terminate-Ack packet may also indicate that the peer is in Closed or Stopped states, and serves to re-synchronize the link configuration.
RUC	receive unknown code	This event occurs when an uninterpretable packet is received from the peer. A Code-Reject packet is sent in response.

(continued)

Table 4.5 continued: PPP Events

Event Label	Event	Description
RXJ+	receive code reject permitted or receive protocol reject	This event occurs when a Code-Reject or a Protocol-Reject packet is received from the peer. The RXJ+ event arises when the rejected value is acceptable, such as a Code-Reject of an extended code, or a Protocol-Reject of an NCR. These are within the scope of normal operation. The implementation MUST stop sending the offending packet type. The RXJ– event arises when the rejected value is catastrophic, such as a Code-Reject of Configure-Request, or a Protocol-Reject of LCP! This event communicates an unrecoverable error that terminates the connection.
RXJ–	receive code reject catastrophic or receive protocol reject	
RXR	receive echo request, receive echo reply, or receive discard request	This event occurs when an Echo-Request, Echo-Reply or Discard-Request packet is received from the peer. The Echo-Reply packet is a response to an Echo-Request packet. There is no reply to an Echo-Reply or Discard-Request packet.

Table 4.6: PPP Actions[8]

Action Label	Action	Definition
tlu	this layer up	This action indicates to the upper layers that the automaton is entering the Opened state. Typically, this action is used by the LCP to signal the Up event to an NCP, Authentication Protocol, or Link Quality Protocol, or MAY be used by an NCP to indicate that the link is available for its network layer traffic.
tld	this layer down	This action indicates to the upper layers that the automaton is leaving the Opened state. Typically, this action is used by the LCP to signal the Down event to an NCP, Authentication Protocol, or Link Quality Protocol, or MAY be used by an NCP to indicate that the link is no longer available for its network layer traffic.
tls	this layer started	This action indicates to the lower layers that the automaton is entering the Starting state, and the lower layer is needed for the link. The lower layer SHOULD respond with an Up event when the lower layer is available. The results of this action are highly implementation dependent.
tlf	this layer finished	This action indicates to the lower layers that the automaton is entering the Initial, Closed or Stopped states, and the lower layer is no longer needed for the link. The lower layer SHOULD respond with a Down event when the lower layer has terminated. Typically, this action MAY be used by the LCP to advance to the Link Dead phase, or MAY be used by an NCP to indicate to the LCP that the link may terminate when there are no other NCPs open. This results of this action are highly implementation dependent.

Table 4.6 continued: PPP Actions

Action Label	Action	Definition
irc	initialize restart count	This action sets the Restart counter to the appropriate value (Max-Terminate or Max-Configure). The counter is decremented for each transmission, including the first.
zrc	zero restart count	This action sets the Restart counter to zero.
scr	send configure request	Configure-Request packet is transmitted. This indicates the desire to open a connection with a specified set of Configuration Options. The Restart timer is started when the Configure-Request packet is transmitted, to guard against packet loss. The Restart counter is decremented each time a Configure-Request is sent.
sca	send configure ack	A Configure-Ack packet is transmitted. This acknowledges the reception of a Configure-Request packet with an acceptable set of Configuration Options.
scn	send configure nak/rej	A Configure-Nak or Configure-Reject packet is transmitted, as appropriate. This negative response reports the reception of a Configure-Request packet with an unacceptable set of Configuration Options, Configure-Nak packets are used to refuse a Configuration Option value, and to suggest a new, acceptable value, Configure-Reject packets are used to refuse all negotiation about a Configuration Option, typically because it is not recognized or implemented. The use of Configure-Nak versus Configure-Reject is more fully described in the chapter on LCP Packet Formats.
str	send terminate request	A Terminate-Request packet is transmitted. This indicates the desire to close a connection. The Restart timer is started when the Terminate-Request pocket is transmitted, to guard against packet loss. The Restart counter is decremented each time a Terminate-Request is sent.
sta	send terminate ack	A Terminate-Ack packet is transmitted. This acknowledges the reception of a Terminate-Request packet or otherwise serves to synchronize the automatons.
scj	send code reject	A Code-Reject packet is transmitted. This indicates the reception of an unknown type of packet.
ser	send echo reply	An Echo-Reply packet is transmitted. This acknowledges the reception of an Echo-Request packet.

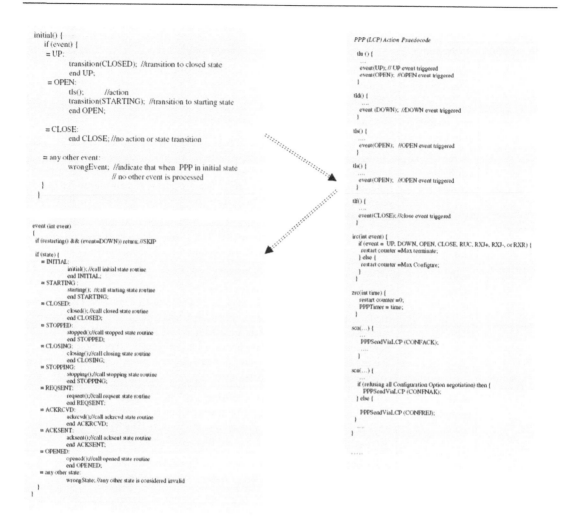

```
initial() {
    if (event() {
        = UP:
            transition(CLOSED); //transition to closed state
            end UP;
        = OPEN:
            tls();      //action
            transition(STARTING); //transition to starting state
            end OPEN;

        = CLOSE:
            end CLOSE; //no action or state transition

        = any other event:
            wrongEvent; //indicate that when PPP in initial state
                        // no other event is processed
    }
}


event (int event)
{
    if (restarting() && (event=DOWN)) return; //SKIP

    if (state) {
        = INITIAL:
            initial(); //call initial state routine
            end INITIAL;
        = STARTING :
            starting(); //call starting state routine
            end STARTING;
        = CLOSED:
            closed(); //call closed state routine
            end CLOSED;
        = STOPPED:
            stopped(); //call stopped state routine
            end STOPPED;
        = CLOSING:
            closing(); //call closing state routine
            end CLOSING;
        = STOPPING:
            stopping(); //call stopping state routine
            end STOPPING;
        = REQSENT:
            reqsent(); //call reqsent state routine
            end REQSENT;
        = ACKRCVD:
            ackrcvd(); //call ackrcvd state routine
            end ACKRCVD;
        = ACKSENT:
            acksent(); //call acksent state routine
            end ACKSENT;
        = OPENED:
            opened(); //call opened state routine
            end OPENED;
        = any other state:
            wrongState; //any other state is considered invalid
    }
}
```

```
PPP (LCP) Action Pseudocode

tlu () {
    ...
    event(UP); // UP event triggered
    event(OPEN);  //OPEN event triggered
}

tld() {
    ...
    event (DOWN); //DOWN event triggered
}

tls() {
    ...
    event(OPEN);  //OPEN event triggered
}

tls() {
    ...
    event(OPEN);  //OPEN event triggered
}

tlf() {
    ...
    event(CLOSE); //close event triggered
}

irc(int event) {
    if (event = UP, DOWN, OPEN, CLOSE, RUC, RXJ+, RXJ-, or RXR) {
        restart counter =Max terminate;
    } else {
        restart counter =Max Configure;
    }
}

zrc(int time) {
    restart counter =0;
    PPPTimer = time;
}

sca(...) {
    ...
    PPPSendViaLCP (CONFACK);
    ...
}

scn(...) {
    ...
    if (refusing all Configuration Option negotiation) then {
        PPPSendViaLCP (CONFNAK);
    } else {
        PPPSendViaLCP (CONFREJ);
    }
    ...
}
    .....
```

Figure 4.29 Initial LCP State

4.5.3 Point-to-Point LCP Pseudocode Example[5]

Initial: PPP link is in the Initial state, the lower layer is unavailable (Down), and no Open event has occurred. The Restart timer is not running in the Initial state.[8]

Starting: The Starting state is the Open counterpart to the Initial state. An administrative Open has been initiated, but the lower layer is still unavailable (Down). The Restart timer is not running in the Starting state. When the lower layer becomes available (Up), a Configure-Request is sent.[8]

```
starting() {
        if (event) {
              = UP:
                      irc(event); //action
                      scr(true); //action
                      transition(REQSENT); //transition to REQSENT state
              end UP;

              = OPEN:
              end OPEN; //no action or state transition

              = CLOSE:
                      tlf(); //action
                      transition(INITIAL); //transition to initial state
              end CLOSE;

              = any other event :
              wrongEvent++; //indicate that when PPP in starting state no other event is processed
        }
}
```

Closed: In the Closed state, the link is available (Up), but no Open has occurred. The Restart timer is not running in the Closed state. Upon reception of Configure-Request packets, a Terminate-Ack is sent. Terminate-Acks are silently discarded to avoid creating a loop.[8]

```
closed (){
        if (event) {
                = DOWN :
                        transition(INITIAL) ; //transition to initial state
                end DOWN;

                = OPEN :
                        irc(event); //action
                        scr(true); //action
                        transition(REQSENT); //transition to REQSENT state
                end OPEN;

                = RCRP, RCRN, RCA, RCN, or RTR:
                        sta(...); //action
                end EVENT;

                = RTA, RXJP, RXR, CLOSE :
                        end EVENT; //no action or state transition

                = RUC:
                        scj(...); //action
                end RUC;

                = RXJN:
                        tlf(); //action
                end RXJN;

                = any other event :
                        wrongEvent; //indicate that when PPP in closed state no other event is processed

                }
        }
```

Stopped: The Stopped state is the Open counterpart to the Closed state. It is entered when the automaton is waiting for a Down event after the This-Layer-Finished action, or after sending a Terminate-Ack. The Restart timer is not running in the Stopped state.[8]

```
stopped (){
        if (event) {
                = DOWN : tls(); //action
                        transition(STARTING) ; //transition to starting state
                end DOWN;

                = OPEN : initializeLink(); //initialize variables
                end OPEN;

                = CLOSE : transition(CLOSED) ; //transition to closed state
                end CLOSE;

                = RCRP : irc(event); //action
                        scr(true); //action
                        sca(...); //action
                        transition(ACKSENT) ; //transition to ACKSENT state
                end RCRP;

                = RCRN : irc(event); //action
                        scr(true); //action
                        scn(...); //action
                        transition(REQSENT) ; //transition to REQSENT state
                end RCRN;

                = RCA ,RCN or RTR : sta(...); //action
                end EVENT;              .

                = RTA, RXJP, or RXR :
                end EVENT;

                = RUC : scj(...); //action
                end RUC;

                = RXJN : tlf(); //action
                end RXJN;

                = any other event :
                        wrongEvent; //indicate that when PPP in stopped state no other event is processed

        }
}
```

Closing: In the Closing state, an attempt is made to terminate the connection. A Terminate-Request has been sent and the Restart timer is running, but a Terminate-Ack has not yet been received. Upon reception of a Terminate-Ack, the Closed state is entered. Upon the expiration of the Restart timer, a new Terminate-Request is transmitted, and the Restart timer is restarted. After the Restart timer has expired Max-Terminate times, the Closed state is entered.[8]

```
closing (){
            if (event) {
                        = DOWN : transition(INITIAL) ; //transition to initial state
            end DOWN;

            = OPEN : transition(STOPPING); //transition to stopping state
                        initializeLink(); //initialize variables
            end OPEN;

            = TOP : str(…); //action
                        initializePPPTimer; //initialize PPP Timer variable
            end TOP;

            = TON : tlf(); //action
                        initializePPPTimer; //initialize PPP Timer variable
                        transition(CLOSED); //transition to CLOSED state
            end TON;

            = RTR : sta(…); //action
            end RTR;

            = CLOSE, RCRP, RCRN, RCA,RCN, RXR, or RXJP:
            end EVENT; //no action or state transition

            = RTA : tlf(); //action
                        transition(CLOSED); //transition to CLOSED stsate
            end RTA;

            = RUC : scj(…); //action
            end RUC;

            = RXJN : tlf(); //action
            end RXJN;

            = any other event :
                        wrongEvent; //indicate that when PPP in closing state no other event is processed
            }
}
```

Stopping: The Stopping state is the Open counterpart to the Closing state. A Terminate-Request has been sent and the Restart timer is running, but a Terminate-Ack has not yet been received.[8]

```
stopping (){
        if (event) {
                    = DOWN : transition(STARTING) ; //transition to STARTING state
                    end DOWN;

                    = OPEN : initializeLink(); //initialize variables
                    end OPEN;

                    = CLOSE : transition(CLOSING); //transition to CLOSE state
                    end CLOSE;

                    = TOP : str(....); //action
                            initialize PPPTimer(); //initialize PPP timer
                    end TOP;

                    = TON : tlf(); //action
                            initialize PPPTimer(); //initialize PPP timer
                            transition(STOPPED); //transition to STOPPED state
                    end TON;

                    = RCRP, RCRN, RCA , RCN, RXJP, RXR : end EVENT; // no action or state transition

                    = RTR : sta(...); //action
                    end RTR;

                    = RTA : tlf(); //action
                            transition(STOPPED); //transition to STOPPED state
                    end RTA;

                    = RUC : scj(...); //action
                    end RUC;

                    = RXJN : tlf(); //action
                            transition(STOPPED); //transition to STOPPED state
                    end RXJN;

                    = any other event : wrongEvent; //indicate that when PPP in stopping state no other event is
                    //processed
        }
}
```

Request-Sent: In the Request-Sent state an attempt is made to configure the connection. A Configure-Request has been sent and the Restart timer is running, but a Configure-Ack has not yet been received nor has one been sent.[8]

```
reqsent (){
        if (event) {
                = DOWN : transition(STARTING); //transition to STARTING state
                end DOWN;

                = OPEN : transition(REQSENT); //transition to REQSENT state
                end OPEN;

                = CLOSE : irc(event); //action
                        str(...); //action
                        transition(CLOSING); //transition to closing state
                end CLOSE;

                = TOP : scr(false); //action
                        initialize PPPTimer(); //initialize PPP timer
                end TOP;

                = TON, RTA, RXJP, or RXR : end EVENT; //no action or state transition

                = RCRP : sca(...); //action
                        if (PAP = Server) {
                                tlu(); //action
                                transition(OPENED); //transition to OPENED state
                        } else { //client
                                transition(ACKSENT); //transition to ACKSENT state
                        }
                end RCRP;

                = RCRN : scn(...); //action
                end RCRN;

                = RCA : if (PAP = Server) {
                        tlu(); //action
                        transition(OPENED); //transition to OPENED state
                } else { //client
                        irc(event); //action
                        transition(ACKRCVD); //transition to ACKRCVD state
                }
                end RCA;

                = RCN : irc(event); //action
                        scr(false); //action
                        transition(REQSENT); //transition to REQSENT state
                end RCN;

                = RTR : sta(...); //action
                end RTR;

                = RUC : scj(..); //action
                break;

                = RXJN : tlf(); //action
                        transition(STOPPED); //transition to STOPPED state
                end RXJN;

                = any other event : wrongEvent; //indicate that when PPP in reqsent state no other event is
                //processed
        }
}
```

Ack-Received: In the Ack-Received state, a Configure-Request has been sent and a Configure-Ack has been received. The Restart timer is still running, since a Configure-Ack has not yet been sent.[8]

```
ackrcvd (){
        if (event) {
                        = DOWN : transition(STARTING); //transition to STARTING state
                        end DOWN;

                        = OPEN, TON, or RXR: end EVENT; //no action or state transition

                        = CLOSE : irc(event); //action
                                str(...); //action
                                transition(CLOSING); //transition to CLOSING state
                        end CLOSE;

                        = TOP : scr(false); //action
                                transition(REQSENT); //transition to REQSENT state
                        end TOP;

                        = RCRP : sca(...); //action
                                tlu(); //action
                                transition(OPENED); //transition to OPENED state
                        end RCRP;

                        = RCRN : scn(...); //action
                        end RCRN;

                        = RCA or RCN : scr(false); //action
                                transition(REQSENT); //transition to REQSENT state
                        end EVENT;

                        = RTR : sta(...); //action
                                transition(REQSENT); //transition to REQSENT state
                        end RTR;
                        = RTA or RXJP : transition(REQSENT); //transition to REQSENT state
                        end EVENT;

                        = RUC : scj(....); //action
                        end RUC;

                        = RXJN : tlf(); //action
                                transition(STOPPED); //event
                        end RXJN;

                        = any other event : wrongEvent; //indicate that when PPP in ackrcvd state no other event is
                        //processed
        }
}
```

Ack-Sent: In the Ack-Sent state, a Configure-Request and a Configure-Ack have both been sent, but a Configure-Ack has not yet been received. The Restart timer is running, since a Configure-Ack has not yet been received.[8]

```
acksent (){
          if (event) {
                    = DOWN : transition(STARTING);
                    end DOWN;

                    = OPEN, RTA, RXJP, TON, or RXR : end EVENT; //no action or state transition
                    = CLOSE : irc(event); //action
                              str(...); //action
                              transition(CLOSING); //transition to CLOSING state
                    end CLOSE;

                    = TOP : scr(false); //action
                              transition(ACKSENT); //transition to ACKSENT state
                    end TOP;

                    = RCRP : sca(...); //action
                    end RCRP;

                    = RCRN : scn(...); //action
                              transition(REQSENT); //transition to REQSENT state
                    end RCRN;

                    = RCA : irc(event); //action
                              tlu(); //action
                              transition(OPENED); //transition to OPENED state
                    end RCA;

                    = RCN : irc(event); //action
                              scr(false); //action
                              transition(ACKSENT); //transition to ACKSENT state
                    end RCN;

                    = RTR : sta(...); //action
                              transition(REQSENT); //transition to REQSENT state
                    end RTR;

                    = RUC : scj(...); //action
                    end RUC;
                    = RXJN : tlf(); //action
                              transition(STOPPED); //transition to STOPPED state
                    end RXJN;

                    = any other event : wrongEvent; //indicate that when PPP in acksent state no other event is
                    //processed
          }
}
```

Opened: In the Opened state, a Configure-Ack has been both sent and received. The Restart timer is not running. When entering the Opened state, the implementation SHOULD signal the upper layers that it is now Up. Conversely, when leaving the Opened state, the implementation SHOULD signal the upper layers that it is now Down.[8]

```
opened (){
        if (event) {
                = DOWN :
                        tld(); //action
                        transition(STARTING); //transition to STARTING state
                end DOWN;

                = OPEN : initializeLink(); //initialize variables
                end OPEN;

                = CLOSE : tld(); //action
                        irc(event); //action
                        str(...); //action
                        transition(CLOSING); //transition to CLOSING state
                end CLOSE;

                = RCRP : tld(); //action
                        scr(true); //action
                        sca(...); //action
                        transition(ACKSENT); //transition to ACKSENT state

                end RCRP;

                = RCRN : tld(); //action
                        scr(true); //action
                        scn(...); //action
                        transition(REQSENT); //transition to RCRN state
                end RCRN;

                = RCA : tld(); //action
                        scr(true); //action
                        transition(REQSENT); //transition to REQSENT state
                end RCA;

                = RCN : tld(); //action
                        scr(true); //action
                        transition(REQSENT); //transition to REQSENT state
                end RCN;

                = RTR : tld(); //action
                        zrc(PPPTimeoutTime); //action
                        sta(...); //action
                        transition(STOPPING); // transition to STOPPING state
                end RTR;

                = RTA : tld(); //action
                        scr(true); //action
                        transition(REQSENT); // transition to REQSENT state
                end RTA;
                = RUC : scj(...); //action
                end RUC;

                = RXJP : end RXJP; //no action or state transition

                = RXJN : tld(); //action
                        irc(event); //action
                        str(...); //action
                        transition(STOPPING); //transition to STOPPING state
                end RXJN;

                = RXR : ser(...); //action
                end RXR;

                = any other event : wrongEvent; //indicate that when PPP in opened state no other event is
                //processed
        }
}
```

4.5.4 Network Layer Middleware[5]

At the network layer, networks can be broken down further into *segments*, smaller sub-networks. Interconnected devices located within the same segment can communicate via their *physical* addresses. Devices located on different segments communicate via a different type of address, referred to as a *network* address. Conversions between a device's physical and network address can occur both within the higher data-link layer, as well as in a network layer protocol. Through the networking address scheme, network layer protocols typically manage:

- data transmitted at the segment level
- datagram traffic
- any routing from the current device to another device.

Like the data-link layer, if the data are meant for the device, then all network layer headers are stripped from the datagram. The remaining data field, called a **packet**, is passed up to the transport layer. If the data are not meant for the device, this layer can also act as a *router* and transmit the data back down the stack to be forwarded to another system.

These same header fields are appended to data coming down from upper layers by the network layer, and then the full network layer datagram is passed to the data-link layer for further processing (see Figure 4.30). Note that the term 'packet' is sometimes used to discuss data transmitted over a network, in general, in addition to data processed at the transport layer.

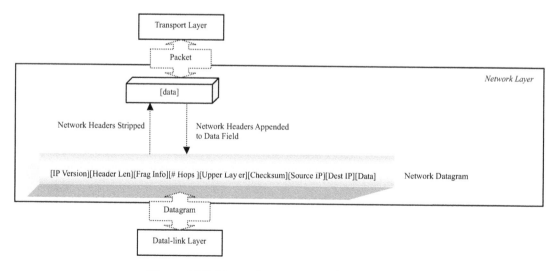

Figure 4.30: Network Layer Data-flow Diagram

4.5.5 Internet Protocol (IP) Example[5]

The networking layer protocol called the Internet Protocol, or IP, is based upon DARPA standard RFC791, and is mainly responsible for implementing addressing and fragmentation functionality (see Figure 4.31).

While the IP layer receives data as packets from upper layers and frames from lower layers, the IP layer actually views and processes data in the form of *datagrams*, whose format is shown in Figure 4.32.

The entire IP datagram is what is received by IP from lower layers. The last field alone within the datagram, the data field, is the *packet* that is sent to upper layers after processing by IP. The remaining fields are stripped or appended, depending on the direction the data are going, to the data field after IP has finished processing. It is these fields that support IP addressing and fragmentation functionality.

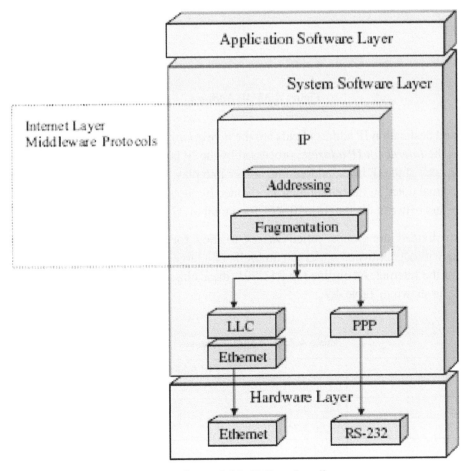

Figure 4.31: IP Functionality

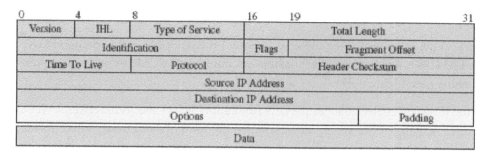

Figure 4.32: IP Datagram[9]

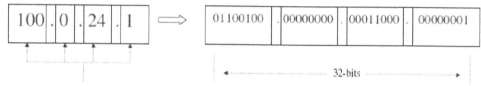

4 sets of 8-bit decimal numbers separated by "dots"

Figure 4.33: IP Address

The source and destination IP address fields are the *networking addresses*, also commonly referred to as the *Internet* or *IP address*, processed by the IP layer. In fact, it is here that one of the main purposes of the IP layer, addressing, comes into play. IP addresses are 32 bits long, in 'dotted-decimal notation', meaning they are divided by 'dots' into four octets (four 8-bit decimal numbers between the ranges of 0–255 for a total of 32 bits), as shown in Figure 4.33.

IP address are divided into groups, called *classes*, to allow for the ability of segments to all communicate without confusion under the umbrella of a larger network, such as the World-Wide-Web, or the Internet. As outlined in RFC791, these classes are organized into ranges of IP addresses, as shown in Table 4.7.

Table 4.7: IP Address Classes[9]

Class	IP Address Range	
A	0.0.0.0	127.255.255.255
B	128.0.0.0	191.255.255.255
C	192.0.0.0	223.255.255.255
D	224.0.0.0	239.255.255.255
E	244.0.0.0	255.255.255.255

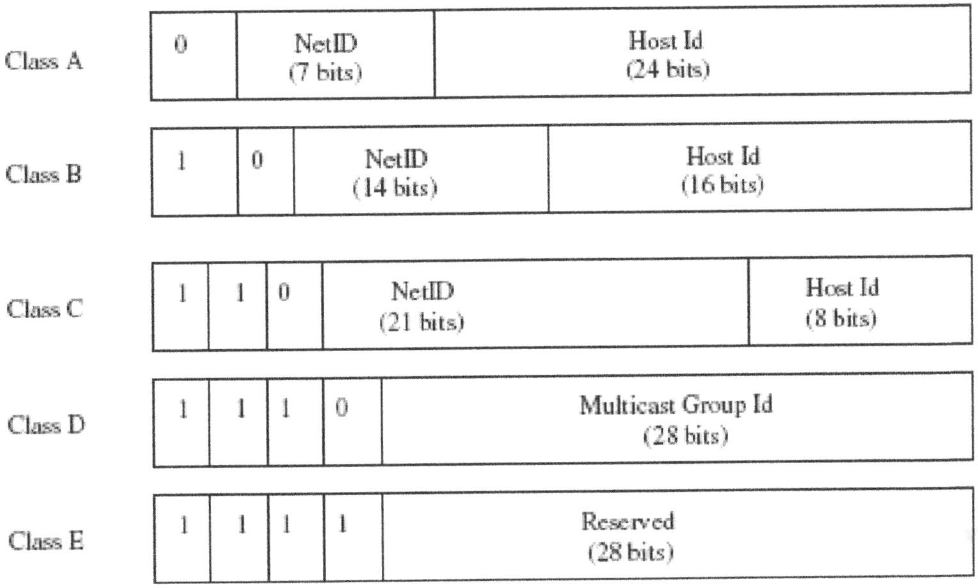

Figure 4.34: IP Classes[9]

The classes (A, B, C, D, and E) are divided according to the value of the first octet in an IP address. If the highest order bit in the octet is a '0', then the IP address is a class 'A' address. If the highest order bit is a '1', then the next bit is checked for a '0' – if it is, then it's a class 'B' address, and so on.

In classes A, B, and C, following the class bit or set of bits is the ***network id***. The network id is unique to each segment or device connected to the Internet, and is assigned by **Internet Network Information Center (InterNIC)**. The ***host id*** portion of an IP address is then left up to the administrators of the device or segment. Class D addresses are assigned for groups of networks or devices, called ***host groups***, and can be assigned by the InterNIC or the IANA (Internet Assigned Numbers Authority). As noted in Figure 4.34, Class E addresses have been reserved for future use.

4.5.6 Internet Protocol (IP) Fragmentation Mechanism[5]

Fragmentation of an IP datagram is done for devices that can only process smaller amounts of networking data at any one time. The IP procedure for fragmenting and reassembling datagrams is a design that supports unpredictability in networking transmissions. This means that IP provides support for a variable number of datagrams containing fragments of data that arrive for reassembly in an arbitrary order, and not necessarily the same order in which they were fragmented. Even fragments of differing datagrams can be handled. In the case of fragmentation, most of the fields in the first 20 bytes of a datagram, called the *header*, are used in the fragmentation and reassembling process.

The *version* field indicates the version of IP being transmitted (i.e., IPv4 is version 4). The *IHL* (internet header length) field is the length of the IP datagram's header. The **total length** field is a 16-bit field in the header which specifies the actual length in octets of the entire datagram including the header, options, padding, and data. The implication behind the size of the total length field is that a datagram can be up to 65 536 (2^{16}) octets in size.

When fragmenting a datagram, the originating device splits a datagram 'N' ways, and copies the contents of the header of the original datagram into all of the smaller datagram headers. The *Internet Identification* (ID) field is used to identify which fragments belong to which datagrams. Under the IP protocol, the data of a larger datagram must be divided into fragments, of which all but the last fragment must be some integral multiple of 8 octet blocks (64 bits) in size.

The *fragment offset* field is a 13-bit field that indicates where in the entire datagram the fragment actually belongs. Data are fragmented into subunits of up to 8192 (2^{13}) fragments of 8 octets (64 bits) each – which is consistent with the total length field being 65 536 octets in size – dividing by 8 for 8 octet groups = 8192. The fragment offset field for the first fragment would be '0', but for other fragments of the same datagram it would be equal to the total length (field) of that datagram fragment plus the number of 8 octet blocks.

The flag fields (shown in Figure 4.35) indicate whether or not a datagram is a fragment of a larger piece. The *MF* (More Fragments) flag of the *flag field* is set to indicate that the

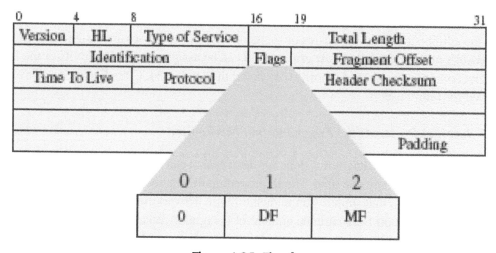

Figure 4.35: Flags[9]

fragment is the last (the end piece) of the datagram. Of course, some systems do not have the capacity to reassemble fragmented datagrams. The *DF* (Don't Fragment) flag of the *flag field* indicates whether or not a device has the resources to assemble fragmented datagrams. It is used by one device's IP layer to inform another that it doesn't have the capacity to reassemble data fragments transmitted to it. Reassembly simply involves taking datagrams with the same ID, source address, destination address, and protocol fields, and using the fragment offset field and MF flags to determine where in the datagram the fragment belongs.

The remaining fields in an IP datagram are summarized as follows:

- Time to live (which indicates the datagram's lifetime)
- Checksum (datagram integrity verification)
- Options field (provides for control functions needed or useful in some situations but unnecessary for the most common communications (i.e., provisions for timestamps, security, and special routing))
- Type of service (used to indicate the quality of the service desired. The type of service is an abstract or generalized set of parameters which characterize the service choices provided in the networks that make up the internet)
- Padding (internet header padding is used to insure that the internet header ends on a 32-bit boundary. The padding is zero)
- Protocol (indicates the next level protocol used in the data portion of the internet datagram. The values for various protocols are specified in 'Assigned Numbers' RFC790, as shown in Table 4.8).

Table 4.8: Flags[9]

Decimal	Octal	Protocol Numbers
0	0	Reserved
1	1	ICMP
2	2	Unassigned
3	3	Gateway-to-Gateway
4	4	CMCC Gateway Monitoring Message
5	5	ST
6	6	TCP
7	7	UCL
8	10	Unassigned
9	11	Secure
10	12	BBN RCC Monitoring
11	13	NVP

(continued)

Table 4.8 continued: Flags[8]

Decimal	Octal	Protocol Numbers
12	14	PUP
13	15	Pluribus
14	16	Telenet
15	17	XNET
16	20	Chaos
17	21	User Datagram
18	22	Multiplexing
19	23	DCN
20	24	TAC Monitoring
21–62	25–76	Unassigned
63	77	Any local network
64	100	SATNET and Backroom EXPAK
65	101	MIT Subnet Support
66–68	102–104	Unassigned
69	105	SATNET Monitoring
70	106	Unassigned
71	107	Internet Packet Core Utility
72–75	110–113	Unassigned
76	114	Backroom SATNET Monitoring
77	115	Unassigned
78	116	WIDEBAND Monitoring
79	117	WIDEBAND EXPAK
80–254	120–376	Unassigned
255	377	Reserved

In Figure 4.36 are open source examples for sending and receiving processing routines for a datagram at the IP layer. Lower layer protocols (i.e., PPP, Ethernet, SLIP, and so on) call some type of 'IPReceive' routine such as the 'void NutIpInput(NUTDEVICE * dev, NETBUF * nb)' in the open source snippet below to indicate to this layer to receive the datagram to disassemble. Higher layer protocols (such as TCP or UDP) call some type of 'IPSend' routine such as the 'int NutIpOutput(u_char proto, u_long dest, NETBUF * nb)' shown in the open source snippet below to transmit the datagram. Within the 'NutIpOutput' below is an example of how an IP header, like that which was shown in Figure 4.32, can be populated.

```
/*
 * Copyright (C) 2001-2007 by egnite Software GmbH. All rights reserved.
 *
 * Redistribution and use in source and binary forms, with or without
 * modification, are permitted provided that the following conditions
 * are met:
 *
 * 1. Redistributions of source code must retain the above copyright
 * notice, this list of conditions and the following disclaimer.
 * 2. Redistributions in binary form must reproduce the above copyright
 * notice, this list of conditions and the following disclaimer in the
 * documentation and/or other materials provided with the distribution.
 * 3. Neither the name of the copyright holders nor the names of
 * contributors may be used to endorse or promote products derived
 * from this software without specific prior written permission.
 *
 * THIS SOFTWARE IS PROVIDED BY EGNITE SOFTWARE GMBH AND CONTRIBUTORS
 * ``AS IS'' AND ANY EXPRESS OR IMPLIED WARRANTIES, INCLUDING, BUT NOT
 * LIMITED TO, THE IMPLIED WARRANTIES OF MERCHANTABILITY AND FITNESS
 * FOR A PARTICULAR PURPOSE ARE DISCLAIMED. IN NO EVENT SHALL EGNITE
 * SOFTWARE GMBH OR CONTRIBUTORS BE LIABLE FOR ANY DIRECT, INDIRECT,
 * INCIDENTAL, SPECIAL, EXEMPLARY, OR CONSEQUENTIAL DAMAGES (INCLUDING,
 * BUT NOT LIMITED TO, PROCUREMENT OF SUBSTITUTE GOODS OR SERVICES; LOSS
 * OF USE, DATA, OR PROFITS; OR BUSINESS INTERRUPTION) HOWEVER CAUSED
 * AND ON ANY THEORY OF LIABILITY, WHETHER IN CONTRACT, STRICT LIABILITY,
 * OR TORT (INCLUDING NEGLIGENCE OR OTHERWISE) ARISING IN ANY WAY OUT OF
 * THE USE OF THIS SOFTWARE, EVEN IF ADVISED OF THE POSSIBILITY OF
 * SUCH DAMAGE.
 *
 * For additional information see http://www.ethernut.de/
 *
 * -
 * Portions Copyright (C) 2000 David J. Hudson <dave@humbug.demon.co.uk>
 *
 * This file is distributed in the hope that it will be useful, but WITHOUT
 * ANY WARRANTY; without even the implied warranty of MERCHANTABILITY or
 * FITNESS FOR A PARTICULAR PURPOSE.
 *
 * You can redistribute this file and/or modify it under the terms of the GNU
 * General Public License (GPL) as published by the Free Software Foundation;
 * either version 2 of the License, or (at your discretion) any later version.
 * See the accompanying file "copying-gpl.txt" for more details.
 *
 * As a special exception to the GPL, permission is granted for additional
 * uses of the text contained in this file. See the accompanying file
 * "copying-liquorice.txt" for details.
 * -
 * Portions Copyright (c) 1983, 1993 by
 * The Regents of the University of California. All rights reserved.
 *
 * Redistribution and use in source and binary forms, with or without
 * modification, are permitted provided that the following conditions
 * are met:
 * 1. Redistributions of source code must retain the above copyright
 * notice, this list of conditions and the following disclaimer.
 * 2. Redistributions in binary form must reproduce the above copyright
 * notice, this list of conditions and the following disclaimer in the
 * documentation and/or other materials provided with the distribution.
 * 3. Neither the name of the University nor the names of its contributors
 * may be used to endorse or promote products derived from this software
 * without specific prior written permission.
 *
 * THIS SOFTWARE IS PROVIDED BY THE REGENTS AND CONTRIBUTORS ``AS IS'' AND
 * ANY EXPRESS OR IMPLIED WARRANTIES, INCLUDING, BUT NOT LIMITED TO, THE
 * IMPLIED WARRANTIES OF MERCHANTABILITY AND FITNESS FOR A PARTICULAR PURPOSE
 * ARE DISCLAIMED. IN NO EVENT SHALL THE REGENTS OR CONTRIBUTORS BE LIABLE
```

Figure 4.36: Open Source Example[6]

```
#include <cfg/ip.h>

#include <net/route.h>
#include <netinet/in.h>
#include <netinet/ip.h>
#include <netinet/icmp.h>
#include <netinet/ip_icmp.h>
#include <netinet/igmp.h>
#include <netinet/udp.h>
#include <sys/socket.h>
#include <arpa/inet.h>

/*!
 * \addtogroup xgIP
 */
/*@{*/

static NutIpFilterFunc NutIpFilter;

/*!
 * \brief Set filter function for incoming IP datagrams.
 *
 * The callbackFunc is called by the IP layer on every incoming IP
 * datagram. Thus it must not block. The implementer returns 0 for
 * allow, -1 for deny.
 *
 * It is recommended to set the filer after DHCP has done its thing,
 * just in case your DHCP server is on a different subnet for example.
 *
 * \param callbackFunc Pointer to callback function to filter IP packets.
 * Set to 0 to disable the filter again.
 */
void NutIpSetInputFilter(NutIpFilterFunc callbackFunc)
{
NutIpFilter = callbackFunc;
}

/*!
 * \brief Process incoming IP datagrams.
 *
 * Datagrams addressed to other destinations and datagrams
 * whose version number is not 4 are silently discarded.
```

Figure 4.36 continued: Open Source Example

```
* \note This routine is called by the Ethernet layer on
* incoming IP datagrams. Applications typically do
* not call this function.
*
* \param dev Identifies the device that received this datagram.
* \param nb The network buffer received.
*/
void NutIpInput(NUTDEVICE * dev, NETBUF * nb)
{
IPHDR *ip;
u_short ip_hdrlen;
u_long dst;
uint_fast8_t bcast;
IFNET *nif;

ip = nb->nb_nw.vp;

/*
* Silently discard datagrams of different IP version as well as
* fragmented or filtered datagrams.
*/
if (ip->ip_v != IPVERSION || /* Version check. */
(ntohs(ip->ip_off) & (IP_MF | IP_OFFMASK)) != 0 || /* Fragmentation. */
(NutIpFilter && NutIpFilter(ip->ip_src))) { /* Filter. */
NutNetBufFree(nb);
return;
}

/*
* IP header length is given in 32-bit fields. Calculate the size in
* bytes and make sure that the header we know will fit in.
*/
ip_hdrlen = ip->ip_hl * 4;
if (ip_hdrlen < sizeof(IPHDR)) {
NutNetBufFree(nb);
return;
}

/*
* No checksum calculation on incoming datagrams!
*/

/*
* Check for broadcast.
*/
dst = ip->ip_dst;
nif = dev->dev_icb;

if (dst == INADDR_BROADCAST ||
(nif->if_local_ip && nif->if_mask != INADDR_BROADCAST && (dst | nif->if_mask) == INADDR_BROADCAST)) {
bcast = 1;
}

/*
* Check for multicast.
*/
else if (IN_MULTICAST(dst)) {
MCASTENTRY *mca;

for (mca = nif->if_mcast; mca; mca = mca->mca_next) {
if (dst == mca->mca_ip) {
break;
}
}
                    if (mca == NULL) {
NutNetBufFree(nb);
                                return;
                }
bcast = 2;
}
```

Figure 4.36 continued: Open Source Example

```
/*
 * Packet is unicast.
 */
else {
bcast = 0;

#ifdef NUTIPCONF_ICMP_ARPMETHOD
/*
 * Silently discard datagrams for other destinations.
 * However, if we haven't got an IP address yet, we
 * allow ICMP datagrams to support dynamic IP ARP method,
 * if this option had been enabled.
 */
if (nif->if_local_ip == 0 && ip->ip_p == IPPROTO_ICMP && (dst & 0xff000000) != 0xff000000 && (dst & 0xff000000) != 0) {
NutNetIfSetup(dev, dst, 0, 0);
}
#endif
if (nif->if_local_ip && (dst == 0 || dst != nif->if_local_ip)) {
NutNetBufFree(nb);
return;
}
}

nb->nb_nw.sz = ip_hdrlen;
nb->nb_tp.vp = ((u_char *) ip) + (ip_hdrlen);
nb->nb_tp.sz = htons(ip->ip_len) - (ip_hdrlen);

switch (ip->ip_p) {
case IPPROTO_ICMP:
NutIcmpInput(dev, nb);
break;
case IPPROTO_UDP:
NutUdpInput(nb, bcast);
break;
case IPPROTO_TCP:
/*
 * Silently discard TCP broadcasts.
 */
if (bcast)
NutNetBufFree(nb);
else
NutTcpInput(nb);
break;
case IPPROTO_IGMP:
NutIgmpInput(dev, nb);
break;
default:
/* Unkown protocol, send ICMP destination (protocol)
 * unreachable message.
 */
if (bcast || !NutIcmpResponse(ICMP_UNREACH, ICMP_UNREACH_PROTOCOL, 0, nb))
NutNetBufFree(nb);
break;
}
}

/*!
 * \brief Send IP datagram.
 *
 * Route an IP datagram to the proper interface.
 *
 * The function will not return until the data has been stored
 * in the network device hardware for transmission. If the
 * device is not ready for transmitting a new packet, the
 * calling thread will be suspended until the device becomes
 * ready again. If the hardware address of the target host needs
 * to be resolved the function will be suspended too.
 *
 * \param proto Protocol type.
 * \param dest Destination IP address. The function will determine
 * the proper network interface by checking the routing
```

Figure 4.36 continued: Open Source Example

```
* table. It will also perform any neccessary hardware
* address resolution.
* \param nb Network buffer structure containing the datagram.
* This buffer will be released if the function returns
* an error.
*
* \return 0 on success, -1 otherwise.
*
* \bug Broadcasts to multiple network devices will fail after the
* first device returns an error.
*/
int NutIpOutput(u_char proto, u_long dest, NETBUF * nb)
{
u_char ha[6];
IPHDR *ip;
NUTDEVICE *dev;
IFNET *nif;
u_long gate;
int rc;

if ((nb = NutNetBufAlloc(nb, NBAF_NETWORK, sizeof(IPHDR))) == 0)
return -1;

/*
* Set those items in the IP header, which are common for
* all interfaces.
*/
ip = nb->nb_nw.vp;
ip->ip_v = 4;
ip->ip_hl = sizeof(IPHDR) / 4;
ip->ip_tos = 0;
ip->ip_len = htons(nb->nb_nw.sz + nb->nb_tp.sz + nb->nb_ap.sz);
ip->ip_off = 0;
if (proto == IPPROTO_IGMP) {
ip->ip_ttl = 1;
} else {
ip->ip_ttl = 0x40;
}
ip->ip_p = proto;
ip->ip_dst = dest;

/*
* Broadcasts are sent on all network interfaces.
*/
if (dest == 0xffffffff) {

memset(ha, 0xff, sizeof(ha));

for (dev = nutDeviceList, rc = 0; dev && rc == 0; dev = dev->dev_next) {
if (dev->dev_type == IFTYP_NET) {

/*
* Set remaining IP header items and calculate the checksum.
*/
nif = dev->dev_icb;
ip->ip_id = htons(nif->if_pkt_id++);
ip->ip_src = nif->if_local_ip;
ip->ip_sum = 0;
ip->ip_sum = NutIpChkSum(0, nb->nb_nw.vp, nb->nb_nw.sz);
/*
* TODO: We must clone the NETBUF!!!
*/
if (nif->if_type == IFT_ETHER)
rc = (*nif->if_output) (dev, ETHERTYPE_IP, ha, nb);
else
rc = (*nif->if_output) (dev, PPP_IP, 0, nb);
}
}
return rc;
}
```

Figure 4.36 continued: Open Source Example

```
/*
 * Get destination's route. This will also return the proper
 * interface.
 */
if ((dev = NutIpRouteQuery(dest, &gate)) == 0) {
NutNetBufFree(nb);
return -1;
}
```

```
/*
 * Set remaining IP header items and calculate the checksum.
 */
nif = dev->dev_icb;
ip->ip_id = htons(nif->if_pkt_id++);
ip->ip_src = nif->if_local_ip;
ip->ip_sum = 0;
ip->ip_sum = NutIpChkSum(0, nb->nb_nw.vp, nb->nb_nw.sz);
```

```
/*
 * On Ethernet we query the MAC address of our next hop,
 * which might be the destination or the gateway to this
 * destination.
 */
if (nif->if_type == IFT_ETHER) {
/*
 * Detect directed broadcasts for the local network. In this
 * case don't send ARP queries, but send directly to MAC broadcast
 * address.
 */
if ((gate == 0) && ((dest | nif->if_mask) == 0xffffffff)) {
memset(ha, 0xff, sizeof(ha));
} else if (NutArpCacheQuery(dev, gate ? gate : dest, ha)) {
/* Note, that a failed ARP request is not considered a
transmission error. It might be caused by a simple
packet loss. */
return 0;
}
return (*nif->if_output) (dev, ETHERTYPE_IP, ha, nb);
} else if (nif->if_type == IFT_PPP)
return (*nif->if_output) (dev, PPP_IP, 0, nb);

NutNetBufFree(nb);
return -1;
}
```

Figure 4.36 continued: Open Source Example

4.5.7 Transport Layer Middleware

Transport layer protocols (see Figure 4.37) are typically responsible for *point-to-point* communication, which means this code is managing, establishing, and closing communication between two specific networked devices. Essentially, this layer is what allows multiple networking applications that reside above the transport layer

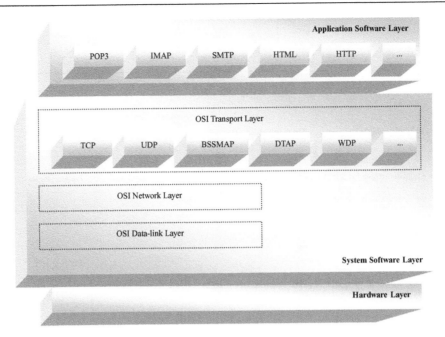

Figure 4.37: Transport Middleware Layer Protocols

to establish client–server, point-to-point communication links to another device via functionality such as:

- flow control that insures packets are transmitted and received at a supportable rate
- insuring packets transmitted have been received and assembled in the correct order
- providing acknowledgments to transmitter upon reception of error-free packet
- requesting re-transmission to transmitter upon reception of defective packet.

As shown in Figure 4.38, generally, data received from the underlying network layer are stripped of the transport header and processed, then transmitted as messages to upper layers. When a transport layer receives a message from an upper layer, the message is processed and a transport header appended to the message before being passed down to underlying layers for further processing for transmission.

The core communication mechanism used when establishing and managing communication between two devices at the transport layer is called a *socket*. Basically, any device that wants to establish a transport layer connection to another device must do so via a socket. So, there is a socket on either end of the point-to-point communication channel for two devices to transmit and receive data. There are several different types of sockets, such as raw, datagram, stream, and sequenced packet for example, depending on the transport layer protocol.

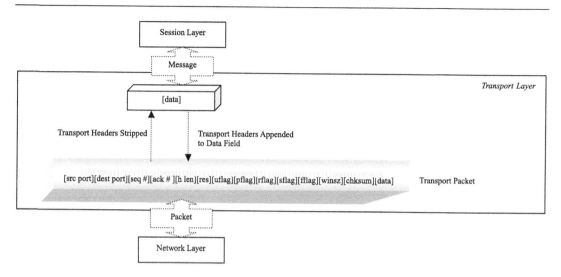

Figure 4.38: Transport Layer Data-flow Diagram

Because one transport layer can manage multiple overlying applications, sockets are bound to *ports* with unique port numbers that have been assigned to each application either by default via industry standard or by the developer. For example, an FTP client being assigned ports 20 or 21, an email/SMTP client being assigned port 23, and an HTTP client being assigned port 80 to name a few. Each device has ports '0' through '65535' available for use, because ports are defined as 16-bit unsigned integers.

As shown in Figure 4.39, in general, transport layer handshaking involves the server waiting for a client-side application to initiate a connection by 'listening' to the relative transport layer socket. Incoming data to the server socket are processed and the IP address, as well as port number, is utilized to determine if the received packet is addressed to an overlying application on the server. Given a successful connection to a client for communication, the server then establishes another independent socket to continue 'listening' for other clients.

4.5.8 Transport Layer Example[5]: User Datagram Protocol (UDP) versus Transmission Control Protocol (TCP)

RFC793 – Transmission Control Protocol (TCP) and RFC768 – User Datagram Protocol (UDP) are two of the more common transport layer (middleware) protocols implemented within an embedded system residing over the networking layer protocol IP (internet protocol). Figure 4.40 is an open source example of UDP functions that utilize lower IP and ICMP middleware layer software.

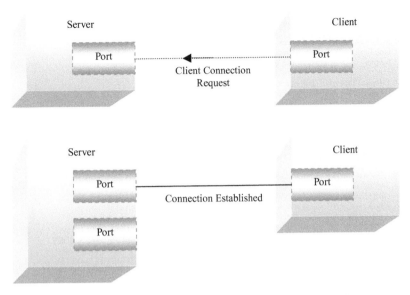

Figure 4.39: Transport Layer Client–Server Handshaking

UDP establishes and dissolves point-to-point *unreliable* connections via a *datagram* socket. This means that the UDP protocol does not provide acknowledgment functionality relative to a UDP packet (see Figure 4.41), and overlying software layers are responsible for managing reliability of transmitted data.

TCP, on the other hand, establishes and dissolves point-to-point *reliable* connections via a datagram socket. Like UDP, TCP transfers and receives data packaged as *segments*, via a socket handling scheme that handles data one message segment at a time. However, TCP provides an acknowledgment at the core of its handshaking scheme and uses a packet structure that differs from UDP (see Figure 4.42).

In addition to the actual data, both UDP and TCP headers contain source and destination port number fields. Both UDP and TCP headers also contain a checksum field to allow both protocols to help insure that data were transmitted without errors. As shown in Table 4.9, TCP headers then provide additional fields to support the additional functionality relative to reliability and handshaking provided by TCP over UDP.

Events are triggered by data within sender and receiver packets, such as user calls (i.e., OPEN, SEND, RECEIVE, CLOSE, ABORT, and STATUS), incoming segments and their relative flags in the case of TCP (SYN, ACK, RST and FIN), and/or timeouts to name a few.

```
/*
 * Copyright I 2001-2003 by egnite Software GmbH. All rights reserved.
 *
 * Redistribution and use in source and binary forms, with or without
 * modification, are permitted provided that the following conditions
 * are met:
 *
 * 1. Redistributions of source code must retain the above copyright
 * notice, this list of conditions and the following disclaimer.
 * 2. Redistributions in binary form must reproduce the above copyright
 * notice, this list of conditions and the following disclaimer in the
 * documentation and/or other materials provided with the distribution.
 * 3. Neither the name of the copyright holders nor the names of
 * contributors may be used to endorse or promote products derived
 * from this software without specific prior written permission.
 *
 * THIS SOFTWARE IS PROVIDED BY EGNITE SOFTWARE GMBH AND CONTRIBUTORS
 * ``AS IS'' AND ANY EXPRESS OR IMPLIED WARRAN TIES, INCLUDING, BUT NOT
 * LIMITED TO, THE IMPLIED WARRANTIES OF MERCHANTABILITY AND FITNESS
 * FOR A PARTICULAR PURPOSE ARE DISCLAIMED. IN NO EVENT SHALL EGNITE
 * SOFTWARE GMBH OR CONTRIBUTORS BE LIABLE FOR ANY DIRECT, INDIRECT,
 * INCIDENTAL, SPECIAL, EXEMPLARY, OR CONSEQUENTIAL DAMAGES (INCLUDING,
 * BUT NOT LIMITED TO, PROCUREMENT OF SUBSTITUTE GOODS OR SERVICES; LOSS
 * OF USE, DATA, OR PROFITS; OR BUSINESS INTERRUPTION) HOWEVER CAUSED
 * AND ON ANY THEORY OF LIABILITY, WHETHER IN CONTRACT, STRICT LI ABILITY,
 * OR TORT (INCLUDING NEGLIGENCE OR OTHERWISE) ARISING IN ANY WAY OUT OF
 * THE USE OF THIS SOFTWARE, EVEN IF ADVISED OF THE POSSIBILITY OF
 * SUCH DAMAGE.
 *
 * For additional information see http://www.ethernut.de/
 *
 * _
 * Portions Copyright I 2000 David J. Hudson <dave@humbug.demon.co.uk>
 *
 * This file is distributed in the hope that it will be useful, but WITHOUT
 * ANY WARRANTY; without even the implied warranty of MERCHANTABILITY or
 * FITNESS FOR A PARTICULAR PURPOSE.
 *
 * You can redistribute this file and/or modify it under the terms of the GNU
 * General Public License (GPL) as published by the Free Software Foundation;
 * either version 2 of the License, or (at your discretion) any later version.
 * See the accompanying file "copying-gpl.txt" for more details.
 *
 * As a special exception to the GPL, permission is granted for additional
 * uses of the text contained in this file. See the accompanying file
 * "copying-liquorice.txt" for details.
 * _
 * Portions Copyright I 1983, 1993 by
 * The Regents of the University of California. All rights reserved.
 *
 * Redistribution and use in source and binary forms, with or without
 * modification, are permitted provided that the following conditions
 * are met:
 * 1. Redistributions of source code must retain the above copyright
 * notice, this list of conditions and the following disclaimer.
 * 2. Redistributions in binary form must reproduce the above copyright
 * notice, this list of conditions and the following disclaimer in the
 * documentation and/or other materials provided with the distribution.
 * 3. Neither the name of the University nor the names of its contributors
 * may be used to endorse or promote products derived from this software
 * without specific prior written permission.
 *
 * THIS SOFTWARE IS PROVIDED BY THE REGENTS AND CONTRIBUTORS ``AS IS '' AND
 * ANY EXPRESS OR IMPLIED WARRANTIES, INCLUDING, BUT NOT LIMITED TO, THE
 * IMPLIED WARRANTIES OF MERCHANTABILITY AND FITNESS FOR A PARTI CULAR PURPOSE
 * ARE DISCLAIMED. IN NO EVENT SHALL THE REGENTS OR CONTRIBUTORS BE LIABLE
```

Figure 4.40: UDP Open Source Example[13]

```
* FOR ANY DIRECT, INDIRECT, INCIDENTAL, SPECIAL, EXEMPLARY, OR CONSEQUENTIAL
* DAMAGES (INCLUDING, BUT NOT LIMITED TO, PROCUREMENT OF SUBSTITUTE GOODS
* OR SERVICES; LOSS OF USE, DATA, OR PROFITS; OR BUSINESS INTERRUPTION)
* HOWEVER CAUSED AND ON ANY THEORY OF LIABILITY, WHETHER IN CONTRACT, STRICT
* LIABILITY, OR TORT (INCLUDING NEGLIGENCE OR OTHERWISE) ARISING IN ANY WAY
* OUT OF THE USE OF THIS SOFTWARE, EVEN I F ADVISED OF THE POSSIBILITY OF
* SUCH DAMAGE.
* -
* Portions Copyright I 1993 by Digital Equipment Corporation.
*
* Permission to use, copy, modify, and distribute this software for any
* purpose with or without fee is hereby granted, provided that the above
* copyright notice and this permission notice appear in all copies, and that
* the name of Digital Equipment Corporation not be used in advertising or
* publicity pertaining to distribution of the document or software without
* specific, written prior permission.
*
* THE SOFTWARE IS PROVIDED  "AS IS" AND DIGITAL EQUIPMENT CORP. DISCLAIMS ALL
* WARRANTIES WITH REGARD TO THIS SOFTWARE, INCLUDING ALL IMPLIED WARRANTIES
* OF MERCHANTABILITY AND FITNESS.  IN NO EVENT SHALL DIGITAL EQUIPMENT
* CORPORATION BE LIABLE FOR ANY SPECIAL, DIRECT, INDIRECT, OR CONSEQUENTIAL
* DAMAGES OR ANY DAMAGES WHATSOEVER RESULTING FROM LOSS OF USE, DATA OR
* PROFITS, WHETHER IN AN ACTION OF CONTRACT, NEGLIGENCE OR OTHER TORTIOUS
* ACTION, ARISING OUT OF OR IN CON NECTION WITH THE USE OR PERFORMANCE OF THIS
* SOFTWARE.
*/

/*
* $Log: udpin.c,v $
* Revision 1.7 2008/04/18 13:13:11 haraldkipp
* Using fast ints.
*
* Revision 1.6 2006/10/08 16:48:22 haraldkipp
* Documentation fixed
*
* Revision 1.5 2005/06/05 16:48:32 haraldkipp
* Additional parameter enables NutUdpInput() to avoid responding to UDP
* broadcasts with ICMP unreachable messages. Fixes bug #1215192.
*
* Revision 1.4 2005/05/26 11:47:24 drsung
* ICMP unreachable will be sent on incoming udp packets with no local peer port.
*
* Revision 1.3 2005/02/02 16:22:35 haraldkipp
* Do not wake up waiting threads if the incoming datagram
* doesn't fit in the buffer.
*
* Revision 1.2 2003/11/24 21:01:04 drsung
* Packet queue added for UDP sockets.
*
* Revision 1.1.1.1 2003/05/09 14:41:45 haraldkipp
* Initial using 3.2.1
*
* Revision 1.10 2003/02/04 18:14:57 harald
* Version 3 released
*
* Revision 1.9 2002/06/26 17:29:36 harald
* First pre-release with 2.4 stack
*
*/

#include <sys/event.h>

#include <netinet/udp.h>
#include <sys/socket.h>
#include <netinet/ip_icmp.h>
#include <netinet/icmp.h>

/*!
 * \addtogroup xgUDP
 */
/*@{*/
```

Figure 4.40 continued: UDP Open Source Example

```
/*!
 * \brief Handle incoming UDP packets.
 *
 * \note This routine is called by the IP layer on
 * incoming UDP packets. Applications typically do
 * not call this function.
 *
 * \param nb Network buffer structure containing the UDP packet.
 * \param bcast Broadcast flag.
 */
/* @@@ 2003-10-24: modified by OS for udp packet queue */
void NutUdpInput(NETBUF * nb, uint_fast8_t bcast)
{
UDPHDR *uh;
UDPSOCKET *sock;

uh = (UDPHDR *) nb->nb_tp.vp;

nb->nb_ap.vp = uh + 1;
nb->nb_ap.sz = nb->nb_tp.sz – sizeof(UDPHDR);
nb->nb_tp.sz = sizeof(UDPHDR);

/*
 * Find a port. If none exists and if this datagram hasn't been
 * broadcasted, return an ICMP unreachable.
 */
if ((sock = NutUdpFindSocket(uh->uh_dport)) == 0) {
if (bcast || NutIcmpResponse(ICMP_UNREACH, ICMP_UNREACH_PORT, 0, nb) == 0) {
        NutNetBufFree(nb);
}
return;
}

/* if buffer size is defined, use packet queue */
if (sock->so_rx_bsz) {
/* New packet fits into the buffer? */
if (sock->so_rx_cnt + nb->nb_ap.sz > sock->so_rx_bsz) {
/* No, so discard it */
NutNetBufFree(nb);
return;
} else {
/* if a first packet is already in the queue, find the end
        —    and add the new packet */
if (sock->so_rx_nb) {
NETBUF *snb;
for (snb = sock->so_rx_nb; snb->nb_next != 0; snb = snb->nb_next);
snb->nb_next = nb;
} else
sock->so_rx_nb = nb;

/* increment input buffer count */
sock->so_rx_cnt += nb->nb_ap.sz;
};
} else { /* no packet queue */
/* if a packet is still buffered, discard it */
if (sock->so_rx_nb) {
NutNetBufFree(sock->so_rx_nb);
}
sock->so_rx_nb = nb;
sock->so_rx_cnt = nb->nb_ap.sz; /* set input buffer count to size of new packet */
};

/* post the event only, if one thread is waiting */
if (sock->so_rx_rdy)
NutEventPost(&sock->so_rx_rdy);
}

/*!
 * \brief Send a UDP packet.
```

Figure 4.40 continued: UDP Open Source Example

```
*
* \param sock Socket descriptor. This pointer must have been
* retrieved by calling NutUdpCreateSocket().
* \param daddr IP address of the remote host in network byte order.
* \param port Remote port number in host byte order.
* \param nb Network buffer structure containing the datagram.
* This buffer will be released if the function returns
* an error.
*
* \note Applications typically do not call this function but
* use the UDP socket interface.
*
* \return 0 on success, -1 otherwise.
*/
int NutUdpOutput(UDPSOCKET * sock, u_long daddr, u_short port, NETBUF * nb)
{
u_long saddr;
u_long csum;
UDPHDR *uh;
NUTDEVICE *dev;
IFNET *nif;

if ((nb = NutNetBufAlloc(nb, NBAF_TRANSPORT, sizeof(UDPHDR))) == 0)
return -1;

uh = nb->nb_tp.vp;
uh->uh_sport = sock->so_local_port;
uh->uh_dport = htons(port);
uh->uh_ulen = htons((u_short)(nb->nb_tp.sz + nb->nb_ap.sz));
```

```
/*
* Get local address for this destination.
*/
if ((dev = NutIpRouteQuery(daddr, &saddr)) != 0) {
nif = dev->dev_icb;
saddr = nif->if_local_ip;
} else
saddr = 0;

uh->uh_sum = 0;
csum = NutIpPseudoChkSumPartial(saddr, daddr, IPPROTO_UDP, uh->uh_ulen);
csum = NutIpChkSumPartial(csum, uh, sizeof(UDPHDR));
uh->uh_sum = NutIpChkSum(csum, nb->nb_ap.vp, nb->nb_ap.sz);

return NutIpOutput(IPPROTO_UDP, daddr, nb);
}
```

Figure 4.40 continued: UDP Open Source Example

UDP and TCP connections then progress from one state to another depending on these events, for example under TCP:

- **LISTEN**, waiting for a connection request
- **ESTABLISHED**, normal and open connection in which data can be received
- **SYN-SENT/SYN-RECEIVED**, synchronize connections reception/transmission of data

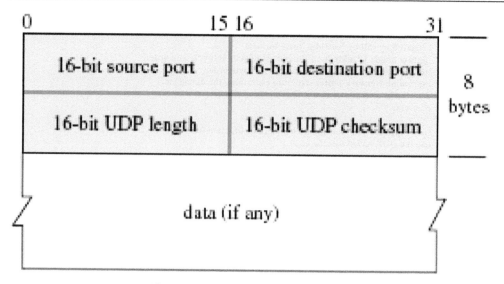

Figure 4.41: UDP Packet Diagram[10]

- **CLOSED**, no connection
- **CLOSING**, waits for a connection termination request acknowledgment
- **CLOSE-WAIT**, waiting for a connection termination request
- **TIME-WAIT**, handshaking delay to allow time for remote connection to process
- **LAST-ACK**, waiting for an acknowledgment of connection termination request

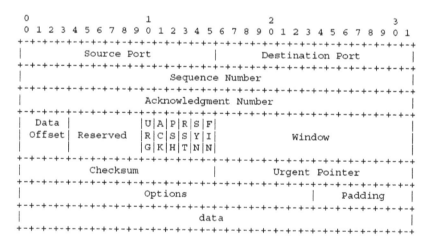

Figure 4.42: TCP Packet Diagram[10]

Table 4.9: Additional TCP Header Fields[5]

TCP Header Field	Description
Acknowledgment Number	TCP handshaking requires that when a TCP connection is established, and acknowledgment is always sent. When an ACK control bit is set, the Acknowledgment Number is the value of the next sequence number the sender of the segment is expecting to receive
Control Bits	
URG	URG: Urgent Pointer field significant
ACK	Acknowledgment field significant
PSH	Push Function
RST	Reset the connection
SYN	Synchronize sequence numbers
FIN	No more data from sender
Data Offset	Contains the location of where data is located within the TCP message segment, after the TCP header
Options	Additional TCP options
End of Option List	Indicates the end of an options list
Maximum Segment Size	Maximum Segment Size
Maximum Segment Size Option Data	This field contains the maximum receive segment size at the TCP which sends this segment
No-Operation	Miscellanous use in options list
Padding	Zeros used to ensure that the TCP header ends, and data start on a 32-bit boundary
Reserved	0 (Reserved)
Sequence Number	When SYN is not present, this field contains the first data octet. Otherwise, this field contains the initial sequence number (ISN) and the first data octet is ISN+1
Urgent Pointer	When the URG control bit is set, this field contains the current value of the urgent pointer which points to the sequence number of the octet following the urgent data
Window	The amount of data the sender of the segment can accept

- **FIN-WAIT-1**, waiting for an acknowledgment or termination request from remote connection
- **FIN-WAIT-2**, waiting for termination request from remote connection.

So as shown in the high-level diagram in Figure 4.43, the handshaking scheme under TCP is based upon connections communicating via these states. The current states are defined by events contained within the content of the transmitted packets.

TCP SYSTEM I.   **TCP SYSTEM II.**

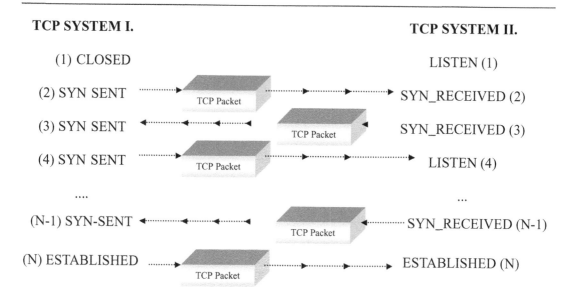

Figure 4.43: High-level TCP States and Handshaking Diagram[5]

4.6 Step 5 Putting it All Together: Tuning the Networking Stack and the Application Requirements

It is important for middleware developers to understand the overall networking requirements of their device and tune networking parameters at all layers of software to real-world performance needs accordingly. Even if the networking components are included as part of a bundle purchased from an off-the-shelf embedded operating system vendor, middleware programmers should not ever *assume* it is configured for their own production-ready requirements. For example, developers that use vxWorks have the option of purchasing an additional tightly networking stack with vxWorks. Access to networking parameters (examples shown in Table 4.10) are provided via the development environment and source code to developers, so that these components can be tuned to the requirements of the device and how it must perform within a network.

So, given the TCP/IP stack parameters shown in Table 4.10 and tuning these – an example to middleware developers is the TCP_MSS_DFLT parameter, which is the TCP Maximum Segment Size (MSS) that can be tuned by analyzing both IP fragmentation as well as managing overhead. The underlying IP stack needs to be considered because TCP segments are repackaged into IP datagrams when data flow down the stack. Thus, the size limitations of the IP datagrams must be taken into account. This is because fragmentation will occur at the IP layer if the TCP segment is too big, resulting in a degradation of performance because more than one datagram must be transmitted at the IP layer for the TCP segment data to be managed successfully.

Table 4.10: Tuning Parameters for Networking Components in vxWorks[12]

Networking Component	Parameter	Description	Value
TCP	TCP_CON_TIMEO_DFLT	Timeout intervals to connect (default 150 = 75 s)	150
	TCP_FLAGS_DFLT	Default value of the TCP flags	(TCP_DO_RFC1323)
	TCP_IDLE_TIMEO_DFLT	Seconds without data before dropping connection	14400
	TCP_MAX_PROBE_DFLT	Number of probes before dropping connection (default 8)	8
	TCP_MSL_CFG	TCP Maximum Segment Lifetime in seconds	30
	TCP_MSS_DFLT	Initial number of bytes for a segment (default 512)	512
	TCP_RAND_FUNC	A random function to use in tcp_init	(FUNCPTR)random
	TCP_RCV_SIZE_DFLT	Number of bytes for incoming TCP data (8192 by default)	8192
	TCP_REXMT_THLD_DFLT	Number of retransmit attempts before error (default 3)	3
	TCP_RND_TRIP_DFLT	Initial value for round-trip-time, in seconds	3
	TCP_SND_SIZE_DFLT	Number of bytes for outgoing TCP data (8192 by default)	8192
UDP	UDP_FLAGS_DFLT	Optional UDP features: default enables checksums	(UDP_DO_CKSUM_SND | UDP_DO_CKSUM_RCV)
	UDP_RCV_SIZE_DFLT	Number of bytes for incoming UDP data (default 41600)	41600
	UDP_SND_SIZE_DFLT	Number of bytes for outgoing UDP data (9216 by default)	9216
	IP_FLAGS_DFLT	Selects otional features of IP layer	(IP_DO_FORWARDING | IP_DO_REDIRECT | IP_DO_CHECKSUM_SND | IP_DO_CHECKSUM_RCV)
IP	IP_FRAG_TTL_DFLT	Number of slow timeouts (2 per second)	60
	IP_QLEN_DFLT	Number of packets stored by receiver	50
	IP_TTL_DFLT	Default TTL value for IP packets	64
	IP_MAX_UNITS	Maximum number of interfaces attached to IP layer	4

Managing the overhead means developers must take into account the TCP and IP headers that are not part of the data being transmitted but must be transmitted along with the data for processing by connected devices. Balancing means doing the full analysis, meaning recognizing that a maximum segment size (MSS) that is lower would reduce fragmentation, but could prove inefficient due to the overhead if it is too low.

Another example for middleware developers relative to tuning for requirements and performance is the TCP window sizes. Under the vxWorks example, the provided TCP/IP implementation includes the TCP socket that receives and sends buffer sizes managed by parameters TCP_RCV_SIZE_DFLT and TCP_SND_SIZE_DFLT. Socket window size is used by TCP to inform connections how much data can be managed at any given time by its sockets. For networking mediums that may require higher window sizes, such as satellite or ATM communication, these values can be tuned accordingly in the project source files. In this example when using this real-world networking stack with vxWorks, the general rules recommended are that these socket buffer sizes should be an even multiple of the maximum segment size (MSS), and three or more times the MSS value. To target networking performance goals, these buffer sizes need to accommodate the *Bandwidth (bytes per second)* × *Round Trip Time (seconds)*.

4.6.1 The Application Requirements

As shown in Figure 4.21 with the OSI model, networking protocols at the application, presentation, and session layers are the protocols that utilize any networking middleware that resides within an embedded device. From the viewpoint of the OSI model, network communication to another device is initiated via the application layer via end-users of the device or end-user network applications. These network applications contain the relevant networking protocols to 'virtually' connect to the networking applications residing in the connected device (see Figure 4.44).

The 'virtual' connection between two networking applications is referred to as a *session*. A *session layer* protocol manages all communication associated with each particular session, such as:

- assigning a port number to each session
- separating and managing the data of independent sessions
- data flow regulation
- error handling
- security for the applications connected.

As shown in Figure 4.45, a message/packet received from the underlying transport layer is stripped of the session layer header for processing, and the remaining data field is transmitted

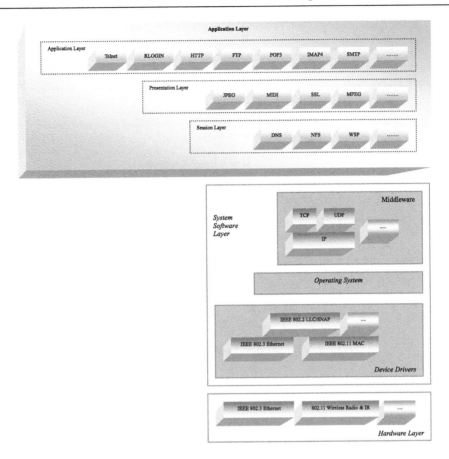

Figure 4.44: Application, Session, and Presentation Layer Protocols

up to the presentation layer protocol. Messages coming down from the presentation layer are processed and appended with a session layer header before being passed down to an underlying layer.

Data coming down from the application layer that requires translation into a generic format for transmission and/or data transmitted from other that requires translation is done via *presentation* layer protocols. In general, this includes data on:

- compression
- decompression
- encryption
- decryption
- protocol conversions
- character conversions.

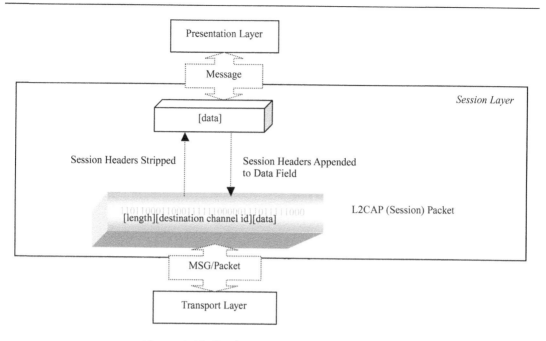

Figure 4.45: Session Layer Data-flow Diagram

In short, data received from the overlying application layer or underlying session layer are translated as required. If data have come from an underlying layer, the presentation layer header is stripped from the data intended for the application layer before being processed and transmitted up the stack. For data coming down from the application layer, after any translation of the data has been completed, a presentation layer header is appended to the data before being transmitted down the stack to the underlying networking protocol (see Figure 4.46).

These higher layer networking protocols can then be implemented as standalone applications with the only responsibility being that of the particular protocol, or within a larger, more-complex device application – as shown with the FTP (File Transfer Protocol) client, SMTP (Simple Mail Transfer Protocol), and Hypertext Transfer Protocol (HTTP) high-level diagram in Figure 4.47.

4.6.2 File Transfer Protocol (FTP) Client Application Example

RFC959, File Transfer Protocol (FTP), is one of the simpler and more common protocols implemented within an embedded system that is used to securely exchange files over a network. The FTP protocol is based on a communication model in which there is an FTP

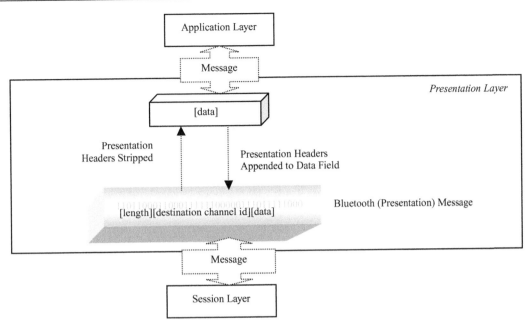

Figure 4.46: Presentation Layer Data-flow Diagram

client, also referred to as a user-protocol interpreter (user PI) that initiates a file transfer, and an FTP *server* or FTP *site* that manages and receives FTP connections. As shown in Figure 4.48, the types of connections that exist between an FTP client and server are:

- *control* connections, which are connections in which commands are transmitted over
- *data* connections, which are connections in which files are transmitted over.

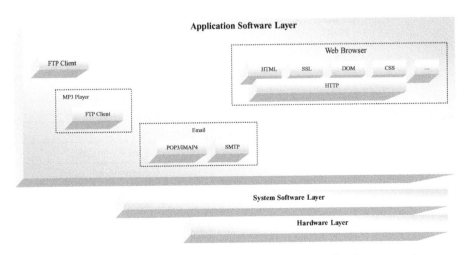

Figure 4.47: FTP, SMTP, and HTTP High-level Application Example

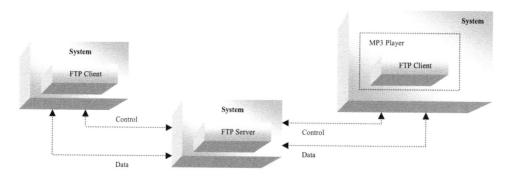

Figure 4.48: FTP Network

FTP clients start FTP sessions by initiating a control connection to a destination system with an FTP server. This FTP control connection is based on a TCP connection to port 21, because FTP requires an underlying transport layer protocol that is a reliable, ordered data stream channel. When FTP client and server communicate over a control connection, they do so via the interchange of commands and reply codes, such as some of the codes shown in Table 4.11.

Figure 4.49 is an open source example of FTP functions, and how this source code utilizes a required underlying networking middleware layer such as TCP socket-related function calls.

Table 4.11: Examples of FTP Commands and Reply Codes[1]

Type	Code	Definition
Command	DELE	Delete. FTP service command
	MODE	Transfer Mode. Transfer parameter command
	PASS	Password. Access control command
	PORT	Data Port. Transfer parameter command
	QUIT	Logout. Access control command
	TYPE	Representation Type. Transfer parameter command
	USER	Username. Access control command
Reply Code	110	Restart marker reply
	120	Service ready in 'x' minutes
	125	Data connection already open
	150	File status OK
	200	Command OK
	202	Command NOT implemented
	211	System Help

```
/*
 * Copyright I 2005 by egnite Software GmbH. All rights reserved.
 *
 * Redistribution and use in source and binary forms, with or without
 * modification, are permitted provided that the following conditions
 * are met:
 *
 * 1. Redistributions of source code must retain the above copyright
 * notice, this list of conditions and the following disclaimer.
 * 2. Redistributions in binary form must reproduce the above copyright
 * notice, this list of conditions and the following disclaimer in the
 * documentation and/or other materials provided with the distribution.
 * 3. Neither the name of the copyright holders nor the names of
 * contributors may be used to endorse or promote products derived
 * from this software without specific prior written permission.
 *
 * THIS SOFTWARE IS PROVIDED BY EGNITE SOFTWARE GMBH AND CONT RIBUTORS
 * ``AS IS" AND ANY EXPRESS OR IMPLIED WARRANTIES, INCLUDING, BUT NOT
 * LIMITED TO, THE IMPLIED WARRANTIES OF MERCHANTABILITY AND FITNESS
 * FOR A PARTICULAR PURPOSE ARE DISCLAIMED. IN NO EVENT SHALL EGNITE
 * SOFTWARE GMBH OR CONTRIBUTORS BE LI ABLE FOR ANY DIRECT, INDIRECT,
 * INCIDENTAL, SPECIAL, EXEMPLARY, OR CONSEQUENTIAL DAMAGES (INCLUDING,
 * BUT NOT LIMITED TO, PROCUREMENT OF SUBSTITUTE GOODS OR SERVICES; LOSS
 * OF USE, DATA, OR PROFITS; OR BUSINESS INTERRUPTION) HOWEVER CAUSED
 * AND ON ANY THEORY OF LIABILITY, WHETHER IN CONTRACT, STRICT LIABILITY,
 * OR TORT (INCLUDING NEGLIGENCE OR OTHERWISE) ARISING IN ANY WAY OUT OF
 * THE USE OF THIS SOFTWARE, EVEN IF ADVISED OF THE POSSIBILITY OF
 * SUCH DAMAGE.
 *
 * For additional information see http://www.ethernut.de/
 *
 */

/*!
 * $Log: ftpserv.c,v $
 * Revision 1.10 2008/01/31 09:38:15 haraldkipp
 * Added return statement in main to avoid warnings with latest GCC.
 *
 * Revision 1.9 2006/09/07 09:00:19 haraldkipp
 * Discovery registration added. Enabled by default on ARM targets only to
 * avoid blowing up AVR code.
 *
 * Revision 1.8 2006/09/05 12:26:35 haraldkipp
 * Added support for SAM9 MMC.
 * DHCP enabled by default.
 *
 * Revision 1.7 2006/08/31 19:15:30 haraldkipp
 * Dummy file system name added to SAM9260 to let it pass the compiler.
 * The application will not yet run on this platform.
 *
 * Revision 1.6 2006/07/26 11:22:55 haraldkipp
 * Added support for AT91SAM7X -EK.
 *
 * Revision 1.4 2006/01/22 17:34:38 haraldkipp
 * Added support for Ethernut 3, PHAT file system and realtime clock.
 *
 * Revision 1.3 2005/04/19 08:51:26 haraldkipp
 * Warn if not Ethernut 2
 *
 * Revision 1.2 2005/02/07 19:05:23 haraldkipp
 * Atmega 103 compile errors fixed
 *
 * Revision 1.1 2005/02/05 20:32:57 haraldkipp
 * First release
 *
 */
```

Figure 4.49: FTP Open Source Example[13]

```
....

/*
 * Baudrate for debug output.
 */
#ifndef DBG_BAUDRATE
#define DBG_BAUDRATE 115200
#endif

/*
 * Wether we should use DHCP.
 */
#define USE_DHCP

/*
 * Wether we should run a discovery responder.
 */
#if defined(__arm__)
#define USE_DISCOVERY
#endif

/*
 * Unique MAC address of the Ethernut Board.
 *
 * Ignored if EEPROM contains a valid configuration.
 */
#define MY_MAC { 0x00, 0x06, 0x98, 0x30, 0x00, 0x35 }

/*
 * Unique IP address of the Ethernut Board.
 *
 * Ignored if DHCP is used.
 */
#define MY_IPADDR "192.168.192.35"

/*
 * IP network mask of the Ethernut Board.
 *
 * Ignored if DHCP is used.
 */
#define MY_IPMASK "255.255.255.0"

/*
 * Gateway IP address for the Ethernut Board.
 *
 * Ignored if DHCP is used.
 */
#define MY_IPGATE "192.168.192.1"

/*
 * NetBIOS name.
 *
 * Use a symbolic name with Win32 Explorer.
 */
//#define MY_WINSNAME "ETHERNUT"

/*
 * FTP port number.
 */
#define FTP_PORTNUM 21

/*
 * FTP timeout.
 *
 * The server will terminate the session, if no new command is received
 * within the specified number of milliseconds.
 */
#define FTPD_TIMEOUT 600000

/*
```

Figure 4.49 continued: FTP Open Source Example

```
 * TCP buffer size.
 */
#define TCPIP_BUFSIZ 5840

/*
 * Maximum segment size.
 *
 * Choose 536 up to 1460. Note, that segment sizes above 536 may result
 * in fragmented packets. Remember, that Ethernut doesn't support TCP
 * fragmentation.
 */
#define TCPIP_MSS 1460

#if defined(ETHERNUT3)

/* Ethernut 3 file system. */
#define FSDEV devPhat0
#define FSDEV_NAME "PHAT0"

/* Ethernut 3 block device interface. */
#define BLKDEV devNplMmc0
#define BLKDEV_NAME "MMC0"

#elif defined(AT91SAM7X_EK)

/* SAM7X-EK file system. */
#define FSDEV devPhat0
#define FSDEV_NAME "PHAT0"

/* SAM7X-EK block device interface. */
#define BLKDEV devAt91SpiMmc0
#define BLKDEV_NAME "MMC0"

#elif defined(AT91SAM9260_EK)

/* SAM9260-EK file system. */
#define FSDEV devPhat0
#define FSDEV_NAME "PHAT0"

/* SAM9260-EK block device interface. */
#define BLKDEV devAt91Mci0
#define BLKDEV_NAME "MCI0"

#elif defined(ETHERNUT2)

/*
 * Ethernut 2 File system
 */
#define FSDEV devPnut
#define FSDEV_NAME "PNUT"

#else

#define FSDEV_NAME "NONE"

#endif

/*! \brief Local timezone, -1 for Central Europe. */
#define MYTZ -1

/*! \brief IP address of the host running a time daemon. */
#define MYTIMED "130.149.17.21"

#ifdef ETHERNUT3
/*! \brief Defined if X1226 RTC is available. */
#define X12RTC_DEV
#endif

/*
 * FTP service.
 *
```

Figure 4.49 continued: FTP Open Source Example

```
* This function waits for client connect, process es the FTP request
* and disconnects. Nut/Net doesn't support a server backlog. If one
* client has established a connection, further connect attempts will
* be rejected.
*
* Some FTP clients, like the Win32 Explorer, open more than one
* connection for background processing. So we run this routine by
* several threads.
*/
void FtpService(void)
{
 TCPSOCKET *sock;

 /*
 * Create a socket.
 */
 if ((sock = NutTcpCreateSocket()) != 0) {

 /*
 * Set specified socket options.
 */
#ifdef TCPIP_MSS
 {
 u_short mss = TCPIP_MSS;
 NutTcpSetSockOpt(sock, TCP_MAXSEG, &mss, sizeof(mss));
 }
#endif
#ifdef FTPD_TIMEOUT
 {
 u_long tmo = FTPD_TIMEOUT;
 NutTcpSetSockOpt(sock, SO_RCVTIMEO, &tmo, sizeof(tmo));
 }
#endif
#ifdef TCPIP_BUFSIZ
 {
 u_short siz = TCPIP_BUFSIZ;
 NutTcpSetSockOpt(sock, SO_RCVBUF, &siz, sizeof(siz));
 }
#endif

 /*
 * Listen on our port. If we return, we got a client.
 */
 printf("\nWaiting for an FTP client...");
 if (NutTcpAccept(sock, FTP_PORTNUM) == 0) {
 printf("%s connected, %u bytes free\n", inet_ntoa(sock->so_remote_addr), (u_int)NutHeapAvailable());
 NutFtpServerSession(sock);
 printf("%s disconnected, %u bytes free\n", inet_ntoa(sock->so_remote_addr), (u_int)NutHeapAvailable());
 } else {
 puts("Accept failed");
 }

 /*
 * Close our socket.
 */
 NutTcpCloseSocket(sock);
 }
}

/*
* FTP service thread.
*/
THREAD(FtpThread, arg)
{
 /* Loop endless for connections. */
 for (;;) {
 FtpService();
 }
}

/*
```

Figure 4.49 continued: FTP Open Source Example

```
 * Assign stdout to the UART device.
 */
void InitDebugDevice(void)
{
u_long baud = DBG_BAUDRATE;

NutRegisterDevice(&DEV_DEBUG, 0, 0);
freopen(DEV_DEBUG_NAME, "w", stdout);
_ioctl(_fileno(stdout), UART_SETSPEED, &baud);
}

/*
 * Setup the 80thernet device. Try DHCP first. If this is
 * the first time boot with empty EEPROM and no DHCP server
 * was found, use hardcoded values.
 */
int InitEthernetDevice(void)
{
u_long ip_addr = inet_addr(MY_IPADDR);
u_long ip_mask = inet_addr(MY_IPMASK);
u_long ip_gate = inet_addr(MY_IPGATE);
u_char mac[6] = MY_MAC;

if (NutRegisterDevice(&DEV_ETHER, 0x8300, 5)) {
puts("No Ethernet Device");
return -1;
}

printf("Configure %s...", DEV_ETHER_NAME);
#ifdef USE_DHCP
if (NutDhcpIfConfig(DEV_ETHER_NAME, 0, 60000) == 0) {
puts("OK");
return 0;
}
printf("initial boot...");
if (NutDhcpIfConfig(DEV_ETHER_NAME, mac, 60000) == 0) {
puts("OK");
return 0;
}
#endif
printf("No DHCP...");
NutNetIfConfig(DEV_ETHER_NAME, mac, ip_addr, ip_mask);
/* Without DHCP we had to set the default gateway manually.*/
if(ip_gate) {
printf("hard coded gate...");
NutIpRouteAdd(0, 0, ip_gate, &DEV_ETHER);
}
puts("OK");

return 0;
}
```

Figure 4.49 continued: FTP Open Source Example

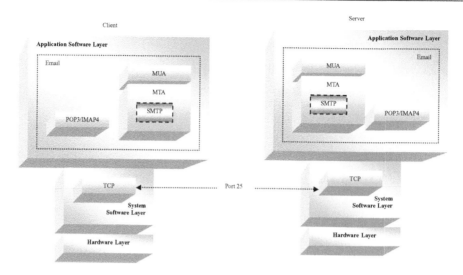

Figure 4.50: RFC2821 Email Model

4.6.3 Simple Mail Transfer Protocol (SMTP) and Email Application Example[5]

RFC2821 for SMTP (Simple Mail Transfer Protocol) is an ASCII-based networking protocol for implementation within electronic mail (email) applications. It is a protocol for reliable and efficient transmission and reception of emails between networked devices. As shown in Figure 4.50, the RFC2821 model reflects an email application with two major elements:

- MUA, a *mail user agent* which is the interface an email application user uses to generate emails
- MTA, the *mail transfer agent* which manages the SMTP communication for exchanging emails between two devices.

Within the MTA, the SMTP protocol dictates that the transmitter of the email is the SMTP *client*, and the receiver of the email is the SMTP *server*. What SMTP requires of the underlying networking middleware is a protocol, such as TCP, that provides a reliable, ordered data stream channel in which SMTP messages can be exchanged. The messages exchanged between SMTP clients and servers have a message format that includes an email **header** (i.e., Reply-To, Date, and From), the **body** of the email (i.e., the content of the email), and the **envelope** (i.e., the addresses of the sender and receiver).

Finally, in order to manage the communication and transmission of messages, the SMTP communication scheme includes the exchange of SMTP commands, such as those shown in Table 4.12.

SMTP defines different buffers that can be implemented on a server to include the various types of data, such as the 'mail-data' buffer to hold the body of an email, a 'forward-path'

Table 4.12: Examples of SMTP Commands and Reply Codes[2]

Type	Code	Definition
Command	HELO	Data object is a fully qualified domain name of the client host, which is how a client identifies itself
	MAIL	Data object is the address of the sender, which identifies the origins of the message
	RCPT	(RECIPIENT) Data object is the address of the recipient, which identifies who the email is for
	RSET	(RESET) Not a data object. Code aborts the current email transaction and allows for any related data to be discarded
	VRFY	(VERIFY) Data object is the email user or mailbox, which allows the SMTP client to verify the recipient's email address without actually transmitting the email to the recipient
Reply Code	211	System Status
	214	Help Message
	220	Service Ready
	221	Service Closing Transmission Channel
	250	Requested Mail Action Completed
	251	User Not Local, Will Forward
	354	Start Mail Input

buffer to hold the addresses of recipients, and 'reverse-path' buffer to hold addresses of senders. This is because data objects that are transmitted can be held pending a confirmation by the sender that the 'end of mail data' has been transmitted by the client device. This 'end of mail data' confirmation (QUIT) is what finalizes a successful email transaction. Finally, because TCP is a reliable byte stream protocol, checksums are usually not needed in an SMTP algorithm to verify the integrity of the data.

Figure 4.51 is an example of SMTP pseudocode implemented in an email application on a client device, and how this source code utilizes an underlying networking middleware layer such as TCP socket-related function calls.

4.6.4 Hypertext Transfer Protocol (HTTP) Cleint and Server Application Example[5]

Based upon several RFC standards, and supported by the World Wide Web (WWW) Consortium, the Hypertext Transfer Protocol (HTTP) 1.1 is the most widely implemented application layer protocol, used to transmit all types of data over the Internet. Under the HTTP protocol, these data (referred to as a *resource*) are identifiable by their *URL* (Uniform

Email Application Task
{
       ...

       Sender = "xx@xx.com";
       Recipient = "yy@yy.com";
       SMTPServer = "smtpserver.xxxx.com"
       SENDER = "tn@xemcoengineering.com",;
       RECIPIENT = "cn@xansa.com";
       CONTENT = "This is a simple e-mail sent by SMTP";
       SMTPSend("Hello!"); // a simple SMTP sample algorithm
       ....

}

SMTPSend (string Subject) {

       TCPSocket s= new TCPSocket(SMTPServer,25); // establishing a TCP connection to port 25 of the

       // destination device
       Timeout = 3 seconds; // timeout for establishing connection 3 seconds
       Transmission Successful = FALSE;
       Time = 0;

       While (time < timeout) {
              read in REPLY;
              If response from recipient then {
                    if REPLY not 220 then {
                          //not willing to accept e-mail
                          close TCP connection
                          time = timeout;
                  } else {
                      transmit to RECIPIENT ("HELO"+hostname); //client identifies itself
                      read in REPLY;
                  } else {
                      transmit to RECIPIENT ("QUIT");
                      read in REPLY;
                      if REPLY not 221 then {
                          //service not closing transmission channel
                          close TCP connection
                          time = timeout;
                    } else {
                      close TCP connection;
                      transmission successful = TRUE;
                      time = timeout;
                  } //end if-then-else "." REPLY not 221
              } //end if-then-else "." REPLY not 250
              } //end if-then-else REPLY not 354
              } // end if-then-else RCPT TO REPLY not 250
              } // end if-then-else MAIL FROM REPLY not 250
              } // end if-then-else HELO REPLY not 250
              } // end if-then-else REPLY not 220
            } else {
                time = time + 1;
            } // end if-then-else response from recipient
       } // end while (time < timeout)
} // end STMPTask

Figure 4.51: SMTP Pseudocode Example[5]

Request Message
<method> <request-URL><version>
<headers>
"/r/n" [blank line]
<body>

Response Message
<version> <status-code><status-phrase>
<headers>
"/r/n" [blank line]
<body>

Figure 4.52: Request and Response Message Formats[11]

Resource Locator). As with the other two networking examples, HTTP is based upon the client–server model that requires its underlying transport protocol to be a reliable, ordered data stream channel, such as TCP. The HTTP transaction starts with the HTTP client opening a connection to an HTTP server by establishing a TCP connection to default port 80 (for example) of the server. The HTTP client then sends a ***request message*** for a particular resource to the HTTP server. The HTTP server responds by sending a ***response message*** to the HTTP client with its requested resource (if available). After the response message is sent, the server closes the connection.

The syntax of request and response messages both have headers that contain message Attribute information that varies according to the message owner, and a body that contains optional data, where the header and body are separated by an empty line. As shown in Figure 4.52, they differ according to the first line of each message – where a request message contains the method (command made by client specifying the action the server needs to perform), the request-URL (address of resource requested), and version (of HTTP) in that order, and the first line of a response message contains the version (of HTTP), the status-code (response code to the client's method), and the status-phrase (readable equivalent of status-code).

Tables 4.13a and 4.13b list the various methods and reply codes that can be implemented in an HTTP server.

Table 4.13a: HTTP Methods[11]

Method	Definition
DELETE	The DELETE method requests that the origin server delete the resource identified by the Request-URI.
GET	The GET method means retrieve whatever information (in the form of an entity) is identified by the Request-URI. The Request-URI refers to a data-producing process, it is the produced data which shall be returned as the entity in the response and not the source of the process, unless that text happens to be the output of the process.

Table 4.13a continued: HTTP Methods

Method	Definition
HEAD	The HEAD method is identical to GET except that the server MUST NOT return a message-body in the response. The metainformation contained in the HTTP headers in response to a HEAD request SHOULD be identical to the information sent in response to a GET request. This method can be used for obtaining metainformation about the entity implied by the request without transferring the entity-body itself. This method is often used for testing hypertext links for validity, accessibility, and recent modification.
OPTIONS	The OPTIONS method represents a request for information about the communication options available on the request/response chain identified by the Request-URI. This method allows the client to determine the options and/or requirements associated with a resource, or the capabilities of a server, without implying a resource action or initiating a resource retrieval.
POST	The POST method is used to request that the destination server accept the entity enclosed in the request as a now subordinate of the resource identified by the Request-URI in the Request-Line. POST is designed to allow a uniform method to cover the following functions: • Annotation of existing resources; • Posting a message to a bulletin board, newsgroup, mailing list, or similar group of articles; • Providing a block of data, such as the result of submitting a form, to a data-handling process; • Extending a database through an append operation.
PUT	The PUT method requests that the enclosed entity be stored under the supplied Request-URI. If the Request-URI refers to an already existing resource, the enclosed entity SHOULD be considered as a modified version of the one residing on the origin server. If the Request-URI does not point to an existing resource, and that URI is capable of being defined as a new resource by the requesting user agent, the origin server can create the resource with that URI.
TRACE	The TRACE method is used to invoke a remote, application-layer loop-back of the request message. TRACE allows the client to see what is being received at the other end of the request chain and use that data for testing or diagnostic information.

Table 4.13b: HTTP Reply Codes[11]

Code	Definition
200	Ok
400	Bad request
404	Not found
501	Not implemented

```
/*
 * Copyright I 2001-2004 by egnite Software GmbH. All rights reserved.
 *
 * Redistribution and use in source and binary forms, with or without
 * modification, are permitted provided that the following conditions
 * are met:
 *
 * 1. Redistributions of source code must retain the above copyright
 * notice, this list of conditions and the following disclaimer.
 * 2. Redistributions in binary form must reproduce the above copyright
 * notice, this list of conditions and the following disclaimer in the
 * documentation and/or other materials provided with the distribution.
 * 3. Neither the name of the copyright holders nor the names of
 * contributors may be used to endorse or promote products derived
 * from this software without specific prior written permission.
 *
 * THIS SOFTWARE IS PROVIDED BY EGNITE SOFTWARE GMBH AND CONTRIBUTORS
 * ``AS IS" AND ANY EXPRESS OR IMPLIED WARRANTIES, INCLUDING, BUT NOT
 * LIMITED TO, THE IMPLIED WARRANTIES OF  MERCHANTABILITY AND FITNESS
 * FOR A PARTICULAR PURPOSE ARE DISCLAIMED. IN NO EVENT SHALL EGNITE
 * SOFTWARE GMBH OR CONTRIBUTORS BE LIABLE FOR ANY DIRECT, INDIRECT,
 * INCIDENTAL, SPECIAL, EXEMPLARY, OR CONSEQUENTIAL DAMAGES (INCLUDING,
 * BUT NOT LIMITED TO, PROCUREMENT OF SUBSTITUTE GOODS OR SERVICES; LOSS
 * OF USE, DATA, OR PROFITS; OR BUSINESS INTERRUPTION) HOWEVER CAUSED
 * AND ON ANY THEORY OF LIABILITY, WHETHER IN CONTRACT, STRICT LIABILITY,
 * OR TORT (INCLUDING NEGLIGENCE OR OTHERWISE) ARISING IN  ANY WAY OUT OF
 * THE USE OF THIS SOFTWARE, EVEN IF ADVISED OF THE POSSIBILITY OF
 * SUCH DAMAGE.
 *
 * For additional information see http://www.ethernut.de/
 *
 */

/*!
 * $Log: httpserv.c,v $
 * Revision 1.19 2008/07/25 10:20:12 olereinhardt
 * Fixed compiler bug for AVR -ICC and added missing PSTR macro around
 * prog_char strings
 *
 * Revision 1.18 2008/07/17 11:56:20 olereinhardt
 * Updated the webserver demo to show new webserver functions (different cgi
 * pathes with I authentication, $QUERY_STRING parameter for ssi
 * included CGIs)
 *
 * Revision 1.17 2008/05/13 19:31:34 thiagocorrea
 * NutHttpSendHeaderBot is marked as deprecated, use NutHttpSendHeaderBottom instead.
 *
 * Revision 1.16 2007/07/17 18:29:30 haraldkipp
 * Server thread names not unique on SAM7X. Fixed by Marti Raudsepp.
 *
 * Revision 1.15 2006/09/07 09:01:36 haraldkipp
 * Discovery registration added.
 * Re-arranged network interface setup to exclude DHCP code from ICCAVR
 * builds and make it work with the demo compiler. Unfinished.
 * Added PHAT file system support. Untested.
 *
 * Revision 1.14 2006/03/02 19:44:03 haraldkipp
 * MMC and PHAT enabled.
 *
 * Revision 1.13 2006/01/11 08:32:57 hwmaier
 * Added explicit type casts to silence a few avr-gcc 3.4.3 warning messages
 *
 * Revision 1.12 2005/11/22 09:14:13 haraldkipp
 * Replaced specific device names by generalized macros.
 *
 * Revision 1.11 2005/10/16 23:22:20 hwmaier
```

Figure 4.53: HTTP Open Source Example[13]

```
* Removed unreferenced nutconfig.h include statement
*
* Revision 1.10 2005/08/05 11:32:50 olereinhardt
* Added SSI and ASP sample
*
* Revision 1.9 2005/04/05 18:04:17 haraldkipp
* Support for ARM7 Wolf Board added.
*
* Revision 1.8 2005/02/23 04:39:26 hwmaier
* no message
*
* Revision 1.7 2005/02/22 02:44:34 hwmaier
* Changes to compile as well for AT90CAN128 device.
*
* Revision 1.6 2004/12/16 10:17:18 haraldkipp
* Added Mikael Adolfsson's excellent parameter parsing routines.
*
* Revision 1.5 2004/03/16 16:48:26 haraldkipp
* Added Jan Dubiec's H8/300 port.
*
* Revision 1.4 2003/11/04 17:46:52 haraldkipp
* Adapted to Ethernut 2
*
* Revision 1.3 2003/09/29 16:33:12 haraldkipp
* Using portable strtok and strtok_r
*
* Revision 1.2 2003/08/07 08:27:58 haraldkipp
* Bugfix, remote not displayed in socket list
*
* Revision 1.1 2003/07/20 15:56:14 haraldkipp
* *** empty log message ***
*
*/

/*!
* \example httpd/httpserv.c
*
* Simple multithreaded HTTP daemon.
*/

/*
* Unique MAC address of the Ethernut Board.
*
* Ignored if EEPROM contains a valid configuration.
*/
#define MY_MAC "\x00\x06\x98\x30\x00\x35"

/*
* Unique IP address of the Ethernut Board.
*
* Ignored if DHCP is used.
*/
#define MY_IPADDR "192.168.192.35"

/*
* IP network mask of the Ethernut Board.
*
* Ignored if DHCP is used.
*/
#define MY_IPMASK "255.255.255.0"

/*
* Gateway IP address for the Ethernut Board.
*
* Ignored if DHCP is used.
*/
#define MY_IPGATE "192.168.192.1"

/* ICCAVR Demo is limited. Try to use the bare minimum. */
#if !defined(__IMAGECRAFT__)
```

Figure 4.53 continued: HTTP Open Source Example

```
/* Wether we should use DHCP. */
#define USE_DHCP
/* Wether we should run a discovery responder. */
#define USE_DISCOVERY
/* Wether to use PHAT file system. */
//#define USE_PHAT

#endif /* __IMAGECRAFT__ */


#ifdef USE_PHAT

#if defined(ETHERNUT3)

/* Ethernut 3 file system. */
#define MY_FSDEV devPhat0
#define MY_FSDEV_NAME "PHAT0"

/* Ethernut 3 block device interface. */
#define MY_BLKDEV devNplMmc0
#define MY_BLKDEV_NAME "MMC0"

#elif defined(AT91SAM7X_EK)

/* SAM7X-EK file system. */
#define MY_FSDEV devPhat0
#define MY_FSDEV_NAME "PHAT0"

/* SAM7X-EK block device interface. */
#define MY_BLKDEV devAt91SpiMmc0
#define MY_BLKDEV_NAME "MMC0"

#elif defined(AT91SAM9260_EK)

/* SAM9260-EK file system. */
#define MY_FSDEV devPhat0
#define MY_FSDEV_NAME "PHAT0"

/* SAM9260-EK block device interface. */
#define MY_BLKDEV devAt91Mci0
#define MY_BLKDEV_NAME "MCI0"

#endif
#endif /* USE_PHAT */

#ifndef MY_FSDEV
#define MY_FSDEV devUrom
#endif

#ifdef MY_FSDEV_NAME
#define MY_HTTPROOT MY_FSDEV_NAME ":/"
#endif


....


/*****************************************************************/
/* ASPCallback */
/* */
/* This routine must have been registered by */
/* NutRegisterAspCallback() and is automatically called by */
/* NutHttpProcessFileRequest() when the server process a page */
/* with an asp function. */
/* */
/* Return 0 on success, -1 otherwise. */
/*****************************************************************/

static int ASPCallback (char *pASPFunction, FILE *stream)
{
```

Figure 4.53 continued: HTTP Open Source Example

```
if (strcmp(pASPFunction, "usr_date") == 0) {
fprintf(stream, "Dummy example: 01.01.2005");
return(0);
}

if (strcmp(pASPFunction, "usr_time") == 0) {
fprintf(stream, "Dummy example: 12:15:02");
return(0);
}

return (-1);
}

/*
* CGI Sample: Show request parameters.
*
* See httpd.h for REQUEST structure.
*
* This routine must have been registered by NutRegisterCgi() and is
* automatically called by NutHttpProcessRequest() when the client
* request the URL 'cgi-bin/test.cgi'.
*/
static int ShowQuery(FILE * stream, REQUEST * req)
{
char *cp;
/*
* This may look a little bit weird if you are not used to C programming
* for flash microcontrollers. The special type 'prog_char' forces the
* string literals to be placed in flash ROM. This saves us a lot of
* precious RAM.
*/
static prog_char head[] = "<HTML><HEAD><TITLE>Parameters</TITLE></HEAD><BODY><H1>Parameters</H1>";
static prog_char foot[] = "</BODY></HTML>";
static prog_char req_fmt[] = "Method: %s<BR>\r\nVersion: HTTP/%d.%d<BR>\r\nContent length: %ld<BR>\r\n";
static prog_char url_fmt[] = "URL: %s<BR>\r\n";
static prog_char query_fmt[] = "Argument: %s<BR>\r\n";
static prog_char type_fmt[] = "Content type: %s<BR>\r\n";
static prog_char cookie_fmt[] = "Cookie: %s<BR>\r\n";
static prog_char auth_fmt[] = "Auth info: %s<BR>\r\n";
static prog_char agent_fmt[] = "User agent: %s<BR>\r\n";

/* These useful API calls create a HTTP response for us. */
NutHttpSendHeaderTop(stream, req, 200, "Ok");
NutHttpSendHeaderBottom(stream, req, html_mt, -1);

/* Send HTML header. */
fputs_P(head, stream);

/*
* Send request parameters.
*/
switch (req->req_method) {
case METHOD_GET:
cp = "GET";
break;
case METHOD_POST:
cp = "POST";
break;
case METHOD_HEAD:
cp = "HEAD";
break;
default:
cp = "UNKNOWN";
break;
}
fprintf_P(stream, req_fmt, cp, req->req_version / 10, req->req_version % 10, req->req_length);
if (req->req_url)
fprintf_P(stream, url_fmt, req->req_url);
if (req->req_query)
fprintf_P(stream, query_fmt, req->req_query);
if (req->req_type)
```

Figure 4.53 continued: HTTP Open Source Example

```
fprintf_P(stream, type_fmt, req->req_type);
if (req->req_cookie)
fprintf_P(stream, cookie_fmt, req->req_cookie);
if (req->req_auth)
fprintf_P(stream, auth_fmt, req->req_auth);
if (req->req_agent)
fprintf_P(stream, agent_fmt, req->req_agent);

/* Send HTML footer and flush output buffer. */
fputs_P(foot, stream);
fflush(stream);

return 0;
}

/*
 * CGI Sample: Show list of threads.
 *
 * This routine must have been registered by NutRegisterCgi() and is
 * automatically called by NutHttpProcessRequest() when the client
 * request the URL 'cgi-bin/threads.cgi'.
 */
static int ShowThreads(FILE * stream, REQUEST * req)
{
static prog_char head[] = "<HTML><HEAD><TITLE>Threads</TITLE></HEAD><BODY><H1>Threads</H1>\r\n"
"<TABLE
BORDER><TR><TH>Handle</TH><TH>Name</TH><TH>Priority</TH><TH>Status</TH><TH>Event<BR>Queue</TH><TH>Timer</TH
><TH>Stack-<BR>pointer</TH><TH>Free<BR>Stack</TH></TR>\r\n";
#if defined(__AVR__)
static prog_char tfmt[] =

"<TR><TD>%04X</TD><TD>%s</TD><TD>%u</TD><TD>%s</TD><TD>%04X</TD><TD>%04X</TD><TD>%04X</TD><TD>%u</T
D><TD>%s</TD></TR>\r\n";
#else
static prog_char tfmt[] =

"<TR><TD>%08lX</TD><TD>%s</TD><TD>%u</TD><TD>%s</TD><TD>%08lX</TD><TD>%08lX</TD><TD>%08lX</TD><TD>%lu
</TD><TD>%s</TD></TR>\r\n";
#endif
static prog_char foot[] = "</TABLE></BODY></HTML>";
static char *thread_states[] = { "TRM", "<FONT COLOR=#CC0000>RUN</FONT>", "<FONT COLOR=#339966>RDY</FONT>", "SLP" };
NUTTHREADINFO *tdp = nutThreadList;

/* Send HTTP response. */
NutHttpSendHeaderTop(stream, req, 200, "Ok");
NutHttpSendHeaderBottom(stream, req, html_mt, -1);

/* Send HTML header. */
fputs_P(head, stream);
for (tdp = nutThreadList; tdp; tdp = tdp->td_next) {
fprintf_P(stream, tfmt, (uptr_t) tdp, tdp->td_priority,
thread_states[tdp->td_state], (uptr_t) tdp->td_queue, (uptr_t) tdp->td_timer,
(uptr_t) tdp->td_sp, (uptr_t) tdp->td_sp – (uptr_t) tdp->td_memory,
*((u_long *) tdp->td_memory) != DEADBEEF ? "Corr" : "OK");
}
fputs_P(foot, stream);
fflush(stream);

return 0;
}

/*
 * CGI Sample: Show list of timers.
 *
 * This routine must have been registered by NutRegisterCgi() and is
 * automatically called by NutHttpProcessRequest() when the client
 * request the URL 'cgi-bin/timers.cgi'.
 */
static int ShowTimers(FILE * stream, REQUEST * req)
{
static prog_char head[] = "<HTML><HEAD><TITLE>Timers</TITLE></HEAD><BODY><H1>Timers</H1>\r\n";
```

Figure 4.53 continued: HTTP Open Source Example

```
static prog_char thead[] =
"<TABLE BORDER><TR><TH>Handle</TH><TH>Countdown</TH><TH>Tick
Reload</TH><TH>Callback<BR>Address</TH><TH>Callback<BR>Argument</TH></TR>\r\n";
#if defined(__AVR__)
static prog_char tfmt[] = "<TR><TD>%04X</TD><TD>%lu</TD><TD>%lu</TD><TD>%04X</TD><TD>%04X</TD></TR>\r\n";
#else
static prog_char tfmt[] = "<TR><TD>%08lX</TD><TD>%lu</TD><TD>%lu</TD><TD>%08lX</TD><TD>%08lX</TD></TR>\r\n";
#endif
static prog_char foot[] = "</TABLE></BODY></HTML>";
NUTTIMERINFO *tnp;
u_long ticks_left;

NutHttpSendHeaderTop(stream, req, 200, "Ok");
NutHttpSendHeaderBottom(stream, req, html_mt, -1);

/* Send HTML header. */
fputs_P(head, stream);
if ((tnp = nutTimerList) != 0) {
fputs_P(thead, stream);
ticks_left = 0;
while (tnp) {
ticks_left += tnp->tn_ticks_left;
fprintf_P(stream, tfmt, (uptr_t) tnp, ticks_left, tnp->tn_ticks, (uptr_t) tnp->tn_callback, (uptr_t) tnp->tn_arg);
tnp = tnp->tn_next;
}
}

fputs_P(foot, stream);
fflush(stream);

return 0;
}

/*
 * CGI Sample: Show list of sockets.
 *
 * This routine must have been registered by NutRegisterCgi() and is
 * automatically called by NutHttpProcessRequest() when the client
 * request the URL 'cgi-bin/sockets.cgi'.
 */
static int ShowSockets(FILE * stream, REQUEST * req)
{
/* String literals are kept in flash ROM. */
static prog_char head[] = "<HTML><HEAD><TITLE>Sockets</TITLE></HEAD>"
"<BODY><H1>Sockets</H1>\r\n"
"<TABLE BORDER><TR><TH>Handle</TH><TH>Type</TH><TH>Local</TH><TH>Remote</TH><TH>Status</TH></TR>\r\n";
#if defined(__AVR__)
static prog_char tfmt1[] = "<TR><TD>%04X</TD><TD>TCP</TD><TD>%s:%u</TD>";
#else
static prog_char tfmt1[] = "<TR><TD>%08lX</TD><TD>TCP</TD><TD>%s:%u</TD>";
#endif
static prog_char tfmt2[] = "<TD>%s:%u</TD><TD>";
static prog_char foot[] = "</TABLE></BODY></HTML>";
static prog_char st_listen[] = "LISTEN";
static prog_char st_synsent[] = "SYNSENT";
static prog_char st_synrcvd[] = "SYNRCVD";
static prog_char st_estab[] = "<FONT COLOR=#CC0000>ESTABL</FONT>";
static prog_char st_finwait1[] = "FINWAIT1";
static prog_char st_finwait2[] = "FINWAIT2";
static prog_char st_closewait[] = "CLOSEWAIT";
static prog_char st_closing[] = "CLOSING";
static prog_char st_lastack[] = "LASTACK";
static prog_char st_timewait[] = "TIMEWAIT";
static prog_char st_closed[] = "CLOSED";
static prog_char st_unknown[] = "UNKNOWN";
prog_char *st_P;
extern TCPSOCKET *tcpSocketList;
TCPSOCKET *ts;

NutHttpSendHeaderTop(stream, req, 200, "Ok");
NutHttpSendHeaderBottom(stream, req, html_mt, -1);
```

Figure 4.53 continued: HTTP Open Source Example

```
/* Send HTML header */
fputs_P(head, stream);
for (ts = tcpSocketList; ts; ts = ts->so_next) {
switch (ts->so_state) {
case TCPS_LISTEN:
st_P = (prog_char *) st_listen;
break;
case TCPS_SYN_SENT:
st_P = (prog_char *) st_synsent;
break;
case TCPS_SYN_RECEIVED:
st_P = (prog_char *) st_synrcvd;
break;
case TCPS_ESTABLISHED:
st_P = (prog_char *) st_estab;
break;
case TCPS_FIN_WAIT_1:
st_P = (prog_char *) st_finwait1;
break;
case TCPS_FIN_WAIT_2:
st_P = (prog_char *) st_finwait2;
break;
case TCPS_CLOSE_WAIT:
st_P = (prog_char *) st_closewait;
break;
case TCPS_CLOSING:
st_P = (prog_char *) st_closing;
break;
case TCPS_LAST_ACK:
st_P = (prog_char *) st_lastack;
break;
case TCPS_TIME_WAIT:
st_P = (prog_char *) st_timewait;
break;
case TCPS_CLOSED:
st_P = (prog_char *) st_closed;
break;
default:
st_P = (prog_char *) st_unknown;
break;
}
/*
* Fixed a bug reported by Zhao Weigang.
*/
fprintf_P(stream, tfmt1, (uptr_t) ts, inet_ntoa(ts->so_local_addr), ntohs(ts->so_local_port));
fprintf_P(stream, tfmt2, inet_ntoa(ts->so_remote_addr), ntohs(ts->so_remote_port));
fputs_P(st_P, stream);
fputs("</TD></TR>\r\n", stream);
fflush(stream);
}

fputs_P(foot, stream);
fflush(stream);

return 0;
}

/*
* CGI Sample: Processing a form.
*
* This routine must have been registered by NutRegisterCgi() and is
* automatically called by NutHttpProcessRequest() when the client
* request the URL 'cgi-bin/form.cgi'.
*
* Thanks to Tom Boettger, who provided this sample for ICCAVR.
*/
int ShowForm(FILE * stream, REQUEST * req)
{
static prog_char html_head[] = "<HTML><BODY><BR><H1>Form Result</H1><BR><BR>";
static prog_char html_body[] = "<BR><BR><p><a href=\"../index.html\">return to main</a></BODY></HTML></p>";
```

Figure 4.53 continued: HTTP Open Source Example

```
NutHttpSendHeaderTop(stream, req, 200, "Ok");
NutHttpSendHeaderBottom(stream, req, html_mt, -1);

/* Send HTML header. */
fputs_P(html_head, stream);

if (req->req_query) {
char *name;
char *value;
int I;
int count;

count = NutHttpGetParameterCount(req);
/* Extract count parameters. */
for (I = 0; I < count; i++) {
name = NutHttpGetParameterName(req, i);
value = NutHttpGetParameterValue(req, i);

/* Send the parameters back to the client. */

#ifdef __IMAGECRAFT__
fprintf(stream, "%s: %s<BR>\r\n", name, value);
#else
fprintf_P(stream, PSTR("%s: %s<BR>\r\n"), name, value);
#endif
}
}

fputs_P(html_body, stream);
fflush(stream);

return 0;
}


/*
 * CGI Sample: Dynamic output cgi included by ssi.shtml file
 *
 * This routine must have been registered by NutRegisterCgi() and is
 * automatically called by NutHttpProcessRequest() when the client
 * request the URL 'cgi-bin/form.cgi'.
 *
 * Thanks to Tom Boettger, who provided this sample for ICCAVR.
 */
int SSIDemoCGI(FILE * stream, REQUEST * req)
{
if (req->req_query) {
char *name;
char *value;
int I;
int count;

count = NutHttpGetParameterCount(req);

/* Extract count parameters. */
#ifdef __IMAGECRAFT__
fprintf(stream, "CGI ssi-demo.cgi called with parameters: These are the parameters\r\n<p>");
#else
fprintf_P(stream, PSTR("CGI ssi-demo.cgi called with parameters: These are the parameters\r\n<p>"));
#endif
for (I = 0; I < count; i++) {
name = NutHttpGetParameterName(req, i);
value = NutHttpGetParameterValue(req, i);

/* Send the parameters back to the client. */

#ifdef __IMAGECRAFT__
fprintf(stream, "%s: %s<BR>\r\n", name, value);
#else
fprintf_P(stream, PSTR("%s: %s<BR>\r\n"), name, value);
#endif
```

Figure 4.53 continued: HTTP Open Source Example

```
}
} else {
time_t now;
tm loc_time;

/* Called without any parameter, show the current time */
now = time(NULL);
localtime_r(&now, &loc_time);
#ifdef __IMAGECRAFT__
fprintf(stream, "CGI ssi-demo.cgi called without any parameter.<br><br>Current time is: %02d.%02d.%04d -- %02d:%02d:%02d<br>\r\n",
loc_time.tm_mday, loc_time.tm_mon+1, loc_time.tm_year+1900, loc_time.tm_hour, loc_time.tm_min, loc_time.tm_sec);
#else
fprintf_P(stream, PSTR("CGI ssi-demo.cgi called without any parameter.<br><br>Current time is: %02d.%02d.%04d --
%02d:%02d:%02d<br>\r\n"),
loc_time.tm_mday, loc_time.tm_mon+1, loc_time.tm_year+1900, loc_time.tm_hour, loc_time.tm_min, loc_time.tm_sec);
#endif
}

fflush(stream);

return 0;
}

/*! \fn Service(void *arg)
 * \brief HTTP service thread.
 *
 * The endless loop in this thread waits for a client connect,
 * processes the HTTP request and disconnects. Nut/Net doesn't
 * support a server backlog. If one client has established a
 * connection, further connect attempts will be rejected.
 * Typically browsers open more  than one connection in order
 * to load images concurrently. So we run this routine by
 * several threads.
 *
 */
THREAD(Service, arg)
{
TCPSOCKET *sock;
FILE *stream;
u_char id = (u_char) ((uptr_t) arg);

/*
 * Now loop endless for connections.
 */
for (;;) {

/*
 * Create a socket.
 */
if ((sock = NutTcpCreateSocket()) == 0) {
printf("[%u] Creating socket failed\n", id);
NutSleep(5000);
continue;
}

/*
 * Listen on port 80. This call will block until we get a connection
 * from a client.
 */
NutTcpAccept(sock, 80);
#if defined(__AVR__)
printf("[%u] Connected, %u bytes free\n", id, NutHeapAvailable());
#else
printf("[%u] Connected, %lu bytes free\n", id, NutHeapAvailable());
#endif

/*
 * Wait until at least 8 kByte of free RAM is availabl e. This will
 * keep the client connected in low memory situations.
 */
#if defined(__AVR__)
```

Figure 4.53 continued: HTTP Open Source Example

```
      while (NutHeapAvailable() < 8192) {
#else
      while (NutHeapAvailable() < 4096) {
#endif
      printf("[%u] Low mem\n", id);
      NutSleep(1000);
      }

      /*
      * Associate a stream with the socket so we can use standard I/O calls.
      */
      if ((stream = _fdopen((int) ((uptr_t) sock), "r+b")) == 0) {
      printf("[%u] Creating stream device failed\n", id);
      } else {
      /*
      * This API call saves us a lot of work. It will parse the
      * client's HTTP request, send any requested file from the
      * registered file system or handle CGI requests by calling
      * our registered CGI routine.
      */
      NutHttpProcessRequest(stream);

      /*
      * Destroy the virtual stream device.
      */
      fclose(stream);
      }

      /*
      * Close our socket.
      */
      NutTcpCloseSocket(sock);
      printf("[%u] Disconnected\n", id);
      }
}
```

Figure 4.53 continued: HTTP Open Source Example

The open source example in Figure 4.53 demonstrates HTTP implemented in a simple web server. The reader can then see an example of how this sample open source code uses underlying TCP (states) in its own HTTP-specific functions.

4.7 Summary

In this chapter, an introduction to core networking concepts and the OSI model was discussed. Moreover, networking middleware was defined as system software that typically resides within the upper data-link layer through to the transport layer in an embedded system. This networking middleware mediates between networking application protocols and the kernel, and/or networking device driver software, as well as mediates and serves different networking application protocols. Finally, underlying networking hardware and system software was explained relative to networking middleware, as well as how to put it all together with networking application layer software. Open source examples were used to help give readers a more clear picture of the implementation of middleware networking protocols from a programmer's perspective within a device, as well as allow the reader to download and utilize these open source examples for themselves.

The next chapter, Chapter 5, introduces database fundamentals relative to their implementation within a middleware layer.

4.8 Problems[5]

1. What is the difference between LANs and WANs?
2. What are the two types of transmission mediums that can connect devices?
3. A. What is the OSI model?
   B. What are the layers of the OSI model?
   C. Give examples of two protocols under each layer.
   D. Where in the Embedded Systems Model does each layer of the OSI model fall? Draw it.
4. A. How does the OSI model compare to the TCP/IP model?
   B. How does the OSI model compare to Bluetooth?
5. Where in the OSI model is networking middleware located?
6. A. Draw the TCP/IP model layers relative to the OSI model.
   B. Which layer would TCP fall under?
7. RS-232 related software is middleware (True/False).
8. PPP manages data as:
   A. Frames.
   B. Datagrams.
   C. Messages.
   D. All of the above.
   E. None of the above.
9. A. Name and describe the four subcomponents that make up PPP software.
   B. What RFCs are associated with each?
10. A. What is the difference between a PPP state and a PPP event?
    B. List and describe three examples of each.
11. A. What is an IP address?
    B. What networking protocol processes IP addresses?
12. Name two examples of application-layer protocols that can either be implemented as stand-alone applications whose sole function is that protocol, or implemented as a sub-component of a larger multifunction application.
13. A. What is the difference between an FTP client and an FTP server?
    B. What type of embedded devices would implement each?

14. SMTP is a protocol that is typically implemented in:
    A. An email application.
    B. A kernel.
    C. A BSP.
    D. Every application.
    E. None of the above.
15. SMTP typically relies on TCP middleware to function (True/False).
16. A. What is HTTP?
    B. What types of applications would incorporate an HTTP client or server?

4.9 End Notes

[1] RFC959 (http://www.freesoft.org/CIE/RFC/959/index.htm).
[2] RFC2821(http://www.freesoft.org/CIE/RFC/2821/index.htm).
[3] Embedded Planet EPC8xx Datasheet.
[4] Embedded Microcomputer Systems, Valvano.
[5] Embedded Systems Architecture, Noergaard – RFC 793. 'Transmission Control Protocol'. DARPA Protocol Specification.
[6] http://www.ethernut.de/en/download/index.html. Open source examples.
[7] VxWorks API Reference Guide: Device Drivers, Version 5.5.
[8] RFC1661 (http://www.freesoft.org/CIE/RFC/1661/index.htm), RFC1334 (http://www.freesoft.org/CIE/RFC/1334/index.htm), RFC1332 (http://www.freesoft.org/CIE/RFC/1332/index.htm)
[9] RFC791 (http://www.freesoft.org/CIE/RFC/791/index.htm).
[10] RFC798 (http://www.freesoft.org/CIE/RFC/798/index.htm).
[11] www.w3.org/Protocols/
[12] WindRiver vxWorks API Documentation and Project.
[13] Egnite Open Source.

CHAPTER 5

File Systems

Chapter Points

- Defines what a file system is and what it manages when utilized as middleware
- Introduces fundamental file system concepts and terminology
- Identifies the major elements of most file system designs

5.1 What is a File System?

File system software provides a scheme to manage data on an embedded computer system. A file system can be accessible and directly utilized by the embedded system's user, as middleware software used by other middleware, as middleware software used by applications in the system to manage data for the application, or some combination of the above. Regardless, a file system manages data by allowing for some combination of the:

- organization
- storage
- creation
- modification
- retrieval

of data from some type of memory medium. Depending on the file system, the memory medium can be volatile RAM, and/or non-volatile memory such as: Flash, CD, tape, floppy disk, and hard disk to name a few. Keep in mind that the file system itself, and the data it manages, may or may not reside on the same device. Meaning, as shown in Figure 5.1, the data the file system manages can be located on some type of hardware storage medium located on the embedded system board or located on some other storage medium accessible to the embedded system (i.e., over a network, on a floppy disk, on a CD, etc.)

Demystifying Embedded Systems Middleware. DOI: 10.1016/B978-0-7506-8455-2.00005-4

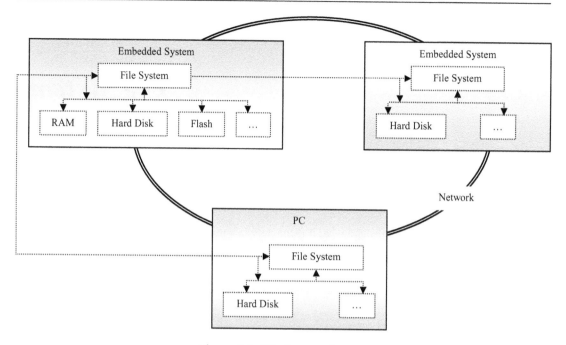

Figure 5.1: File System Access

5.2 How Does a File System Manage Data?

As implied in its name, a *file system* manages data in a fundamental element called a *file*. A file is simply a set of data that has been grouped together and assigned a unique 'name'. To maintain its relevance in the embedded device, a file system then must have a reliable and efficient scheme to create filenames, process filenames, and locate the files this metadata represents on the storage medium.

Real-world Advice

Know Your Standards!

File systems will adhere to standards for everything from naming scheme and convention (i.e., characters, size, encoding, etc.) to I/O APIs. For example, some implementations provide a standard asynchronous I/O API to interface to files located on the device that adheres to the international standard *IEEE 1003.1 POSIX* (portable operating system interface for computing environments), regardless of the underlying file system on the device. This asynchronous I/O API is a standard interface that is utilized by any embedded application to allow for simpler and faster portability of applications across different platforms that provide an application interface that adheres to this specification.

So, keep in mind when trying to understand a particular file system implementation that it may adhere to proprietary standards, industry specifications, or some combination of both.

The type of data contained in files is typically **NOT** constrained by the file system, meaning that as far as a file system is concerned, files can contain any kind of data or some combination of different types of data, such as graphics, source code, and/or document text to name a few. However, while the type of data within a file may not be relevant to a file system, whether or not data bits need to be structured in a particular way within a file can vary from file system to file system. Supported file structure types can range from unstructured, commonly referred to as *raw*, to rigidly *structured* data files of a particular size and format. For example, with file systems that support raw files, the file system essentially views data within a file as data bit streams comprised of 0s and 1s that can be freely accessed in any form and/or order by other users and/or software using the file system (see Figure 5.2). In short, a file system needs to support the structure of the data within a file in order for that particular file to be compatible with the file system.

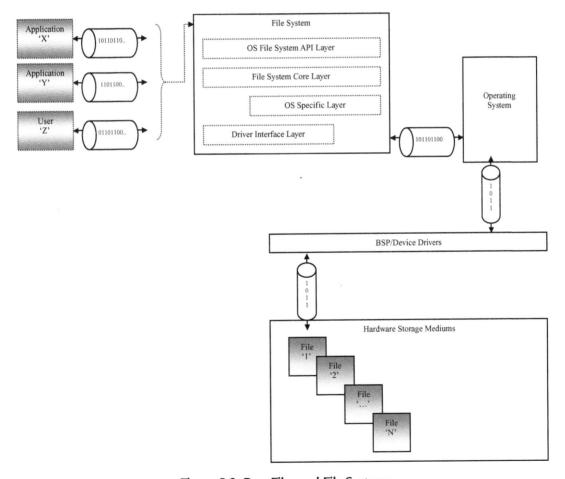

Figure 5.2: Raw Files and File Systems

The first steps to understanding the fundamentals and ultimately any file system implementation are:

Step 1. Understand what the purpose of the file system is within the system, and simply keep this in mind regardless of how complex a particular file system implementation is. As introduced at the start of this chapter, the purpose of a file system is to manage data stored on some type of storage medium located within the embedded device and/or some remotely accessible storage medium.

Step 2. Understand the APIs that are provided by a file system in support of a file system's inherent purpose. These APIs can, of course, differ from file system to file system – but in general include some combination of:
- Naming and creating files
- Configuring files
- Removing files
- Opening and closing files
- Writing to and reading from files
- Creating and configuring *directories* for groups of files
- Removing directories
- Reading directories
- Additional/extended functions
- File system creation, mounting, and unmounting
- Symbolic, hard, and/or dynamic links
- Journaling/atomic transactions

Step 3. Using the Embedded Systems Model, define and understand all required architecture components that underlie the file system, including:

Step 3.1. Know your file system-specific standards (see Chapter 3).

Step 3.2. Understanding the hardware (see Chapter 2). If the reader comprehends the hardware, it is easier to understand why a particular file system implements functionality in a certain way relative to the storage medium, as well as the hardware requirements of a particular file system implementation.

Step 3.3. Define and understand the specific underlying system software components, such as the available device drivers supporting the storage medium(s) and the operating system API (see Chapter 2).

Step 4. Define the particular file system architecture model based on an understanding of the generic file system model, and then define and understand what type of functionality and data exists at each layer. This includes file-system-specific data, such as data structures and the functions included at each layer. This step will be addressed in detail in a later section.

5.3 File System Data and the File System Reference Model

At the file system level, there are two general types of data:

- *User Content Data*. The data *files* that belong to the users and/or other software using the file system. As discussed at the start of this chapter, a file system typically does *not* constrain the type of content that can be in a file.
- *File-system-specific Data*. This includes data structures and metadata that are specific to that particular file system. Essentially, it is all the data and functionality in the file system implementation itself.

The key to understanding a file system implementation is by keeping in mind that 'all' the concepts and features provided by a file system are in support of the fundamental abstraction, the *file* containing user content data – and 'everything' that falls under file-system-specific data builds upon and revolves around this fundamental file system abstraction.

The components that make up a file system implementation can widely vary between designs from different vendors. However, to simplify understanding of all file system implementations it is useful to visualize that, at the highest level, all file systems contain some combination of the four components shown in the 'General File System Model' in Figure 5.3, specifically:

- a *File System Operation API* layer which contains the libraries with the defined file-level operational APIs that file system users, other middleware and applications can use to create, access, and manage files
- the *File System Core* layer manages file system data objects, metadata, and RAM usage by the file system. This layer is responsible for data management and the translation between the file system's view of the storage medium to how data are actually accessed through the device driver interface (i.e., blocks in Flash, sectors on a hard disk, etc.) and the operating system's file system interface
- the *OS Specific* layer is the interface to the embedded system's operating system
- the *Driver Interface* layer which is the interface to the hardware storage medium device drivers.

Remember!

The Model versus Real-world File System Implementations
Remember that what is shown in Figure 5.3 is a reference model, meaning some file systems may have a subset of these components, others have merged/split some of the functionality of various layers into fewer/more components, and/or may have additional components. However, this model is a powerful tool that the reader can use to understand the fundamentals of just about any file system implemented in an embedded system on the field today.

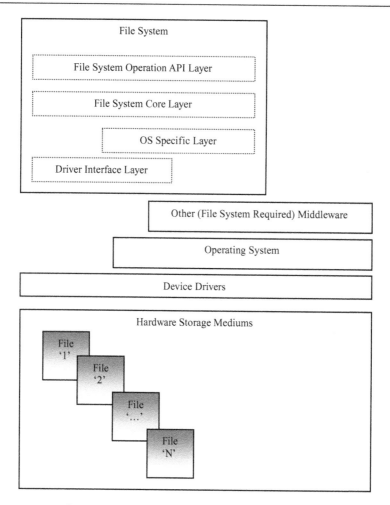

Figure 5.3: General File System Reference Model

These file system components work in conjunction with and interface to applications, other middleware, the embedded system's operating system and/or device drivers to provide file system functionality to higher layers of software. The next several sections will outline these layers in more detail.

5.3.1 Driver Interface Layer

As introduced in Chapter 2, the hardware storage medium(s) that the file system(s) interfaces (interface) to all require a device driver library to allow access to the hardware by other software components like the file system. Any file system code that utilizes these device drivers directly falls under the file system's ***device driver layer***. Figures 5.4a and 5.4b show that what specific file system components exist at the driver interface layer and how they are

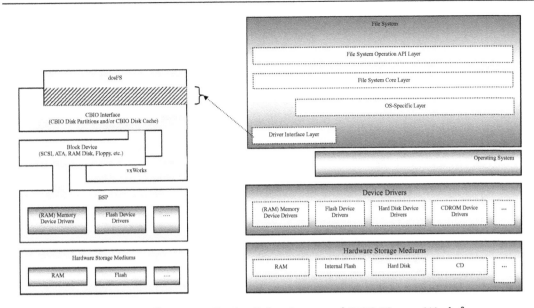

Figure 5.4a: File System Device Driver Layer and DOS FS on vxWorks[8]

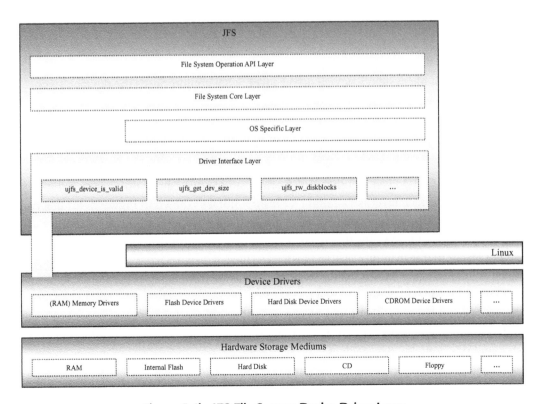

Figure 5.4b: JFS File System Device Driver Layer

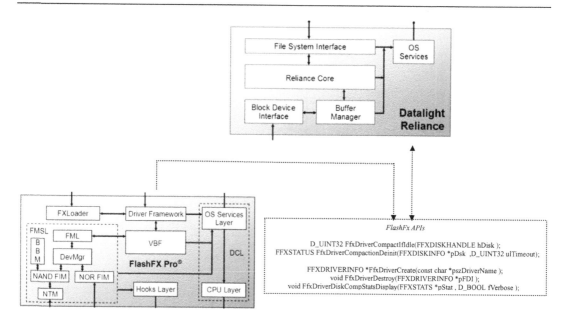

Figure 5.4c: Datalight's FlashFx High-level Diagram[14]

integrated into the device will vary depending on the underlying system software. In other words, relative to a file system's device driver layer, what compromises the device driver library will determine what and how hardware is accessible to the file system. The Figure 5.4a example is with a file system ported on a version of vxWorks that includes the CBIO interface, an underlying middleware component in itself. Any file system code that utilizes CBIO functions accessing block devices directly would fall under the device interface layer.

Like WindRiver's CBIO layer, another real-world example that can be utilized by a file system's driver interface layer is Datalight's FlashFx library (shown in Figure 5.4c) that can underlie FAT or Reliance embedded file systems. As its name implies, FlashFx (and libraries like it) is created for file systems that reside on Flash memory for the purpose of allowing overlying layers to transparently utilize Flash as a (block) disk device would be used. Flash memory is used in many embedded designs because aside from being programmable at run-time, Flash is considered competitive in terms of power requirements, size, amount of storage space, and price relative to other types of non-volatile memory. Libraries such as FlashFx also provide a simpler abstraction layer for overlying software to use that work around some of Flash memory's complexities, such as:

- supporting the different types of Flash requires different types of special programming schemes. This can include having to erase on a sector-by-sector basis, manage and optimizing timing for reads, writes, and erases, as well as requirements relative to used Flash only allowing write operations after a prior erase operation

• Flash memory lifetime is limited by a finite number of write and erase cycles, so any scheme that optimizes and limits the access of Flash helps insure that the Flash part will not wear out before the end of the device's lifecycle

• Flash memory types differ in terms of reliability. They can contain pre-existing defective blocks, and/or defective blocks can develop over time that require some type of scheme to manage bad blocks and protect data.

It is then important for middleware developers to understand the overall requirements of their device, and tune the associated parameters to real-world performance needs accordingly. For example, developers that use vxWorks have the option of using the FlashFx library with vxWorks with the Reliance file system or some other FAT file system. Access to parameters (examples shown in Table 5.1) is provided via the development environment and source

Table 5.1: Examples of Datalight's FlashFx Tuning Parameters[14]

FlashFx Parameter	Description
FFXCONF_(Flash Type)	At least one Flash type must be enabled that defines the type of Flash technologies that the driver will support.
FFXCONF_NANDSUPPORT	NAND Flash Support.
FFXCONF_NORSUPPORT	NOR Flash Support.
FFXCONF_ISWFSUPPORT	Intel Sibley Wireless Flash (ISWF) support.
FFXCONF_BBMSUPPORT	Bad Block management (BBM) support.
FFXCONF_(File System Type)	The types of file systems that will be overlying FlashFX.
FFXCONF_RELIANCESUPPORT	Reliance File System.
FFXCONF_FATSUPPORT	FAT File System.
FFXCONF_READAHEADENABLED	Disables/Enables the FlashFX adaptive read ahead feature.
FFX_MAX_DEVICES	The maximum number of devices which needs to be supported.
FFX_DEVn_FIMS (n = 0 ...max devs)	The FIMs (Flash Interface Modules) which will be associated with the device.
FFX_DEVn_NTMS (n = 0 ...max devs)	If a NAND-type of FIM is used, then a list of NTMs (NAND Technology Modules) associated with the device needs to be specified.
FFX_DEVn_SETTINGS (n = 0 ...max devs)	*UnchachedAddress* = base address of the Flash array. *ReservedLo, ReservedHi* = the amount of Flash at the start and end of the Flash array which FlashFX does not access. *MaxArraySize* = maximum amount of Flash to use in the Flash array.
FFX_DEVn_BBMFORMAT (n = 0 ...max devs)	BBM (Bad Block Management) format settings for the device.

code to developers, so that these components can be tuned to the functional and performance requirements for instance.

The example shown in Figure 5.4d is the JFS file system open source with functions that utilize device driver-level functionality.

5.3.2 OS Specific Layer

File system code that falls under the file system's *OS Specific layer* (see Figure 5.5a):

1. makes any OS kernel API calls, such as the Linux calls in the JFS source code example shown in Figure 5.5b.
2. utilizes the functionality provided by the OS interfaces in support of the file system. For example, in order to manage data files and directories a file system will store

```
/*  << devices.c >>
 *   Copyright (c) International Business Machines Corp., 2000                    - 2003
 *
 *   This program is free software;              you can redistribute it and/or modify
 *   it under the terms of the GNU General Public License as published                    by
 *   the Free Software Foundation; either version 2 of the License, or
 *   (at your option) any later version.
 *
 *   This program is distributed in the hope that it will be useful,
 *   but WITHOUT ANY WARRANTY;    without even the implied warranty of
 *   MERCHANTABILITY or FITNESS FOR A PARTICULAR PURPOSE.         See
 *   the GNU General Public License for more details.
 *
 *   You should have received a copy of the GNU General Public License
 *   along with this program;          if not, write to the Free Software
 *   Foundation, Inc        ., 59 Temple Place, Suite 330, Boston, MA 02111            - 1307 USA
 */
#include     <config.h>
#include     <errno.h>
#include     <fcntl.h>
#include     <unistd.h>
#include     <string.h>
#include     <stdio.h>
#include     <stdlib.h>

#ifdef    HAVE_SYS_MOUNT_H
#include     <sys/mount.h>
#endif

#include     <sys/types.h>
#include     <sys/stat.h>
#include     <sys/ioctl.h>
#if   defined   (__DragonFly__)
#include     <machine/param.h>
#include     <sys/diskslice.h>
#include     <sys/disklabel.h>
#endif

#include     "jfs_types.h"
#include     "jfs_filsys.h"
#include     "devices.h"
#include     "debug.h"
```

Figure 5.4d: JFS File System Device Driver Layer Function Code

```
/*
 * NAME: ujfs_get_dev_size
 *
 * FUNCTION: Uses the device driver interface to determine the raw capacity of
 *    the specified device.
 *
 * PRE   CONDITIONS:
 *
 * POST CONDITIONS:
 *
 * PARAMETERS:
 *    device    - device
 *    size    - filled in with size of device; not modified if failure occurs
 *
 * NOTES:
 *
 * DATA STRUCTURES:
 *
 * RETURNS: 0 if successful; anything else indicates failures
 */

int   ujfs_get_dev_size(FILE *device, int64_t *size)
{
          off_t Starting_Position;                /* position within file/device upon
                                                   * entry to this function. */
          off_t Current_Position = 16777215;      /* position we are attempting
                                                   * to read from. */
          off_t    Last_Valid_Position = 0;       /* Last position we could successfully
                                                   * read from. */
          off_t First_Invalid_Position;           /* first invalid position we attempted
                                                   * to read from/seek to. */

          int   Seek_Result;        /* value returned by lseek. */
          size_t Read_Resul      t = 0;           /* value returned by read. */
          int   rc;
          struct    stat stat_data;
          int   devfd = fileno(device);

          rc = fstat(devfd, &stat_data);
          if  (!rc && S_ISREG(stat_data.st_mode)) {
                    /* This is a regular file.             */
                    *size = (int64_t) ((stat_data.st_size / 1024) *              1024);
                    return    NO_ERROR;

          }
#ifdef    BLKGETSIZE64
          {
                    uint64_t sz;
                    if  (ioctl(devfd, BLKGETSIZE64, &sz) >= 0) {
                              *size = sz;
                              return    0;
                    }
          }
#endif
#ifdef    BLKGETSIZE
          {
                    unsigned    long  num_sectors = 0;

                    if  (ioctl(devfd, BLKGETSIZE, &num_sectors) >              = 0) {
                              /* for now, keep size as multiple of 1024, *
                               * not 512, so eliminate any odd sector.                    */
                              *size = PBSIZE * (int64_t) ((num_sectors / 2) * 2);
                              return    NO_ERROR;

                    }
          }
#endif
```

Figure 5.4d continued: JFS File System Device Driver Layer Function Code

```
#if    defined   (__DragonFly__)
            {
                    struct    diskslices dss;
                    struct    disklabel dl;
                    struct    diskslice *sliceinfo;
                    int    slice;
                    dev_t dev = stat_data.st_rdev;

                    rc = ioctl(devfd, DIOCGSLICEINFO, &dss);
                    if  (rc < 0)
                            return      - 1;

                    slice = dkslice(dev);
                    sliceinfo = &dss.dss_slices[slice];

                    DBG_TRACE(("ujfs_get_device_size: slice = %d                \ n" , slice));

                    if (sliceinfo) {
                            if  (slice == WHOLE_DISK_SLICE || slice == 0) {
                                    *size = (int64_t) sliceinfo        - >ds_size * dss.dss_secsize;
                                    DBG_TRACE(("ujfs_get_device_size: slice represents disk            \ n" ));
                            } else  {
                                    if  (sliceinfo      - >ds_label) {
                                            DBG_TRACE(("ujfs_get_device_size: slice has disklabel              \ n" ));
                                            rc = ioctl(devfd, DIOCGDINFO, &dl);
                                            if  (!rc) {
                                                    *size = (int64_t) dl.d_secperunit *
dss.dss_secsize;
                                            } else  {
                                                    return    ( - 1);
                                            }
                                    }
                            }
                    } else  {
                            return    ( - 1);
                    }

                    DBG_TRACE(("ujfs_get_device_size: size in bytes = %ld            \ n" , *size));
                    DBG_TRACE(("ujfs_get_device_size: size in megabytes = %ld          \ n" ,
                            *size / (1024 * 1024)));

                    return    0;
            }
#endif
            /*
             * If the ioctl above fails or is undefined, use a binary search to
             * find the last byte in the partition.           This works because an lseek to
             * a position within the partition does not return an error while an
             * lseek to a position beyond the end of the            partition does.     Note that
             * some SCSI drivers may log an 'access beyond end of device' error
             * message.
             */

            /* Save the starting position so that we can restore it when we are
             * done! */
            Starting_Position = ftello(device);
            if  (Starting_Position          < 0)
                    return      ERROR_SEEK;

            /*
             * Find a position beyond the end of the partition.            We will start by
             * attempting to seek to and read the 16777216th byte in the partition.
             * We start here because a JFS partition must be at least this big.              If
             * it is not, then we can not format it as JFS.
             */
            do  {
                    /* Seek to the location we wish to test. */
                    Seek_Result = fseeko(device, Current_Position, SEEK_SET);
                    if  (Seek_Result == 0) {
                            /* Can we read from this location? */
                            Read_Result = fgetc(d        evice);
                            if  (Read_Result != EOF) {
                                    /* The current test position is valid.            Save it
                                     * for future reference. */
                                    Last_Valid_Position = Current_Position;

                                    /* Lets calculate the next location to test. */
                                    Current_Position =          ((Current_Position + 1) * 2)
                                                    - 1;

                            }
                    }
            } while   ((Seek_Result == 0) && (Read_Result == 1));

            /*
             * We have exited the while loop, which means that Current Position is
             * beyond the end of the partition or is unreadable due to a hardware
             * problem (bad block).           Since the odds of hitting a bad block are very
             * low, we will ignore that condition for now.              If time becomes
             * available, then this issue can be revisited.
             */
```

Figure 5.4d continued: JFS File System Device Driver Layer Function Code

```
                    /* Is the drive greater than 16MB? */
                    if  (Last_Valid_Position == 0) {
                               /* Determine if drive is readable at all.                    If it is, the drive
                                * is too small.          If not, it could be a newly created partion,
                                * so we need to issue a different error message                   */
                               *size = 0;                    /* Indicates not readable at all */
                               Seek_Result = fseeko(device, Last_Valid_Position, SEEK_SET);
                               if  (Seek_Result == 0) {
                                          /* Can we read from this location? */
                                          Read_Result =      fgetc(device);
                                          if  (Read_Result != EOF)
                                                     /* non   - zero indicates readable, but too small */
                                                     *size = 1;
                               }
                               goto  restore;
                    }

        /*  The drive is larger than 16MB.             Now we must find out exactly how         large.
            * We now have a point within the partition and one beyond it.                    The end
            * of the partition must lie between the two.              We will use a binary
            * search to find it.*/

            /* Setup for the binary search. */
            First_Invalid_Position = Current_Position;
            Current_Po    sition = Last_Valid_Position +
                                          ((Current_Position          - Last_Valid_Position) / 2);

            /*
             * Iterate until the difference between the last valid position and the
             * first invalid position is 2 or less.
             */
            while   ((First_Invalid_Position          - Last_Valid_Position) > 2) {
                               /* Seek to the location we wish to test. */
                               Seek_Result = fseeko(device, Current_Position, SEEK_SET);
                               if  (Seek_Result == 0) {
                                          /* Can we read from this location? */
                                          Read_Result =      fgetc(device);
                                          if  (Read_Result != EOF) {
                                                     /* The current test position is valid.
                                                      * Save it for future reference. */
                                                     Last_Valid_Position = Current_Position;

                                                     /*
                                                      * Lets calculate the next location to test. It
                                                      * should be half way          between the current test
                                                      * position and the first invalid position that
                                                      * we know of.
                                                      */
                                                     Current_Position = Current_Position +
                                                                              ((First_Invalid_Position
                                                                              Last_Valid_Position) / 2);

                                          }
                               } else
                                          Read_Result = 0;

                               if  (Read_Result != 1) {
                                          /* Out test position is beyond the end of the partition.
                                           * It becomes our first known invalid position. */
                                          First_Invalid_Position = Current_Position;

                                          /* Our new test position should be half way between our
                                           * last kno     wn valid position and our current test
                                           * position. */
                                          Current_Position =
                                          Last_Valid_Position +
                                          ((Current_Position          - Last_Valid_Position) / 2);
                               }
            }

            /*
             * The size of the drive should be Last_Valid_Position + 1 as
             * Last_Valid_Position is an offset from the beginning of the partition.
             */
            *size = Last_Valid_Position + 1;

restore:
            /* Restore the original position. */
            if  (fseeko(device, Starting_Position, SEEK_SET) != 0)
                       return   ERROR_SEEK;

            return   NO_ERROR;
}
.........
```

Figure 5.4d continued: JFS File System Device Driver Layer Function Code

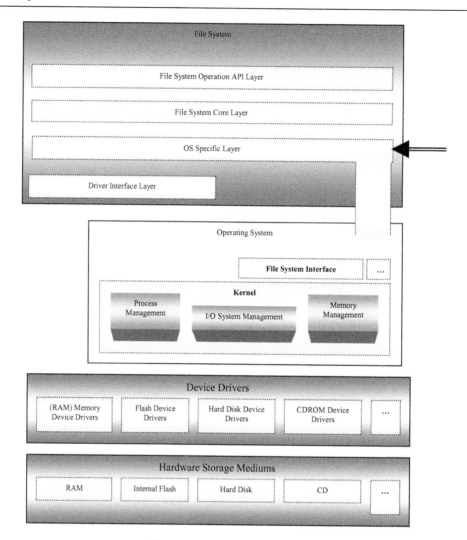

Figure 5.5a: OS Specific Layer

```
 *   the GNU General Public License for more details.
 *   You should have received a copy of the GNU General Public License
 *   along with this program; if not, write to the Free Software
 *   Foundation, Inc., 59 Temple Place, Suite 330, Boston, MA 02111-1307 USA
 */

#include <linux/config.h>
#include <linux/fs.h>
#include <linux/module.h>
#include <linux/blkdev.h>
#include <linux/completion.h>
#include <asm/uaccess.h>
#include "jfs_incore.h"
#include "jfs_filsys.h"
#include "jfs_metapage.h"
#include "jfs_superblock.h"
#include "jfs_dmap.h"
#include "jfs_imap.h"
#include "jfs_debug.h"

.....

static int __init init_jfs_fs(void)
{
        int rc;

        printk("JFS development version: $Name: v1_1_7 $\n");

        jfs_inode_cachep =
            kmem_cache_create("jfs_ip", sizeof (struct jfs_inode_info), 0, 0,init_once, NULL);
        if (jfs_inode_cachep == NULL)
                return -ENOMEM;

        /*Metapage initialization */
        rc = metapage_init();
        if (rc) {
                jfs_err("metapage_init failed w/rc = %d", rc);
                goto free_slab;
        }

        /* Transaction Manager initialization */
        rc = txInit();
        if (rc) {
                jfs_err("txInit failed w/rc = %d", rc);
                goto free_metapage;
        }

        /* I/O completion thread (endio) */
        jfsIOthread = kernel_thread(jfsIOWait, 0,CLONE_FS | CLONE_FILES | CLONE_SIGHAND);
        if (jfsIOthread < 0) {
                jfs_err("init_jfs_fs: fork failed w/rc = %d", jfsIOthread);
                goto end_txmngr;
        }
        wait_for_completion(&jfsIOwait);        /* Wait until thread starts */

        jfsCommitThread = kernel_thread(jfs_lazycommit, 0,CLONE_FS | CLONE_FILES | CLONE_SIGHAND);
        if (jfsCommitThread < 0) {
                jfs_err("init_jfs_fs: fork failed w/rc = %d", jfsCommitThread);
                goto kill_iotask;
        }
        wait_for_completion(&jfsIOwait);        /* Wait until thread starts */

        jfsSyncThread = kernel_thread(jfs_sync, 0,CLONE_FS | CLONE_FILES | CLONE_SIGHAND);
        if (jfsSyncThread < 0) {
                jfs_err("init_jfs_fs: fork failed w/rc = %d", jfsSyncThread);
                goto kill_committask;
        }
        wait_for_completion(&jfsIOwait);        /* Wait until thread starts */

#ifdef PROC_FS_JFS
        jfs_proc_init();
#endif

        return register_filesystem(&jfs_fs_type);

kill_committask:
        jfs_stop_threads = 1;

        wake_up(&jfs_commit_thread_wait);
        wait_for_completion(&jfsIOwait);        /* Wait for thread exit */
kill_iotask:
        jfs_stop_threads = 1;

        wake_up(&jfs_IO_thread_wait);
        wait_for_completion(&jfsIOwait);        /* Wait for thread exit */
end_txmngr:
        txExit();
free_metapage:
        metapage_exit();
free_slab:
        kmem_cache_destroy(jfs_inode_cachep);
        return rc;
}

.......
```

Figure 5.5b: JFS Source Example Utilizing Linux Kernel Calls

information, a.k.a. metadata, about each particular file and directory it is responsible for in some type of data structure typically provided by an operating system's interface API. The file system itself may then derive its own data structure(s) from the OS provided structure to be used internally, and in conjunction with the data structure provided by the operating system. Metadata stored in these data structures will vary from file system to file system depending on the requirements of the embedded device, but generally includes such data as:

- location of file or directory on hardware storage medium
- the size of the file or directory
- the type of file
- the date the file or directory was created and/or last modified
- the file or directory owner
- file or directory permissions, such as read-only, read-write, shared, etc.

to name a few. While the semantics will vary as to what this directory/file descriptor data structure is called in a particular file system implementation, its purpose and the general type of data it contains are consistent with other file systems. Figures 5.5c and 5.5d

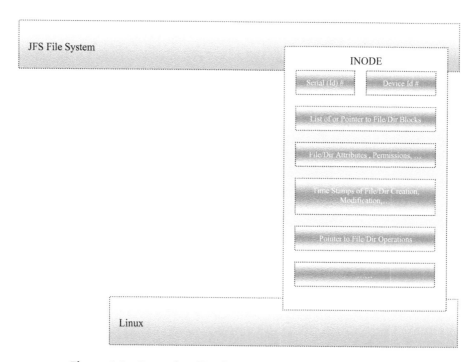

Figure 5.5c: Example of Inode Data Structure Block Diagram

```
/* <<file.c in JFS>>
 * Copyright (c) International Business Machines Corp., 2000-2002
 * Portions Copyright (c) Christoph Hellwig, 2001-2002
 *
 * This program is free software; you can redistribute it and/or
 * modify it under the terms of the GNU General Public License as
 * published by the Free Software Foundation; either version 2 of
 * the License, or (at your option) any later version.
 *
 * This program is distributed in the hope that it will be useful,
 * but WITHOUT ANY WARRANTY; without even the implied warranty of
 * MERCHANTABILITY or FITNESS FOR A PARTICULAR PURPOSE.  See
 * the GNU General Public License for more details.
 *
 * You should have received a copy of the GNU General Public
 * License along with this program; if not, write to the Free
 * Software Foundation, Inc., 59 Temple Place, Suite 330, Boston,
 * MA 02111-1307 USA
 */
static int jfs_open(struct inode *inode, struct file *file)
{
int rc;

if ((rc = generic_file_open(inode, file)))
        return rc;

/*
 * We attempt to allow only one "active" file open per aggregate
 * group.  Otherwise, appending to files in parallel can cause
 * fragmentation within the files.
 *
 * If the file is empty, it was probably just created and going
 * to be written to.  If it has a size, we'll hold off until the
 * file is actually grown.
 */
if (S_ISREG(inode->i_mode) && file->f_mode & FMODE_WRITE &&
                            (inode->i_size == 0))
{
        struct jfs_inode_info *ji = JFS_IP(inode);
        spin_lock_irq(&ji->ag_lock);
        if (ji->active_ag == -1) {
          ji->active_ag = ji->agno;
          atomic_inc(&JFS_SBI(inode->i_sb)->
                     bmap->db_active[ji->agno]);

        }
        spin_unlock_irq(&ji->ag_lock);
}

return 0;
}
```

```
/* << fs.h in Linux>>
 *  This program is free software; you can redistribute
 *  it and/or modify it under the terms of the GNU General
 *  Public License as published by the Free Software Foundation;
 *  either version 2 of the License, or (at your option)
 *  any later version.
 *
 *  This program is distributed in the hope that it will be useful,
 *  but WITHOUT ANY WARRANTY; without even the implied warranty of
 *  MERCHANTABILITY or FITNESS FOR A PARTICULAR PURPOSE.  See
 *  the GNU General Public License for more details.
 *
 *  You should have received a copy of the GNU General Public License
 *  along with this program; if not, write to the Free Software
 *  Foundation, Inc., 59 Temple Place, Suite 330,
 *  Boston, MA 02111-1307 USA
 */

struct inode {
  ......

  unsigned long              i_inostamp;//inode fileset stamp
  unsigned long              i_inodenumber; //inode serial number
  unsigned long              i_devnumber; //device id number

  umode_t                    i_mode; //attributes,formats,permissions,...
  nlink_t                    i_nlink;//number of object links
  uid_t                      i_uid;//file owner's user id
  gid_t                      i_gid;//file owner's group id
  kdev_t                     i_rdev;//special file device id
  loff_t                     i_size; //inode inline data size

  time_t                     i_atime;//timestamp of data last accessed
  time_t                     i_mtime; //timestamp of data last modified
  time_t                     i_ctime; //timestamp of data last changed
  time_t                     i_otime; //timestamp of file/dir creation

  unsigned long              i_blksize;//block size
  unsigned long              i_numblocks;//number of allocated blocks
  struct semaphore           i_devicesem; //device locking semaphore
  struct inode_operations    *i_op; //file inode  struct file_operations
  struct super_block         *i_sb; //JFS-private superblock info

  struct address_space       *i_mapping;//address space operations
  struct dquot               *i_dquot[MAXQUOTAS];//disk quota options
  struct block_device        *i_bdev;//pointer to block device
  struct char_device         *i_cdev; //pointer to character device

  unsigned long              i_dnotify_mask; // directory notify events
  struct dnotify_struct      *i_dnotify; // for directory notifications

  unsigned long              i_state;//data related inode  unsigned char

  ......
```

```
                    #define di_xtroot
                    u._file._u2._xtroot
                    #define di_dxd
                    u._file._u2._special._d
        xd
                    #define di_btroot
                    di_xtroot
                    #define di_inlinedata
                    u._file._u2._special._u
                    #define di_rdev
                    u._file._u2._special._u
        ._rdev
                    #define di_fastsymlink
                    u._file._u2._special._u
        ._fastsymlink
                    #define di_inlineea
        u._file._u2._special._inlineea
        } u;
};
......
```

Figure 5.5d: Inode Data Structure JFS and Linux Inode Source Code Example

show a block diagram and sample code of a directory/file descriptor data structure in a Linux-supported implementation, commonly referred to as an ***inode***, containing metadata type fields.

It is because of a directory/file descriptor data structure that a file system is able to create the *illusion* that a file is a contiguous entity to file system users and applications, even if that is not how the file is stored in the storage medium. Remember that, at the hardware level, a file system views the storage device as broken down into smaller-sized addressable storage units.

Depending on the size of a file, the data within a file can comprise one or more of these addressable storage units. Moreover, these units may or may not be contiguous, thus the need to track the units that comprise a file in a data structure like a directory/file descriptor data structure. Then, as shown in Figure 5.5e, a file system utilizes a directory/file descriptor data structure in order to translate to and from the physical data addresses in order to locate and manage the data unit(s) that comprise a file.

5.3.3 File System Core Layer

At the heart of any file system's core layer (see Figure 5.6a) are the directory/file descriptor data structures utilized to manage the data. This means the functionality included at this level revolves around these data structures, and at a minimum includes some combination of:

- directory and file descriptor data structure management
- data storage management
- directory management.

5.3.4 Directory and File Descriptor Data Structure Management

The file system core layer includes functionality that manages the set of directory/file descriptor data structures that represent the various files and directories accessible to the file system, such as the creation of a descriptor when a file or directory is created, and/or the management of the file system's *control block* (shown in Figure 5.6a). The control block is an allocated portion of the storage medium for file system-related information storage and retrieval to/from RAM. JFS, for instance, has a relative control block on the storage medium it supports, commonly referred to as the *superblock* in this and some other file system implementations. The JFS source code example in Figure 5.6b shows an inode operations library for managing inodes, as well as code to manage inode-related data.

File system implementations, also, may include with their directory/file descriptor data structure management scheme some additional **log management** functionality. These logs track file

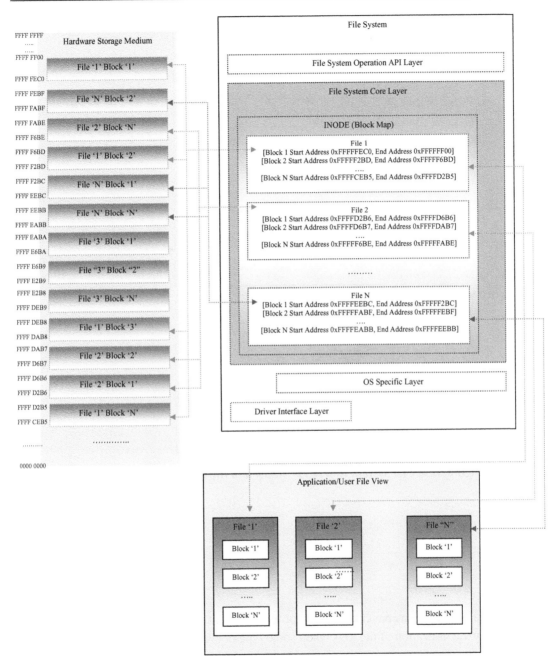

Figure 5.5e: General Directory/File Descriptor Data Structure Block General Translation Example

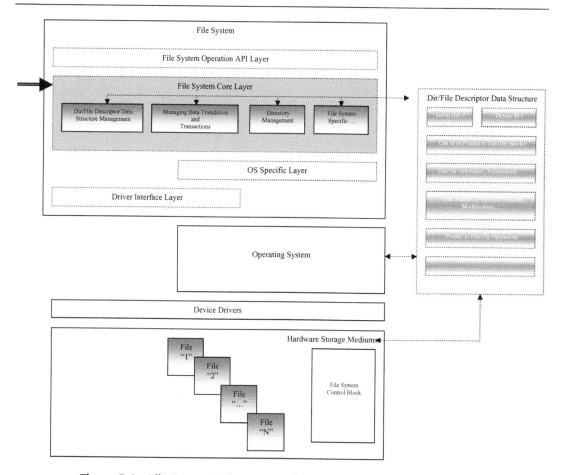

Figure 5.6a: File System Reference Model and the File System Core Layer

system operations and data changes to allow for improvement of file system data integrity and recoverability via utilization of the logs when some type of system failure has occurred. Log management in these file systems is typically implemented in support of what are commonly referred to as *(atomic) transactional* and/or *journaling* file systems, where by definition these file systems are intended to be more reliable. Figure 5.6c shows a systems-level example of a transactional file system (TRFS) implemented in a vxWorks-based system, whereas Figures 5.6d and 5.6e show examples of IBM's JFS (journaled file system) log management library.

```
/* <inode.c>
 * Copyright (c) International Business Machines Corp., 2000-2002
 * Portions Copyright (c) Christoph Hellwig, 2001-2002
 *
 * This program is free software; you can redistribute it and/or modify
 * it under the terms of the GNU General Public License as published by
 * the Free Software Foundation; either version 2 of the License, or
 * (at your option) any later version.
 *
 * This program is distributed in the hope that it will be useful,
 * but WITHOUT ANY WARRANTY; without even the implied warranty of
 * MERCHANTABILITY or FITNESS FOR A PARTICULAR PURPOSE. See
 * the GNU General Public License for more details.
 *
 * You should have received a copy of the GNU General Public License
 * along with this program; if not, write to the Free Software
 * Foundation, Inc., 59 Temple Place, Suite 330, Boston, MA 02111-1307 USA
 */

#include <linux/fs.h>
#include "jfs_incore.h"
#include "jfs_filsys.h"
#include "jfs_dmap.h"
#include "jfs_imap.h"
#include "jfs_extent.h"
#include "jfs_unicode.h"
#include "jfs_debug.h"


......

void jfs_clear_inode(struct inode *inode)
{
        struct jfs_inode_info *ji = JFS_IP(inode);

        if (is_bad_inode(inode))
                /*
                 * We free the fs-dependent structure before making the
                 * inode bad
                 */
                return;

        jfs_info("jfs_clear_inode called ip = 0x%p", inode);

        spin_lock_irq(&ji->ag_lock);
        if (ji->active_ag != -1) {
                struct bmap *bmap = JFS_SBI(inode->i_sb)->bmap;
                atomic_dec(&bmap->db_active[ji->active_ag]);
                ji->active_ag = -1;
        }
        spin_unlock_irq(&ji->ag_lock);

        ASSERT(list_empty(&ji->anon_inode_list));

        if (ji->atlhead) {
                jfs_err("jfs_clear_inode: inode %p has anonymous tlocks",
                        inode);
                jfs_err("i_state = 0x%lx, cflag = 0x%lx", inode->i_state,
                        ji->cflag);
        }

        free_jfs_inode(inode);
}
```

Figure 5.6b: Example of JFS Inode Operations

```
void jfs_read_inode(struct inode *inode)
{
        int rc;

        rc = alloc_jfs_inode(inode);
        if (rc) {
                jfs_warn("In jfs_read_inode, alloc_jfs_inode failed");
                goto bad_inode;
        }
        jfs_info("In jfs_read_inode, inode = 0x%p", inode);

        if (diRead(inode))
                goto bad_inode_free;

        if (S_ISREG(inode->i_mode)) {
                inode->i_op = &jfs_file_inode_operations;
                inode->i_fop = &jfs_file_operations;
                inode->i_mapping->a_ops = &jfs_aops;
        } else if (S_ISDIR(inode->i_mode)) {
                inode->i_op = &jfs_dir_inode_operations;
                inode->i_fop = &jfs_dir_operations;
                inode->i_mapping->a_ops = &jfs_aops;
                inode->i_mapping->gfp_mask = GFP_NOFS;
        } else if (S_ISLNK(inode->i_mode)) {
                if (inode->i_size >= IDATASIZE) {
                        inode->i_op = &page_symlink_inode_operations;
                        inode->i_mapping->a_ops = &jfs_aops;
                } else
                        inode->i_op = &jfs_symlink_inode_operations;
        } else {
                inode->i_op = &jfs_special_inode_operations;
                init_special_inode(inode, inode->i_mode,
                                        kdev_t_to_nr(inode->i_rdev));
        }

        return;

 bad_inode_free:
        free_jfs_inode(inode);
 bad_inode:
        make_bad_inode(inode);
}

/* This define is from fs/open.c */
#define special_file(m) (S_ISCHR(m)||S_ISBLK(m)||S_ISFIFO(m)||S_ISSOCK(m))

/*
 * Workhorse of both fsync & write_inode
 */
int jfs_commit_inode(struct inode *inode, int wait)
{
        int rc = 0;
        tid_t tid;
        static int noisy = 5;

        jfs_info("In jfs_commit_inode, inode = 0x%p", inode);

        /*
         * Don't commit if inode has been committed since last being
         * marked dirty, or if it has been deleted.
         */
        if (test_cflag(COMMIT_Nolink, inode) ||
            !test_cflag(COMMIT_Dirty, inode))
                return 0;

        if (isReadOnly(inode)) {
                /* kernel allows writes to devices on read-only
                 * partitions and may think inode is dirty
                 */
                if (!special_file(inode->i_mode) && noisy) {
                        jfs_err("jfs_commit_inode(0x%p) called on "
                                        "read-only volume", inode);
                        jfs_err("Is remount racy?");
                        noisy--;
                }
                return 0;
        }

        tid = txBegin(inode->i_sb, COMMIT_INODE);
        down(&JFS_IP(inode)->commit_sem);
        rc = txCommit(tid, 1, &inode, wait ? COMMIT_SYNC : 0);
        txEnd(tid);
        up(&JFS_IP(inode)->commit_sem);
        return rc;
}
```

Figure 5.6b continued: Example of JFS Inode Operations

```
void jfs_write_inode(struct inode *inode, int wait)
{
        if (test_cflag(COMMIT_Nolink, inode))
                return;
        /*
         * If COMMIT_DIRTY is not set, the inode isn't really dirty.
         * It has been committed since the last change, but was still
         * on the dirty inode list.
         */
        if (!test_cflag(COMMIT_Dirty, inode)) {
                /* Make sure committed changes hit the disk */
                jfs_flush_journal(JFS_SBI(inode->i_sb)->log, wait);
                return;
        }

        if (jfs_commit_inode(inode, wait)) {
                jfs_err("jfs_write_inode: jfs_commit_inode failed!");
        }
}

void jfs_delete_inode(struct inode *inode)
{
        jfs_info("In jfs_delete_inode, inode = 0x%p", inode);

        if (test_cflag(COMMIT_Freewmap, inode))
                freeZeroLink(inode);

        diFree(inode);

        clear_inode(inode);
}

void jfs_dirty_inode(struct inode *inode)
{
        static int noisy = 5;

        if (isReadOnly(inode)) {
                if (!special_file(inode->i_mode) && noisy) {
                        /* kernel allows writes to devices on read-only
                         * partitions and may try to mark inode dirty
                         */
                        jfs_err("jfs_dirty_inode called on read-only volume");
                        jfs_err("Is remount racy?");
                        noisy--;
                }
                return;
        }

        set_cflag(COMMIT_Dirty, inode);
}
......
```

Figure 5.6b continued: Example of JFS Inode Operations

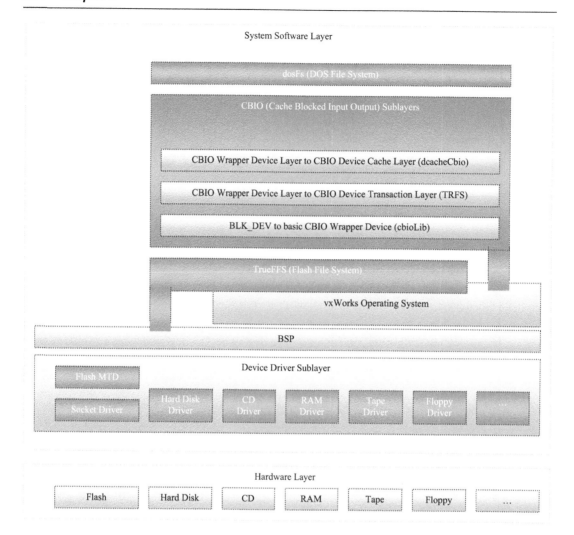

Figure 5.6c: Example of Transactional File System (TRFS) and vxWorks

```
/*
                 * update transaction lsn;
                 */
                else {
                        /* inherit oldest/smallest lsn of page */
                        logdiff(diffp, mp->lsn, log);
                        logdiff(difft, tblk->lsn, log);
                        if (diffp < difft) {
                                /* update tblock lsn with page lsn */
                                tblk->lsn = mp->lsn;

                                /* move tblock after page on logsynclist */
                                list_del(&tblk->synclist);
                                list_add(&tblk->synclist, &mp->synclist);
                        }
                }

        LOGSYNC_UNLOCK(log);

        /*
         *      write the log record
         */
    writeRecord:
        lsn = lmWriteRecord(log, tblk, lrd, tlck);

        /*
         * forward log syncpt if log reached next syncpt trigger
         */
        logdiff(diffp, lsn, log);
        if (diffp >= log->nextsync)
                lsn = lmLogSync(log, 0);

        /* update end-of-log lsn */
        log->lsn = lsn;

        LOG_UNLOCK(log);

        /* return end-of-log address */
        return lsn;
}


/*
 * NAME:        lmWriteRecord()
 *
 * FUNCTION:        move the log record to current log page
 *
 * PARAMETER:       cd          - commit descriptor
 *
 * RETURN:          end-of-log address
 *
 * serialization: LOG_LOCK() held on entry/exit
 */
static int
lmWriteRecord(struct jfs_log * log, struct tblock * tblk, struct lrd * lrd,
              struct tlock * tlck)
{
        int lsn = 0;                    /* end-of-log address */
        struct lbuf *bp;        /* dst log page buffer */
        struct logpage *lp; /* dst log page */
        caddr_t dst;                    /* destination address in log page */
        int dstoffset;                  /* end-of-log offset in log page */
        int freespace;                  /* free space in log page */
        caddr_t p;                      /* src meta-data page */
        caddr_t src;
        int srclen;
        int nbytes;                     /* number of bytes to move */
        int i;
        int len;
        struct linelock *linelock;
        struct lv *lv;
        struct lvd *lvd;
        int l2linesize;

        len = 0;

        /* retrieve destination log page to write */
        bp = (struct lbuf *) log->bp;
        lp = (struct logpage *) bp->l_ldata;
        dstoffset = log->eor;

        /* any log data to write ? */
        if (tlck == NULL)
                goto moveLrd;

        /*
         *      move log record data
         */
        /* retrieve source meta-data page to log */
        if (tlck->flag & tlckPAGELOCK) {
                p = (caddr_t) (tlck->mp->data);
                linelock = (struct linelock *) & tlck->lock;
        }
```

Figure 5.6d: Example of JFS Log Manager Utilized for Journaling

```
                   * update transaction lsn;
                   */
            else {
                       /* inherit older relost lsn of page */
                    logdiff(diffp, mp->lsn, log);
                    logdiff(difft, tblk->lsn, log);
                    if (diffp < difft) {
                            /* update tblock lsn with older lsn */
                            tblk->lsn = mp->lsn;

                            /* move tblock after page on logsynclist */
                            list_del(&tblk->synclist);
                            list_add(&tblk->synclist, &mp->synclist);
                    }
            }

            LOGSYNC_UNLOCK(log);

            /*
             *      write log record
             */
    writeRecord:
            lsn = lmWriteRecord(log, tblk, lrd, tlck);

            /*
             * forward log syncpt if log reached next syncpt trigger
             */
            logdiff(diffp, lsn, log);
            if (diffp >= log->nextsync)
                    lsn = lmLogSync(log, 0);

            /* update end-of-log lsn */
            log->lsn = lsn;

            LOG_UNLOCK(log);

            /* return end-of-log address */
            return lsn;
}


/*
 * NAME:    lmWriteRecord()
 *
 * FUNCTION:        move log record to current log page
 *
 * PARAMETER:       cd              - commit descriptor
 *
 * RETURN:          end-of-log address
 *
 * serialization: LOG_LOCK() held on entry/exit
 */
static int
lmWriteRecord(struct jfs_log * log, struct tblock * tblk, struct lrd * lrd,
              struct tlck * tlck)
{
        int lsn = 0;                     /* end-of-log address */
        struct lbuf *bp;        /* dst log page buffer */
        struct logpage *lp;     /* dst log page */
        caddr_t dst;                     /* destination address in log page */
        int dstoffset;                   /* end-of-log offset in log page */
        int freespace;                   /* free space in log page */
        caddr_t p;                       /* src meta-data page */
        caddr_t src;
        int srclen;
        int nbytes;                      /* number of bytes to move */
        int i;
        int len;
        struct linelock *linelock;
        struct lv *lv;
        struct lvd *lvd;
        int l2linesize;

        len = 0;

        /* retrieve destination log page to write */
        bp = (struct lbuf *) log->bp;
        lp = (struct logpage *) bp->l_ldata;
        dstoffset = log->eor;

        /* any log data to write ? */
        if (tlck == NULL)
                goto moveLrd;

        /*
         *      move log record data
         */
        /* retrieve source meta-data page to log */
        if (tlck->flag & tlckPAGELOCK) {
                p = (caddr_t) (tlck->mp->data);
                linelock = (struct linelock *) & tlck->lock;
        }
```

Figure 5.6d continued: Example of JFS Log Manager Utilized for Journaling

```
        * Ensure that inode isn't reused before
        * lazy commit thread finishes processing
        */
    if (tblk->xflag & (COMMIT_CREATE | COMMIT_DELETE)) {
            atomic_inc(&tblk->ip->i_count);
            /*
             * Avoid a race deadlock.
             *
             * If the inode is locked, we may be blocked in
             * txCommit_inode.  If so, we don't want the
             * lazy_commit thread doing the last iput() on the inode
             * since that may block on the locked inode.  Instead,
             * commit the transaction synchronously, so the last iput
             * will be done by the calling thread (or later)
             */
            if (tblk->ip->i_state & I_LOCK)
                    tblk->xflag &= ~COMMIT_LAZY;
    }

    ASSERT((!(tblk->xflag & COMMIT_DELETE)) ||
            ((tblk->ip->i_nlink == 0) &&
            !test_cflag(COMMIT_Nolink, tblk->ip)));

    /*
     *      write COMMIT log record
     */
    lrd->type = cpu_to_le16(LOG_COMMIT);
    lrd->length = 0;
    lsn = lmLog(log, tblk, lrd, NULL);

    lmGroupCommit(log, tblk);

    /*
     *      - transaction is now committed -
     */

    /*
     * force pages in careful update
     * (imap addressing structure update)
     */
    if (flag & COMMIT_FORCE)
            txForce(tblk);

    /*
     *      update allocation map.
     *
     * update inode allocation map and inode:
     * free pager lock on memory object of inode if any.
     * update  block allocation map.
     *
     * txUpdateMap() resets XAD_NEW in XAD.
     */
    if (tblk->xflag & COMMIT_FORCE)
            txUpdateMap(tblk);

    /*
     *      free transaction locks and pageout/free pages
     */
    txRelease(tblk);

    if ((tblk->flag & tblkGC_LAZY) == 0)
            txUnlock(tblk);

    /*
     *      reset in-memory object state
     */
    for (k = 0; k < cd.nip; k++) {
            ip = cd.iplist[k];
            jfs_ip = JFS_IP(ip);

            /*
             * reset in-memory inode state
             */
            jfs_ip->bxflag = 0;
            jfs_ip->blid = 0;
    }

out:
    if (rc != 0)
            txAbort(tid, 1);

TheEnd:
    jfs_info("txCommit: tid = %d, returning %d", tid, rc);
    return rc;
}
```

Figure 5.6d continued: Example of JFS Log Manager Utilized for Journaling

```
/* jfs_txnmgr.c: transaction manager
 * Copyright (C) International Business Machines Corp., 2000-2004
 * Portions Copyright (C) Christoph Hellwig, 2001-2002
 *
 * This program is free software; you can redistribute it and/or modify
 * it under the terms of the GNU General Public License as published by
 * the Free Software Foundation; either version 2 of the License, or
 * (at your option) any later version.
 *
 * This program is distributed in the hope that it will be useful,
 * but WITHOUT ANY WARRANTY; without even the implied warranty of
 * MERCHANTABILITY or FITNESS FOR A PARTICULAR PURPOSE. See
 * the GNU General Public License for more details.
 *
 * You should have received a copy of the GNU General Public License
 * along with this program; if not, write to the Free Software
 * Foundation, Inc., 59 Temple Place, Suite 330, Boston, MA 02111-1307 USA
 * notes:
 * transaction starts with txBegin() and ends with txCommit()
 * or txAbort().
 * transaction commit management
 * -----------------------------
 */
......

/*
 * NAME: txCommit()
 *
 * FUNCTION: commit the changes to the objects specified in
 * clist. For journalled segments only the
 * changes of the caller are committed, ie by tid.
 * for non-journalled segments the data are flushed to
 * disk and then the change to the disk inode and indirect
 * blocks committed (so blocks newly allocated to the
 * segment will be made a part of the segment atomically).
 *
 * all of the segments specified in clist must be in
 * one file system. no more than 6 segments are needed
 * to handle all unix svcs.
 *
 * if the i_nlink field (i.e. disk inode link count)
 * is zero, and the type of inode is a regular file or
 * directory, or symbolic link , the inode is truncated
 * to zero length. the truncation is committed but the
 * VM resources are unaffected until it is closed (see
 * iput and iclose).
 *
 * PARAMETER:
 *
 * RETURN:
 *
 * serialization:
 * on entry the inode lock on each segment is assumed
 * to be held.
 *
 * i/o error:
 */
int txCommit(tid_t tid,              /* transaction identifier */
        int nip,             /* number of inodes to commit */
        struct inode **iplist,        /* list of inode to commit */
        int flag)
{
        int rc = 0;
        struct commit cd;
        struct jfs_log *log;
        struct tblock *tblk;
        struct lrd *lrd;
        int lsn;
        struct inode *ip;
        struct jfs_inode_info *jfs_ip;
        int k, n;
        ino_t top;
        struct super_block *sb;

        jfs_info("txCommit, tid = %d, flag = %d", tid, flag);
        /* is read-only file system ? */
        if (isReadOnly(iplist[0])) {
                rc = -EROFS;
                goto TheEnd;
        }

        sb = cd.sb = iplist[0]->i_sb;
        cd.tid = tid;

        if (tid == 0)
                tid = txBegin(sb, 0);
        tblk = tid_to_tblock(tid);
```

Figure 5.6e: Example of JFS Transaction Manager Using JFS Log Manager for Journaling

```
/*
        * initialize commit structure
        */
       log = JFS_SBI(sb)->log;
       cd.log = log;

       /* initialize log record descriptor in commit */
       lrd = &cd.lrd;
       lrd->logtid = cpu_to_le32(tblk->logtid);
       lrd->backchain = 0;

       tblk->xflag |= flag;

       if ((flag & (COMMIT_FORCE | COMMIT_SYNC)) == 0)
               tblk->xflag |= COMMIT_LAZY;
       /*
        * prepare non-journaled objects for commit
        *
        * flush data pages of non-journaled file
        * to prevent the file getting non-initialized disk blocks
        * in case of crash.
        * (new blocks - )
        */
       cd.iplist = iplist;
       cd.nip = nip;

       /*
        * acquire transaction lock on (on-disk) inodes
        *
        * update on-disk inode from in-memory inode
        * acquiring transaction locks for AFTER records
        * on the on-disk inode of file object
        *
        * sort the inodes array by inode number in descending order
        * to prevent deadlock when acquiring transaction lock
        * of on-disk inodes on multiple on-disk inode pages by
        * multiple concurrent transactions
        */
       for (k = 0; k < cd.nip; k++) {
               top = (cd.iplist[k])->i_ino;
               for (n = k + 1; n < cd.nip; n++) {
                       ip = cd.iplist[n];
                       if (ip->i_ino > top) {
                               top = ip->i_ino;
                               cd.iplist[n] = cd.iplist[k];
                               cd.iplist[k] = ip;
                       }
               }

               ip = cd.iplist[k];
               jfs_ip = JFS_IP(ip);

               if (test_and_clear_cflag(COMMIT_Syncdata, ip) &&
                ((tblk->flag && COMMIT_DELETE) == 0))
                       fsync_inode_data_buffers(ip);

               /*
                * Mark inode as not dirty. It will still be on the dirty
                * inode list, but we'll know not to commit it again unless
                * it gets marked dirty again
                */
               clear_cflag(COMMIT_Dirty, ip);

               /* inherit anonymous tlock(s) of inode */
               if (jfs_ip->atlhead) {
                       lid_to_tlock(jfs_ip->atltail)->next = tblk->next;
                       tblk->next = jfs_ip->atlhead;
                       if (!tblk->last)
                               tblk->last = jfs_ip->atltail;
                       jfs_ip->atlhead = jfs_ip->atltail = 0;
                       TXN_LOCK();
                       list_del_init(&jfs_ip->anon_inode_list);
                       TXN_UNLOCK();
               }

               /*
                * acquire transaction lock on on-disk inode page
                * (become first tlock of the tblk's tlock list)
                */
               if (((rc = diWrite(tid, ip))))
                       goto out;
       }

       /*
        * write log records from transaction locks
        *
        * txUpdateMap() resets XAD_NEW in XAD.
        */
       if ((rc = txLog(log, tblk, &cd)))
               goto TheEnd;
```

Figure 5.6e continued: Example of JFS Transaction Manager Using JFS Log Manager for Journaling

```
/*
                 * Ensure that inode isn't reused before
                 * lazy commit thread finishes processing
                 */
                if (tblk->xflag & (COMMIT_CREATE | COMMIT_DELETE)) {
                        atomic_inc(&tblk->ip->i_count);
                        /*
                         * Avoid a rare deadlock
                         *
                         * If the inode is locked, we may be blocked in
                         * jfs_commit_inode. If so, we don't want the
                         * lazy_commit thread doing the last iput() on the inode
                         * since that may block on the locked inode. Instead,
                         * commit the transaction synchronously, so the last iput
                         * will be done by the calling thread (or later)
                         */
                        if (tblk->ip->i_state & I_LOCK)
                                tblk->xflag &= ~COMMIT_LAZY;
                }

                ASSERT((!(tblk->xflag & COMMIT_DELETE)) ||
                 ((tblk->ip->i_nlink == 0) &&
                        !test_cflag(COMMIT_Nolink, tblk->ip)));

                /*
                 * write COMMIT log record
                 */
                lrd->type = cpu_to_le16(LOG_COMMIT);
                lrd->length = 0;
                lsn = lmLog(log, tblk, lrd, NULL);

                lmGroupCommit(log, tblk);

                /*
                 * - transaction is now committed -
                 */

                /*
                 * force pages in careful update
                 * (imap addressing structure update)
                 */
                if (flag & COMMIT_FORCE)
                        txForce(tblk);

                /*
                 * update allocation map.
                 *
                 * update inode allocation map and inode:
                 * free pager lock on memory object of inode if any.
                 * update block allocation map.
                 *
                 * txUpdateMap() resets XAD_NEW in XAD.
                 */
                if (tblk->xflag & COMMIT_FORCE)
                        txUpdateMap(tblk);

                /*
                 * free transaction locks and pageout/free pages
                 */
                txRelease(tblk);

                if ((tblk->flag & tblkGC_LAZY) == 0)
                        txUnlock(tblk);


                /*
                 * reset in-memory object state
                 */
                for (k = 0; k < cd.nip; k++) {
                        ip = cd.iplist[k];
                        jfs_ip = JFS_IP(ip);

                        /*
                         * reset in-memory inode state
                         */
                        jfs_ip->bxflag = 0;
                        jfs_ip->blid = 0;
                }

out:
        if (rc != 0)
                txAbort(tid, 1);

TheEnd:
        jfs_info("txCommit: tid = %d, returning %d", tid, rc);
        return rc;
}
```

Figure 5.6e continued: Example of JFS Transaction Manager Using JFS Log Manager for Journaling

5.3.5 Data Storage Management

At the core of a file system's data management scheme is the ability to locate and manage the data blocks belonging to each file located on the hardware storage medium(s). The file descriptor data structure records the blocks that are associated with a particular file, as well as where to locate these blocks in some type of ***block map*** (see Figure 5.6f).

While how a file descriptor data structure records the block data information in its block map will differ between file systems, the most common algorithms include one or some combination of:

- **Direct Addressing**, where the block map contains a list of the data block addresses that make up the file.
- **Indirect Addressing**, where the block map contains a pointer to another block, referred to as the *indirect block*. The indirect block then contains a list of the data block addresses that make up the file. This allows for a file system to support larger file sizes over direct addressing without having to dramatically increase the size of the file descriptor data structure.
- **Double-indirect Addressing**, where the block map contains a pointer to another block, referred to as the *double-indirect block*. The double-indirect block then contains a list of *indirect blocks*. Each indirect block then contains a list of the data block addresses that make up the file. As with indirect addressing, double-indirect addressing allows for a file system to support larger file sizes over direct, as well as over indirect, addressing.
- **Extent-based Addressing**, where the block map is an *extent list* made up of addresses that each represent a range of blocks (data blocks, indirect blocks, and/or double-indirect blocks). An address in the extent list represents the starting address of a set of blocks, as well as the number of consecutive blocks in the set in addition to the first block.

Shown in Figure 5.6g is a sample inode that contains the field that supports JFS, which uses extent-based addressing in its data management scheme. Figure 5.6h is a JFS sample inode initialization code which demonstrates some usage by JFS of its extent-based addressing algorithm.

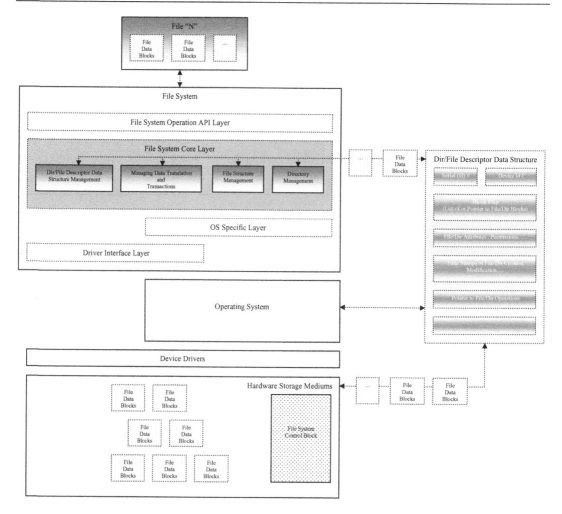

Figure 5.6f: Management of File Data

```
/*   <<jfs_dinode.h>>
 *   Copyright (c) International Business Machines Corp., 2000-2001
 *
 *.  This program is free software;  you can redistribute it and/or modify
 *   it under the terms of the GNU General Public License as published by
 *   the Free Software Foundation; either version 2 of the License, or
 *   (at your option) any later version.
 *
 *   This program is distributed in the hope that it will be useful,
 *   but WITHOUT ANY WARRANTY;  without even the implied warranty of
 *   MERCHANTABILITY or FITNESS FOR A PARTICULAR PURPOSE.  See
 *   the GNU General Public License for more details.
 *
 *   You should have received a copy of the GNU General Public License
 *   along with this program;  if not, write to the Free Software
 *   Foundation, Inc., 59 Temple Place, Suite 330, Boston, MA 02111-1307 USA
 */

/*
 *      on-disk inode : 512 bytes
 *
 * note: align 64-bit fields on 8-byte boundary.
 */
struct dinode {
        /*
         *      I. base area (128 bytes)
         *      ------------------------
         *
         * define generic/POSIX attributes
         */
        u32 di_inostamp;    /* 4: stamp to show inode belongs to fileset */
        s32 di_fileset;             /* 4: fileset number */
        u32 di_number;              /* 4: inode number, aka file serial number */
        u32 di_gen;                 /* 4: inode generation number */

        pxd_t di_ixpxd;             /* 8: inode extent descriptor */

        s64 di_size;                /* 8: size */
        s64 di_nblocks;             /* 8: number of blocks allocated */

        u32 di_nlink;               /* 4: number of links to the object */

        u32 di_uid;                 /* 4: user id of owner */
        u32 di_gid;                 /* 4: group id of owner */

        u32 di_mode;                /* 4: attribute, format and permission */

        struct timestruc_t di_atime;  /* 8: time last data accessed */
        struct timestruc_t di_ctime;  /* 8: time last status changed */
        struct timestruc_t di_mtime;  /* 8: time last data modified */
        struct timestruc_t di_otime;  /* 8: time created */

        dxd_t di_acl;               /* 16: acl descriptor */

        dxd_t di_ea;                /* 16: ea descriptor */

        u32 di_next_index;  /* 4: Next available dir_table index */

        s32 di_acltype;             /* 4: Type of ACL */

        /*
         *      Extension Areas.
         *
         *      Historically, the inode was partitioned into 4 128-byte areas,
         *      the last 3 being defined as unions which could have multiple
         *      uses.  The first 96 bytes had been completely unused until
         *      an index table was added to the directory.  It is now more
         *      useful to describe the last 3/4 of the inode as a single
         *      union.  We would probably be better off redesigning the
         *      entire structure from scratch, but we don't want to break
         *      commonality with OS/2's JFS at this time.
         */
        union {
                struct {
                        /*
                         * This table contains the information needed to
                         * find a directory entry from a 32-bit index.
                         * If the index is small enough, the table is inline,
                         * otherwise, an x-tree root overlays this table
                         */
                        struct dir_table_slot _table[12]; /* 96: inline */

                        dtroot_t _dtroot;               /* 288: dtree root */
                } _dir;                                 /* (384) */
        #define di_dirtable u._dir._table
        #define di_dtroot   u._dir._dtroot
        #define di_parent       di_dtroot.header.idotdot
        #define di_DASD             di_dtroot.header.DASD
```

Figure 5.6g: Example Inode and Extent Addressing

```
/*
 * NAME: init_inode
 *
 * FUNCTION: initialize inode fields for an inode with a single extent or inline
 *           data or a directory inode
 *
 * PARAMETERS:
 *     new_inode       - Pointer to inode to be initialized
 *     fileset_num     - Fileset number for inode
 *     inode_num       - Inode number of inode
 *     num_blocks      - Number of aggregate blocks allocated to inode
 *     size            - Size in bytes allocated to inode
 *     first_block     - Offset of first block of inode's extent
 *     mode            - Mode for inode
 *     inode_type      - Indicates the type of inode to be initialized.
 *                       Currently supported types are inline data, extents,
 *                       and no data.  The other parameters to this function
 *                       will provide the necessary information.
 *     inoext_address  - Address of inode extent containing this inode
 *     aggr_block_size - Aggregate block size
 *     inostamp        - Stamp used to identify inode as belonging to fileset
 *
 * RETURNS: None
 */

void init_inode(struct dinode *new_inode,
                int fileset_num,
                unsigned inode_num,
                int64_t num_blocks,
                int64_t size,
                int64_t first_block,
                mode_t mode,
                ino_data_type inode_type,
                int64_t inoext_address,
                int aggr_block_size,
                unsigned inostamp)
{
        /*
         * initialize inode with where this stuff lives
         */
        new_inode->di_inostamp = inostamp;
        new_inode->di_fileset = fileset_num;
        new_inode->di_number = inode_num;
        new_inode->di_gen = 1;
        PXDaddress(&(new_inode->di_ixpxd), inoext_address);
        PXDlength(&(new_inode->di_ixpxd), INODE_EXTENT_SIZE / aggr_block_size);
        new_inode->di_mode = mode;
        new_inode->di_nblocks = num_blocks;
        new_inode->di_size = size;
        new_inode->di_nlink = 1;
        new_inode->di_next_index = 2;

        switch (inode_type) {
        case inline_data:
                new_inode->di_dxd.flag = DXD_INLINE;
                DXDlength(&(new_inode->di_dxd),
                          sizeof (struct dinode) - offsetof(struct dinode,
                                                            di_inlinedata));
                DXDaddress(&(new_inode->di_dxd), 0);
                break;
        case extent_data:
        case max_extent_data:
                ((xtpage_t *) & (new_inode->di_DASD))->header.flag =
                    DXD_INDEX | BT_ROOT | BT_LEAF;
                /*
                 * Since this is the root, we don't actually use the next and
                 * prev entries.  Set to 0 in case we decide to use this space
                 * for something in the future.
                 */
                ((xtpage_t *) & (new_inode->di_DASD))->header.next = 0;
                ((xtpage_t *) & (new_inode->di_DASD))->header.prev = 0;
                ((xtpage_t *) & (new_inode->di_DASD))->header.nextindex =
                    XTENTRYSTART + 1;
                ((xtpage_t *) & (new_inode->di_DASD))->header.maxentry =
                    XTROOTMAXSLOT;
                ((xtpage_t *) & (new_inode->di_DASD))->xad[XTENTRYSTART].flag =
                    0;
                ((xtpage_t *) & (new_inode->di_DASD))->xad[XTENTRYSTART].rsvrd =
                    0;
                XADoffset(&((xtpage_t *) & (new_inode->di_DASD))->
                          xad[XTENTRYSTART], 0);
                XADlength(&((xtpage_t *) & (new_inode->di_DASD))->
                          xad[XTENTRYSTART], num_blocks);
                XADaddress(&((xtpage_t *) & (new_inode->di_DASD))->
                          xad[XTENTRYSTART], first_block);
                break;
```

Figure 5.6g continued: Example Inode and Extent Addressing

```
/* <inode.c>
 * Copyright (c) International Business Machines Corp., 2000-2002
 *
 * This program is free software; you can redistribute it and/or modify
 * it under the terms of the GNU General Public License as published by
 * the Free Software Foundation; either version 2 of the License, or
 * (at your option) any later version.
 *
 * This program is distributed in the hope that it will be useful,
 * but WITHOUT ANY WARRANTY; without even the implied warranty of
 * MERCHANTABILITY or FITNESS FOR A PARTICULAR PURPOSE. See
 * the GNU General Public License for more details.
 *
 * You should have received a copy of the GNU General Public License
 * along with this program; if not, write to the Free Software
 * Foundation, Inc., 59 Temple Place, Suite 330, Boston, MA 02111-1307 USA
 */

/*
 * NAME: init_inode
 *
 * FUNCTION: Initialize inode fields for an inode with a single extent or inline
 * data or a directory inode
 *
 * PARAMETERS:
 * new_inode - Pointer to inode to be initialized
 * fileset_num - Fileset number for inode
 * inode_num - Inode number of inode
 * num_blocks - Number of aggregate blocks allocated to inode
 * size - Size in bytes allocated to inode
 * first_block - Offset of first block of inode's extent
 * mode - Mode for inode
 * inode_type - Indicates the type of inode to be initialized.
 * Currently supported types are inline data, extents,
 * and no data. The other parameters to this function
 * will provide the necessary information.
 * inoext_address - Address of inode extent containing this inode
 * aggr_block_size - Aggregate block size
 * inostamp - Stamp used to identify inode as belonging to fileset
 *
 * RETURNS: None
 */
```

Figure 5.6h: JFS Source Code and Extent Addressing

```
void init_inode(struct dinode *new_inode,
                int fileset_num,
                unsigned inode_num,
                int64_t num_blocks,
                int64_t size,
                int64_t first_block,
                mode_t mode,
                ino_data_type inode_type,
                int64_t inoext_address,
                int aggr_block_size,
                unsigned inostamp)
{
        /*
         * Initialize inode with where this stuff lives
         */
        new_inode->di_inostamp = inostamp;
        new_inode->di_fileset = fileset_num;
        new_inode->di_number = inode_num;
        new_inode->di_gen = 1;
        PXDaddress(&(new_inode->di_ixpxd), inoext_address);
        PXDlength(&(new_inode->di_ixpxd), INODE_EXTENT_SIZE / aggr_block_size);
        new_inode->di_mode = mode;
        new_inode->di_nblocks = num_blocks;
        new_inode->di_size = size;
        new_inode->di_nlink = 1;
        new_inode->di_next_index = 2;

        switch (inode_type) {
        case inline_data:
                new_inode->di_dxd.flag = DXD_INLINE;
                DXDlength(&(new_inode->di_dxd),
                            sizeof (struct dinode) - offsetof(struct dinode,
                                                        di_inlinedata));
                DXDaddress(&(new_inode->di_dxd), 0);
                break;
        case extent_data:
        case max_extent_data:
                ((xtpage_t *) & (new_inode->di_DASD))->header.flag =
                DXD_INDEX | BT_ROOT | BT_LEAF;
                /*
                 * Since this is the root, we don't actually use the next and
                 * prev entries. Set to 0 in case we decide to use this space
                 * for something in the future.
                 */
                ((xtpage_t *) & (new_inode->di_DASD))->header.next = 0;
                ((xtpage_t *) & (new_inode->di_DASD))->header.prev = 0;
                ((xtpage_t *) & (new_inode->di_DASD))->header.nextindex =
                XTENTRYSTART + 1;
                ((xtpage_t *) & (new_inode->di_DASD))->header.maxentry =
                XTROOTMAXSLOT;
                ((xtpage_t *) & (new_inode->di_DASD))->xad[XTENTRYSTART].flag =
                0;
                ((xtpage_t *) & (new_inode->di_DASD))->xad[XTENTRYSTART].rsvrd =
                0;
                XADoffset(&((xtpage_t *) & (new_inode->di_DASD))->
                            xad[XTENTRYSTART], 0);
                XADlength(&((xtpage_t *) & (new_inode->di_DASD))->
                            xad[XTENTRYSTART], num_blocks);
                XADaddress(&((xtpage_t *) & (new_inode->di_DASD))->
                            xad[XTENTRYSTART], first_block);
                break;
```

Figure 5.6h continued: JFS Source Code and Extent Addressing

```
        case no_data:
                /*
                 * No data to be filled in here, don't do anything
                 */
                ((xtpage_t *) & (new_inode->di_DASD))->header.flag =
                DXD_INDEX | BT_ROOT | BT_LEAF;
                /*
                 * Since this is the root, we don't actually use the next and
                 * prev entries. Set to 0 in case we decide to use this space
                 * for something in the future.
                 */
                ((xtpage_t *) & (new_inode->di_DASD))->header.next = 0;
                ((xtpage_t *) & (new_inode->di_DASD))->header.prev = 0;
                ((xtpage_t *) & (new_inode->di_DASD))->header.nextindex =
                XTENTRYSTART;
                ((xtpage_t *) & (new_inode->di_DASD))->header.maxentry =
                XTROOTMAXSLOT;
                ((xtpage_t *) & (new_inode->di_DASD))->xad[XTENTRYSTART].flag =
                0;
                ((xtpage_t *) & (new_inode->di_DASD))->xad[XTENTRYSTART].rsvrd =
                0;
                break;
        default:
                DBG_ERROR(("Internal error: %s(%d): Unrecognized inode data type %d\n",
                        __FILE__, __LINE__, inode_type))
                break;
        }

        new_inode->di_atime.tv_sec = new_inode->di_ctime.tv_sec =
         new_inode->di_mtime.tv_sec = new_inode->di_otime.tv_sec =
         (unsigned) time(NULL);
        return;
}
```

Figure 5.6h continued: JFS Source Code and Extent Addressing

5.3.6 Directory Management

A *directory* is a mechanism in file systems that allows for one or more files and/or directories
to be grouped under a single name. Essentially the same descriptor data structure used to
represent files in a file system is typically used to represent a directory, where the directory
descriptor data structure is responsible for storing the list of other directory and/or file
descriptor data structures that are assigned to it.

There are several schemes utilized in different file system designs for how directories keep
track of their file and subdirectory names, including: *linear,* where file and subdirectory
names are managed as a linear list within the directory descriptor data structure; *B-Tree*
(i.e., B-Tree, B+Tree, B*Tree), which are hierarchical 'tree' data structures where file
and subdirectory names are inserted/deleted sorted nodes (parent and/or child); and *hash
table* data structures, where file and directory names are sorted and used as keys for faster
retrieval – just to name a few.

Figure 5.6i shows an external inode with fields utilized in the directory management scheme
sample code shown in Figure 5.6j.

```
/*    << fs.h in Linux>>
 *
 *    This program is free software;  you can redistribute
 *    it and/or modify it under the terms of the GNU General
 *    Public License as published by the Free Software Foundation;
 *    either version 2 of the License, or (at your option)
 *    any later version.
 *
 *    This program is distributed in the hope that it will be useful,
 *    but WITHOUT ANY WARRANTY;  without even the implied warranty of
 *    MERCHANTABILITY or FITNESS FOR A PARTICULAR PURPOSE.  See
 *    the GNU General Public License for more details.
 *
 *    You should have received a copy of the GNU General Public License
 *    along with this program;  if not, write to the Free Software
 *    Foundation, Inc., 59 Temple Place, Suite 330,
 *    Boston, MA 02111-1307 USA
 */


struct inode {

  ......

  unsigned long                    i_inostamp;//inode fileset stamp
  unsigned long                    i_inodenumber; //inode serial number
  unsigned long                    i_devnumber;//device id number

  umode_t                          i_mode; //attributes,formats,permissions,..
  nlink_t                          i_nlink;//number of object links
  uid_t                            i_uid;//file owner's user id
  gid_t                            i_gid;//file owner's group id
  kdev_t                           i_rdev;//special file device id
  loff_t                           i_size; //inode inline data size

  time_t                           i_atime;//timestamp of data last accessed
  time_t                           i_mtime; //timestamp of data last modified
  time_t                           i_ctime; //timestamp of data last changed
  time_t                           i_otime; //timestamp of file/dir creation

  unsigned long                    i_blksize;//block size
  unsigned long                    i_blocks;//number of allocated blocks
  struct semaphore                 i_dev_sem; //device locking semaphore
  struct inode_operations          *i_op; //file inode
  struct file_operations           *i_fop;//jfs dir or file operations
  struct super_block               *i_sb; //JFS-private superblock info

  struct address_space             *i_mapping;//address space operations
  struct dquot                     *i_dquot[MAXQUOTAS];//disk quota options
  struct block_device              *i_bdev;//pointer to block device
  struct char_device               *i_cdev; //pointer to character device

  unsigned long                    i_dnotify_mask; // directory notify events
  struct dnotify_struct            *i_dnotify; // for directory notifications

  unsigned long                    i_state;//data related inode
  unsigned char                    i_sock; //socket number

  ......
};
```

Figure 5.6i: External Linux Inode Sample Source Code

```
/*
                                 * xtTruncate isn't guaranteed to fully truncate
                                 * the xtree.  The caller needs to check i_size
                                 * after committing the transaction to see if
                                 * additional truncation is needed.  The
                                 * COMMIT_Stale flag tells caller that we
                                 * initiated the truncation.
                                 */
                                xtTruncate(tid, ip, 0, COMMIT_PWMAP);
                                set_cflag(COMMIT_Stale, ip);

                                tblk->xflag = xflag_save;
                        } else
                                ip->i_size = 1;

                        jfs_ip->next_index = 2;
                } else
                        ip->i_size = IDATASIZE;

        /*
         * acquire a transaction lock on the root
         *
         * action: directory initialization;
         */
        tlck = txLock(tid, ip, (struct metapage *) & jfs_ip->bxflag,
                        tlckDTREE | tlckENTRY | tlckBTROOT);
        dtlck = (struct dt_lock *) & tlck->lock;

        /* linelock root */
        ASSERT(dtlck->index == 0);
        lv = & dtlck->lv[0];
        lv->offset = 0;
        lv->length = DTROOTMAXSLOT;
        dtlck->index++;

        p = &jfs_ip->i_dtroot;

        p->header.flag = DXD_INDEX | BT_ROOT | BT_LEAF;

        p->header.nextindex = 0;

        /* init freelist */
        fsi = 1;
        f = &p->slot[fsi];

        /* init data area of root */
        for (fsi++; fsi < DTROOTMAXSLOT; f++, fsi++)
                f->next = fsi;
        f->next = -1;

        p->header.freelist = 1;
        p->header.freecnt = 8;

        /* init '..' entry */
        p->header.idotdot = cpu_to_le32(idotdot);

        return;
}
......



/*
 *      add_index()
 *
 *      Adds an entry to the directory index table.  This is used to provide
 *      each directory entry with a persistent index in which to resume
 *      directory traversals
 */
static u32 add_index(tid_t tid, struct inode *ip, s64 bn, int slot)
{
        struct super_block *sb = ip->i_sb;
        struct jfs_sb_info *sbi = JFS_SBI(sb);
        struct jfs_inode_info *jfs_ip = JFS_IP(ip);
        u64 blkno;
        struct dir_table_slot *dirtab_slot;
        u32 index;
        struct linelock *llck;
        struct lv *lv;
        struct metapage *mp;
        s64 offset;
        uint page_offset;
        struct tlock *tlck;
        s64 xaddr;

        ASSERT(DO_INDEX(ip));
```

Figure 5.6j: JFS B+Tree Directory Scheme Sample Source Code

```
if (jfs_ip->next_index < 2) {
        jfs_warn("add_index: next_index = %d.  Resetting!",
                    jfs_ip->next_index);
        jfs_ip->next_index = 2;
}

index = jfs_ip->next_index++;

if (index <= MAX_INLINE_DIRTABLE_ENTRY) {
        /*
         * i_size reflects size of inode table, or 8 bytes per entry.
         */
        ip->i_size = (loff_t) (index - 1) << 3;

        /*
         * dir table fits inline within inode
         */
        dirtab_slot = &jfs_ip->i_dirtable[index-2];
        dirtab_slot->flag = DIR_INDEX_VALID;
        dirtab_slot->slot = slot;
        DTSaddress(dirtab_slot, bn);

        set_cflag(COMMIT_Dirtable, ip);

        return index;
}
if (index == (MAX_INLINE_DIRTABLE_ENTRY + 1)) {
        struct dir_table_slot temp_table[12];

        /*
         * It's time to move the inline table to an external
         * page and begin to build the x-tree
         */
        if (dbAlloc(ip, 0, sbi->nbperpage, &xaddr))
                goto clean_up;        /* No space */

        /*
         * Save the table, we're going to overwrite it with the
         * xtree root
         */
        memcpy(temp_table, &jfs_ip->i_dirtable, sizeof(temp_table));

        /*
         * Initialize empty x-tree
         */
        xtInitRoot(tid, ip);

        /*
         * Allocate the first block & add it to the xtree
         */
        if (xtInsert(tid, ip, 0, 0, sbi->nbperpage, &xaddr, 0)) {
                /* This really shouldn't fail */
                jfs_warn("add_index: xtInsert failed!");
                memcpy(&jfs_ip->i_dirtable, temp_table,
                        sizeof (temp_table));
                goto clean_up;
        }
        ip->i_size = PSIZE;
        ip->i_blocks += LBLK2PBLK(sb, sbi->nbperpage);

        if ((mp = get_index_page(ip, 0)) == 0) {
                jfs_err("add_index: get_metapage failed!");
                xtTruncate(tid, ip, 0, COMMIT_PWMAP);
                memcpy(&jfs_ip->i_dirtable, temp_table,
                        sizeof (temp_table));
                goto clean_up;
        }
        tlck = txLock(tid, ip, mp, tlckDATA);
        llck = (struct linelock *) & tlck->lock;
        ASSERT(llck->index == 0);
        lv = &llck->lv[0];

        lv->offset = 0;
        lv->length = 6;        /* tlckDATA slot size is 16 bytes */
        llck->index++;

        memcpy(mp->data, temp_table, sizeof(temp_table));

        mark_metapage_dirty(mp);
        release_metapage(mp);

        /*
         * Logging is now disabled in xtree routines
         */
        clear_cflag(COMMIT_Dirtable, ip);
}
```

Figure 5.6j continued: JFS B+Tree Directory Scheme Sample Source Code

5.3.7 Impact of File System Core on Embedded Device

What most differentiates the behavior and performance of one file system over another are the elements that make up a file system's core layer, specifically the directory and file descriptor data structure, data storage management, and directory management schemes implemented within the file system design (see Figure 5.7). In the case of a directory and file descriptor data structure design, for example, the maximum file size that can be managed via a file system is determined by the scheme in which this data structure tracks the data. Furthermore, given the ability to support larger file sizes, a file system that implements an inefficient scheme may take longer to navigate the data structure to track down data within a large file. This also holds true for how a directory (data structure) stores file names and any subdirectory information – tracking down a file or subdirectory may take longer if an inefficient scheme is implemented to traverse the data structure.

Depending on the file system, the less a file system has to access the hardware storage medium to retrieve and/or write file system data blocks, the more efficiently it can perform. So, file systems can have an advantage over other file systems on performance with a storage management scheme that:

- does as much as possible in (faster) RAM before storing any data back on a (slower) hardware storage medium. A drawback is hardware storage medium is not always in sync with the current state of the file system if system failure occurs, thus making recovery of the file system more difficult and decreasing file system reliability
- supports larger block sizes. A drawback is if the entire block is not utilized then storage medium space usage is not optimal
- is able to store data blocks compactly and contiguously on the storage medium. A drawback is that compaction algorithms that resolve fragmentation issues are more complex to implement over creating larger block sizes, for example.

While a file system can have an advantage, the less it accesses the hardware storage medium over other file systems, there are other file systems that implement schemes based on constant storage medium access in order to make the system more reliable, which in some embedded designs with high reliability requirements would provide an advantage. These file systems, commonly referred to as journaled or (atomic) transactional file systems, log file system transactions in some manner to be utilized in a file system recovery in case of some type of system failure. Drawbacks of a journaled/(atomic) transactional file system will depend on its internal design, such as if logging data locks up the file system in any way and how logged data are written/retrieved to/from the storage medium, for example.

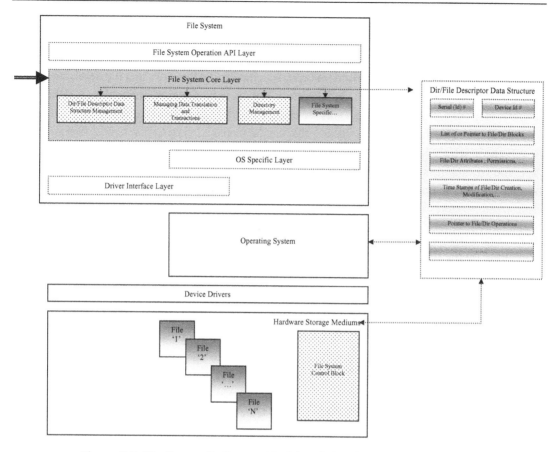

Figure 5.7: File System Reference Model and the File System Core Layer

5.3.8 File System Operation API Layer

While file systems can vary on the API functionality provided in the *File System Operation API Layer* (shown in Figure 5.8a), and/or how these operations are implemented, file systems all provide some universally similar file system operations. As introduced in Section 5.1, examples of these operations include:

- **Creating and Configuring Files**, given a directory name and a valid new file name within the size and character type restrictions provided by the file system, a file descriptor data structure is created for each new file, and relevant fields filled (i.e., size, permissions, etc.). The file descriptor data structure is then added to the directory's descriptor data structure.
- **Renaming Files**, given a directory name, the old file name, and a new file name – if the new file name does not already exist as an entry in the directory's descriptor data structure and if there is no other software/user accessing the file, then the old file name is updated to the new file name in some manner.

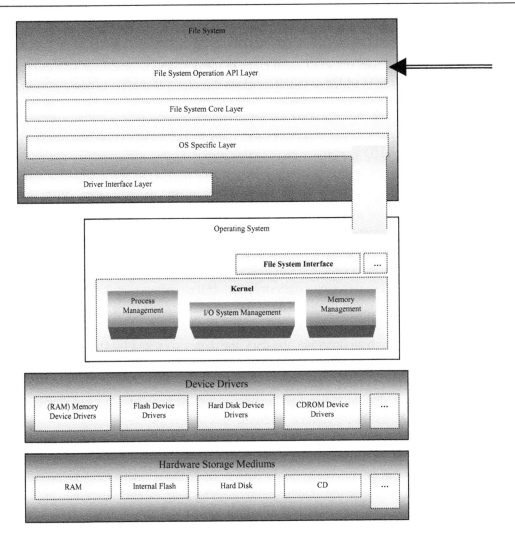

Figure 5.8a: General Embedded System File System Reference Model

- **Copying or Moving Files**, given a source directory name, a destination directory name, and the file name – if the file name exists as an entry in the source directory's descriptor data structure and it does not exist as an entry in the destination directory's descriptor data structure, the file is added to the destination directory. If the file is being moved, it is then removed from the source directory.
- **Removing Files**, given the directory name and file name, the file system first finds the directory's descriptor data structure and looks up the name of the file to retrieve the serial number (id) of the file's descriptor data structure. If the attributes in the file's descriptor's data structure are verified to insure that the file can be deleted by the requesting software/user, and if there is no other software/user accessing the file, the file system frees the file's resources in some manner, including removing any references to the file from its directory's descriptor data structure.
- **Opening Files**, given the directory name and file name, the file system first finds the directory's descriptor data structure and looks up the name of the file to retrieve the serial number (id) of the file's descriptor data structure. If the attributes in the file descriptor's data structure are verified to insure that the file can be opened by the requesting software/user, then I/O operations are allowed to be performed on the file.
- **Writing to Files**, given an open file, the data, the data's size, and location in the file the data are to be stored at – the file descriptor data structure relevant field is modified according to the file system's data storage management scheme (i.e., direct addressing, indirect addressing, double-indirect addressing, extent addressing, etc.) and then the data are stored on to the hardware storage medium.
- **Reading from Files**, given an open file, the data, the data's size, and location in the file the data is stored at – the file descriptor data structure relevant field is used to locate the desired data according to the file system's data storage management scheme (i.e., direct addressing, indirect addressing, double-indirect addressing, extent addressing, etc.) and then the data are loaded from the hardware storage medium.
- **Creating Directories**, given a new directory name – a directory descriptor data structure is created for each new directory, and relevant fields filled (i.e., permissions, flags, etc.). The directory descriptor data structure is then added to the parent directory's descriptor data structure.
- **Removing Directories**, given a parent directory name and the name of the directory to be removed – the parent directory's descriptor data structure is used to look up the name of the directory to be deleted to retrieve the serial number (id) of its descriptor data structure. If the attributes in the directory's descriptor data structure are verified to insure that the directory can be deleted by the requesting software/user, and if there is no other software/user accessing any contents of the directory, the file system frees the directory resources in some manner, including removing any references to the directory from its parent directory's descriptor data structure.

- **Reading Directories**, given a directory name – the directory's descriptor data structure is utilized to display its contents (file names and subdirectories).
- **Additional/Extended Functions**
  - *Creating and Initializing the File System,* where provided parameters and assigned hardware storage medium block(s), sector(s), or volume(s) are used to create and initialize a new file system. In general, this includes allocating a file system control block(s) on the storage medium block(s), sector(s), or volume(s), creating any necessary directory/file descriptor data structures, and creating an empty root directory on the assigned storage medium block(s), sector(s), or volume(s).
  - *File System Verification*, where an unmounted file system is checked to determine if it is 'clean', a.k.a. if its metadata information is up to date and no data corruption has been found. If a file system is 'dirty', the verification process has uncovered inconsistent and/or corrupted data.
  - *Mounting the File System*, where the hardware storage medium is accessed to retrieve and load file system metadata from the file system's control block into RAM. The file system and respective hardware storage medium block(s), sector(s), or volume(s) are then ready for access and use.
  - *Unmounting the File System,* a proper shutdown of the file system where the hardware storage medium block(s), sector(s), or volume(s) are put in a 'clean' state by copying the latest file system metadata in RAM back to the file system's control block on the hardware storage medium.
  - Symbolic, Hard, and/or Dynamic links.

Figure 5.8b shows examples of APIs available under vxWorks, and Figures 5.8c, 5.8d, and 5.8e show examples of how various directory and file operations are implemented in the open source JFS implementation. While the internal source code of how operations are implemented will differ between file systems, many file systems have 'similar' operations as those shown in these examples and can give the reader a feel for what to expect.

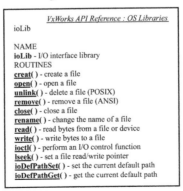

```
VxWorks API Reference : OS Libraries
ioLib

NAME
ioLib - I/O interface library
ROUTINES
creat( ) - create a file
open( ) - open a file
unlink( ) - delete a file (POSIX)
remove( ) - remove a file (ANSI)
close( ) - close a file
rename( ) - change the name of a file
read( ) - read bytes from a file or device
write( ) - write bytes to a file
ioctl( ) - perform an I/O control function
lseek( ) - set a file read/write pointer
ioDefPathSet( ) - set the current default path
ioDefPathGet( ) - get the current default path
```

Figure 5.8b: Example vxWorks Operations[13]

chdir( ) - set the current default path
getcwd( ) - get the current default path (POSIX)
getwd( ) - get the current default path
ioGlobalStdSet( ) - set the file descriptor for global standard input/output/error
ioGlobalStdGet( ) - get the file descriptor for global standard input/output/error
ioTaskStdSet( ) - set the file descriptor for task standard input/output/error
ioTaskStdGet( ) - get the file descriptor for task standard input/output/error
isatty( ) - return whether the underlying driver is a tty device

DESCRIPTION
This library contains the interface to the basic I/O system. It includes: Interfaces to the seven basic driver-provided functions: **creat( )**, **remove( )**, **open( )**, **close( )**, **read( )**, **write( )**, and **ioctl( )**. Interfaces to several file system functions, including **rename( )** and **lseek( )**. Routines to set and get the current working directory. Routines to assign task and global standard file descriptors.

FILE DESCRIPTORS
At the basic I/O level, files are referred to by a file descriptor. A file descriptor is a small integer returned by a call to **open( )** or **creat( )**. The other basic I/O calls take a file descriptor as a parameter to specify the intended file. Three file descriptors are reserved and have special meanings:
  0 **(STD_IN)** - standard input
  1 **(STD_OUT)** - standard output
  2 **(STD_ERR)** - standard error output
VxWorks allows two levels of redirection. First, there is a global assignment of the three standard file descriptors. By default, new tasks use this global assignment. The global assignment of the three standard file descriptors is controlled by the routines **ioGlobalStdSet( )** and **ioGlobalStdGet( )**. Second, individual tasks may override the global assignment of these file descriptors with their own assignments that apply only to that task. The assignment of task-specific standard file descriptors is controlled by the routines **ioTaskStdSet( )** and **ioTaskStdGet( )**.

INCLUDE FILES: **ioLib.h**
SEE ALSO: **iosLib**, ansiStdio,

creat( )
NAME
creat( ) - create a file
SYNOPSIS
int creat
(
const char * name, /* name of the file to create */
int flag /* O_RDONLY, O_WRONLY, or O_RDWR */
)
DESCRIPTION
This routine creates a file called *name* and opens it with a specified *flag*. This routine determines on which device to create the file; it then calls the create routine of the device driver to do most of the work. Therefore, much of what transpires is device/driver-dependent. The parameter *flag* is set to **O_RDONLY** (0), **O_WRONLY** (1), or **O_RDWR** (2) for the duration of time the file is open. To create NFS files with a UNIX chmod-type file mode, call **open( )** with the file mode specified in the third argument.
NOTE
For more information about situations when there are no file descriptors available, see the manual entry for **iosInit( )**.
RETURNS
A file descriptor number, or ERROR if a filename is not specified, the device does not exist, no file descriptors are available, or the driver returns ERROR.
SEE ALSO
ioLib, **open( )**

open( )
NAME
open( ) - open a file
SYNOPSIS
int open (
const char * name, /* name of the file to open */
int flags, /* O_RDONLY, O_WRONLY, O_RDWR, or O_CREAT */
int mode /* mode of file to create (UNIX chmod style) */)
DESCRIPTION
This routine opens a file for reading, writing, or updating, and returns a file descriptor for that file. The arguments to **open( )** are the filename and the type of access:

O_RDONLY (0)   (or READ)   - open for reading only.

O_WRONLY (1)   (or WRITE)   - open for writing only.

O_RDWR (2)   (or UPDATE)   - open for reading and writing.

O_CREAT (0x0200)   - create a file.

In general, **open( )** can only open pre-existing devices and files. However, for NFS network devices only, files can also be created with **open( )** by performing a logical OR operation with **O_CREAT** and the *flags* argument. In this case, the file is created with a UNIX chmod-style file mode, as indicated with *mode*. For example: fd = open ("/usr/myFile", O_CREAT | O_RDWR, 0644); Only the NFS driver uses the *mode* argument.

NOTE: For more information about situations when there are no file descriptors available, see the manual entry for **iosInit( )**.

RETURNS: A file descriptor number, or ERROR if a file name is not specified, the device does not exist, no file descriptors are available, or the driver returns ERROR.

Figure 5.8b continued: Example vxWorks Operations

ERRNO
ELOOP
SEE ALSO
ioLib, **creat( )**
        VARARGS2

---

unlink( )
NAME
unlink( ) - delete a file (POSIX)
SYNOPSIS
STATUS unlink
(char * name /* name of the file to remove */)
DESCRIPTION
This routine deletes a specified file. It performs the same function as **remove( )** and is provided for POSIX compatibility.
RETURNS
OK if there is no delete routine for the device or the driver returns OK; ERROR if there is no such device or the driver returns ERROR.
SEE ALSO
ioLib, **remove( )**

---

remove( )
NAME
remove( ) - remove a file (ANSI)
SYNOPSIS
STATUS remove
(const char * name /* name of the file to remove */)
DESCRIPTION
This routine deletes a specified file. It calls the driver for the particular device on which the file is located to do the work.
RETURNS
OK if there is no delete routine for the device or the driver returns OK; ERROR if there is no such device or the driver returns ERROR.
SEE ALSO
ioLib, *American National Standard for Information Systems - Programming Language - C, ANSI X3.159-1989: Input/Output* (**stdio.h**),

---

close( )
NAME
close( ) - close a file
SYNOPSIS
STATUS close
(int fd /* file descriptor to close */)
DESCRIPTION
This routine closes the specified file and frees the file descriptor. It calls the device driver to do the work.
RETURNS
The status of the driver close routine, or ERROR if the file descriptor is invalid.
SEE ALSO
ioLib

---

rename( )
NAME
rename( ) - change the name of a file
SYNOPSIS
int rename
(const char * oldname, /* name of file to rename */
const char * newname /* name with which to rename file */)
DESCRIPTION

       This routine changes the name of a file from *oldfile* to *newfile*.

Figure 5.8b continued: Example vxWorks Operations

NOTE
Only certain devices support **rename( )**. To confirm that your device supports it, consult the respective **xxDrv** or xxFs listings to verify that ioctl FIORENAME exists. For example, dosFs and rt11Fs support **rename( )**, but **netDrv** and **nfsDrv** do not.
RETURNS
OK, or ERROR if the file could not be opened or renamed.
SEE ALSO
ioLib

read( )
NAME
read( ) - read bytes from a file or device
SYNOPSIS
int read
(
int fd, /* file descriptor from which to read */
char * buffer, /* pointer to buffer to receive bytes */
size_t maxbytes /* max no. of bytes to read into buffer */
)
DESCRIPTION
This routine reads a number of bytes (less than or equal to *maxbytes*) from a specified file descriptor and places them in *buffer*. It calls the device driver to do the work.
RETURNS
The number of bytes read (between 1 and *maxbytes*, 0 if end of file), or ERROR if the file descriptor does not exist, the driver does not have a read routines, or the driver returns ERROR. If the driver does not have a read routine, errno is set to ENOTSUP.
SEE ALSO
ioLib

write( )
NAME
write( ) - write bytes to a file
SYNOPSIS
int write
(
int fd, /* file descriptor on which to write */
char * buffer, /* buffer containing bytes to be written */
size_t nbytes /* number of bytes to write */
)
DESCRIPTION
This routine writes *nbytes* bytes from *buffer* to a specified file descriptor *fd*. It calls the device driver to do the work.
RETURNS
The number of bytes written (if not equal to *nbytes*, an error has occurred), or ERROR if the file descriptor does not exist, the driver does not have a write routine, or the driver returns ERROR. If the driver does not have a write routine, errno is set to ENOTSUP.
SEE ALSO
ioLib

ioctl( )
NAME
ioctl( ) - perform an I/O control function
SYNOPSIS
int ioctl
(
int fd, /* file descriptor */
int function, /* function code */
int arg /* arbitrary argument */
)

Figure 5.8b continued: Example vxWorks Operations

DESCRIPTION
This routine performs an I/O control function on a device. The control functions used by VxWorks device drivers are defined in the header file **ioLib.h**. Most requests are passed on to the driver for handling. Since the availability of **ioctl( )** functions is driver-specific, these functions are discussed separately in **tyLib**, **pipeDrv**, **nfsDrv**, **dosFsLib**, **rt11FsLib**, and **rawFsLib**.
The following example renames the file or directory to the string "newname":
ioctl (fd, FIORENAME, "newname");
Note that the function FIOGETNAME is handled by the I/O interface level and is not passed on to the device driver itself. Thus this function code value should not be used by customer-written drivers.

RETURNS
The return value of the driver, or ERROR if the file descriptor does not exist.
SEE ALSO
ioLib, **tyLib**, **pipeDrv**, **nfsDrv**, **dosFsLib**, **rt11FsLib**, **rawFsLib**, *VxWorks Programmer's Guide: I/O System, Local File Systems*
VARARGS2

OS Libraries : Routines

lseek( )
NAME
lseek( ) - set a file read/write pointer
SYNOPSIS
int lseek
(
int fd, /* file descriptor */
long offset, /* new byte offset to seek to */
int whence /* relative file position */
)
DESCRIPTION
This routine sets the file read/write pointer of file *fd* to *offset*. The argument *whence*, which affects the file position pointer, has three values:

SEEK_SET (0)   - set to *offset*

SEEK_CUR (1)   - set to current position plus *offset*

SEEK_END (2)   - set to the size of the file plus *offset*

This routine calls **ioctl( )** with functions FIOWHERE, FIONREAD, and FIOSEEK.
RETURNS
The new offset from the beginning of the file, or ERROR.
ARGSUSED
SEE ALSO **ioLib**

OS Libraries : Routines

ioDefPathSet( )
NAME
ioDefPathSet( ) - set the current default path
SYNOPSIS
STATUS ioDefPathSet
(
char * name /* name of the new default device and path */
)
DESCRIPTION
This routine sets the default I/O path. All relative pathnames specified to the I/O system will be prepended with this pathname. This pathname must be an absolute pathname, i.e., *name* must begin with an existing device name.
RETURNS
OK, or ERROR if the first component of the pathname is not an existing device.
SEE ALSO
ioLib, **ioDefPathGet( )**, **chdir( )**, **getcwd( )**

OS Libraries : Routines

ioDefPathGet( )
NAME
ioDefPathGet( ) - get the current default path
SYNOPSIS
void ioDefPathGet
(
char * pathname /* where to return the name */
)

Figure 5.8b continued: Example vxWorks Operations

DESCRIPTION
This routine copies the name of the current default path to *pathname*. The parameter *pathname* should be **MAX_FILENAME_LENGTH** characters long.
RETURNS
N/A
SEE ALSO
ioLib, **ioDefPathSet**( ), **chdir**( ), **getcwd**( )

---

chdir( )
NAME
chdir( ) - set the current default path
SYNOPSIS
STATUS chdir
(
 char * pathname /* name of the new default path */
)
DESCRIPTION
This routine sets the default I/O path. All relative pathnames specified to the I/O system will be prepended with this pathname. This pathname must be an absolute pathname, i.e., *name* must begin with an existing device name.
RETURNS
OK, or ERROR if the first component of the pathname is not an existing device.
SEE ALSO
ioLib, **ioDefPathSet**( ), **ioDefPathGet**( ), **getcwd**( )

---

getcwd( )
NAME
getcwd( ) - get the current default path (POSIX)
SYNOPSIS
char *getcwd
(
 char * buffer, /* where to return the pathname */
 int size /* size in bytes of buffer */
)
DESCRIPTION
This routine copies the name of the current default path to *buffer*. It provides the same functionality as **ioDefPathGet**( ) and is provided for POSIX compatibility.
RETURNS
A pointer to the supplied buffer, or NULL if *size* is too small to hold the current default path.
SEE ALSO
ioLib, **ioDefPathSet**( ), **ioDefPathGet**( ), **chdir**( )

---

getwd( )
NAME
getwd( ) - get the current default path
SYNOPSIS
char *getwd
(
 char * pathname /* where to return the pathname */
)
DESCRIPTION
This routine copies the name of the current default path to *pathname*. It provides the same functionality as **ioDefPathGet**( ) and **getcwd**( ). It is provided for compatibility with some older UNIX systems.
The parameter *pathname* should be **MAX_FILENAME_LENGTH** characters long.
RETURNS
A pointer to the resulting path name.
SEE ALSO

ioLib

Figure 5.8b continued: Example vxWorks Operations

ioGlobalStdSet( )
NAME
ioGlobalStdSet( ) - set the file descriptor for global standard input/output/error
SYNOPSIS
void ioGlobalStdSet
(
int stdFd, /* std input (0), output (1), or error (2) */
int newFd /* new underlying file descriptor */
)
DESCRIPTION
This routine changes the assignment of a specified global standard file descriptor *stdFd* (0, 1, or, 2) to the specified underlying file descriptor *newFd*. *newFd* should be a file descriptor open to the desired device or file. All tasks will use this new assignment when doing I/O to *stdFd*, unless they have specified a task-specific standard file descriptor (see **ioTaskStdSet( )**). If *stdFd* is not 0, 1, or 2, this routine has no effect.
RETURNS
N/A
SEE ALSO
ioLib, **ioGlobalStdGet( )**, **ioTaskStdSet( )**

ioGlobalStdGet( )
NAME
ioGlobalStdGet( ) - get the file descriptor for global standard input/output/error
SYNOPSIS
int ioGlobalStdGet
(
int stdFd /* std input (0), output (1), or error (2) */
)
DESCRIPTION
This routine returns the current underlying file descriptor for global standard input, output, and error.
RETURNS
The underlying global file descriptor, or ERROR if *stdFd* is not 0, 1, or 2.
SEE ALSO
ioLib, **ioGlobalStdSet( )**, **ioTaskStdGet( )**

ioTaskStdSet( )
NAME
ioTaskStdSet( ) - set the file descriptor for task standard input/output/error
SYNOPSIS
void ioTaskStdSet
(
int taskId, /* task whose std fd is to be set (0 = self) */
int stdFd, /* std input (0), output (1), or error (2) */
int newFd /* new underlying file descriptor */
)
DESCRIPTION
This routine changes the assignment of a specified task-specific standard file descriptor *stdFd* (0, 1, or, 2) to the specified underlying file descriptor*newFd*. *newFd* should be a file descriptor open to the desired device or file. The calling task will use this new assignment when doing I/O to *stdFd*, instead of the system-wide global assignment which is used by default. If *stdFd* is not 0, 1, or 2, this routine has no effect.
NOTE
This routine has no effect if it is called at interrupt level.
RETURNS
N/A
SEE ALSO
ioLib, **ioGlobalStdGet( )**, **ioTaskStdGet( )**

ioTaskStdGet( )
NAME

    ioTaskStdGet( ) - get the file descriptor for task standard input/output/error

Figure 5.8b continued: Example vxWorks Operations

SYNOPSIS
int ioTaskStdGet
(
int taskId, /* ID of desired task (0 = self) */
int stdFd /* std input (0), output (1), or error (2) */
)
DESCRIPTION
This routine returns the current underlying file descriptor for task-specific standard input, output, and error.
RETURNS
The underlying file descriptor, or ERROR if *stdFd* is not 0, 1, or 2, or the routine is called at interrupt level.
SEE ALSO
ioLib, **ioGlobalStdGet( )**, **ioTaskStdSet( )**

OS Libraries : Routines

isatty( )
NAME
isatty( ) - return whether the underlying driver is a tty device
SYNOPSIS
BOOL isatty
(
int fd /* file descriptor to check */
)
DESCRIPTION
This routine simply invokes the **ioctl( )** function FIOISATTY on the specified file descriptor.
RETURNS
TRUE, or FALSE if the driver does not indicate a tty device.
SEE ALSO

ioLib

Figure 5.8b continued: Example vxWorks Operations

```
/*
 * Copyright (C) International Business Machines Corp., 2000-2003
 * Portions Copyright (C) Christoph Hellwig, 2001-2002
 *
 * This program is free software;  you can redistribute it and/or modify
 * it under the terms of the GNU General Public License as published by
 * the Free Software Foundation; either version 2 of the License, or
 * (at your option) any later version.
 *
 * This program is distributed in the hope that it will be useful,
 * but WITHOUT ANY WARRANTY;  without even the implied warranty of
 * MERCHANTABILITY or FITNESS FOR A PARTICULAR PURPOSE.  See
 * the GNU General Public License for more details.
 *
 * You should have received a copy of the GNU General Public License
 * along with this program;  if not, write to the Free Software
 * Foundation, Inc., 59 Temple Place, Suite 330, Boston, MA 02111-1307 USA
 */
struct inode_operations jfs_dir_inode_operations = {
        .create         = jfs_create,
        .lookup         = jfs_lookup,
        .link           = jfs_link,
        .unlink         = jfs_unlink,
        .symlink        = jfs_symlink,


        .mkdir          = jfs_mkdir,
        .rmdir          = jfs_rmdir,
        .mknod          = jfs_mknod,
        .rename         = jfs_rename,
#ifdef JFS_XATTR
        .setxattr       = jfs_setxattr,
        .getxattr       = jfs_getxattr,
        .listxattr      = jfs_listxattr,
        .removexattr    = jfs_removexattr,
#endif
};


struct file_operations jfs_dir_operations = {
        .read           = generic_read_dir,
        .readdir        = jfs_readdir,
        .fsync          = jfs_fsync,
};
```

(comments)

Filename : < namei.c >

//jfs_create - creates regular file in parent directory
//jfs_lookup - search for filename or directory
//jfs_link - create object link in parent directory
//jfs_unlink - remove object link from parent directory
//jfs_symlink - creates symbolic link to object in
            directory
//jfs_mkdir - create child directory in parent directory
//jfs_rmdir - remove link to child directory
//jfs_mknod - create a special file/device
//jfs_rename - rename file or directory

//jfs_setxattr - set extended file/dir attributes
//jfs_getxattr - get extended file/dir attribute
//jfs_listxattr - list extended file/dir attributes
//jfs_removexattr - remove extended file/dir attribute

//generic_read_dir - (not available)
//jfs_readdir - sequentially read directory entries from offset
//jfs_fsync - data synchronization with storage medium

Figure 5.8c: JFS File System Directory Operations

```
/*
 *   Copyright (C) International Business Machines Corp., 2000-2003
 *   Portions Copyright (C) Christoph Hellwig, 2001-2002
 *
 *   This program is free software;  you can redistribute it and/or modify
 *   it under the terms of the GNU General Public License as published by
 *   the Free Software Foundation; either version 2 of the License, or
 *   (at your option) any later version.
 *
 *   This program is distributed in the hope that it will be useful,
 *   but WITHOUT ANY WARRANTY;  without even the implied warranty of
 *   MERCHANTABILITY or FITNESS FOR A PARTICULAR PURPOSE.  See
 *   the GNU General Public License for more details.
 *
 *   You should have received a copy of the GNU General Public License
 *   along with this program;  if not, write to the Free Software
 *   Foundation, Inc., 59 Temple Place, Suite 330, Boston, MA 02111-1307 USA
 */

struct inode_operations jfs_file_inode_operations = {
        .truncate    = jfs_truncate,
#ifdef JFS_XATTR
        .setxattr    = jfs_setxattr,
        .getxattr    = jfs_getxattr,
        .listxattr   = jfs_listxattr,
        .removexattr        = jfs_removexattr,
#endif
};

struct file_operations jfs_file_operations = {
        .open        = jfs_open,
        .llseek      = generic_file_llseek,
        .write       = generic_file_write,
        .read        = generic_file_read,
        .mmap        = generic_file_mmap,
        .fsync       = jfs_fsync,
        .release     = jfs_release,
};

struct inode_operations jfs_special_inode_operations = {
#ifdef JFS_XATTR
        .setxattr    = jfs_setxattr,
        .getxattr    = jfs_getxattr,
        .listxattr   = jfs_listxattr,
        .removexattr        = jfs_removexattr,
#endif
};
```

(comments)

Filename : < file.c >

//jfs_truncate - truncate a file

//jfs_setxattr - set extended file/dir attributes
//jfs_getxattr - get extended file/dir attribute
//jfs_listxattr - list extended file/dir attributes
//jfs_removexattr - remove extended file/dir attribute

//jfs_open - open a file
//generic_file_llseek - (not available)
//generic_file_write - (not available)
//generic_file_read - (not available)
//generic_file_mmap - (not available)
//jfs_fsync - data synchronization with storage medium
//jfs_release - file release

//jfs_setxattr - set extended file/dir attributes
//jfs_getxattr - get extended file/dir attribute
//jfs_listxattr - list extended file/dir attributes
//jfs_removexattr - remove extended file/dir attribute

Figure 5.8d: JFS File System File Operations

```
/* <namei.c>
 * Copyright (C) International Business Machines Corp., 2000-2003
 * Portions Copyright (C) Christoph Hellwig, 2001-2002
 *
 * This program is free software; you can redistribute it and/or modify
 * it under the terms of the GNU General Public License as published by
 * the Free Software Foundation; either version 2 of the License, or
 * (at your option) any later version.
 *
 * This program is distributed in the hope that it will be useful,
 * but WITHOUT ANY WARRANTY; without even the implied warranty of
 * MERCHANTABILITY or FITNESS FOR A PARTICULAR PURPOSE. See
 * the GNU General Public License for more details.
 *
 * You should have received a copy of the GNU General Public License
 * along with this program; if not, write to the Free Software
 * Foundation, Inc., 59 Temple Place, Suite 330, Boston, MA 02111-1307 USA
 */
```

Figure 5.8e: JFS Operations (Function) Source Code

```
#include <linux/fs.h>
#include "jfs_incore.h"
#include "jfs_superblock.h"
#include "jfs_inode.h"
#include "jfs_dinode.h"
#include "jfs_dmap.h"
#include "jfs_unicode.h"
#include "jfs_metapage.h"
#include "jfs_xattr.h"
#include "jfs_debug.h"

........

/*
 * NAME:   jfs_create(dip, dentry, mode)
 *
 * FUNCTION:        create a regular file in the parent directory <dip>
 *                  with name = <from dentry> and mode = <mode>
 *
 * PARAMETER:       dip      - parent directory vnode
 *                  dentry   - dentry of new file
 *                  mode     - create mode (rwxrwxrwx).
 *
 * RETURN:          Errors from subroutines
 *
 */
static int jfs_create(struct inode *dip, struct dentry *dentry, int mode)
{
        int rc = 0;
        tid_t tid;                       /* transaction id */
        struct inode *ip = NULL;         /* child directory inode */
        ino_t ino;
        struct component_name dname;     /* child directory name */
        struct btstack btstack;
        struct inode *iplist[2];
        struct tblock *tblk;

        jfs_info("jfs_create: dip:0x%p name:%s", dip, dentry->d_name.name);

        /*
         * search parent directory for entry/freespace
         * (dtSearch() returns parent directory page pinned)
         */
        if ((rc = get_UCSname(&dname, dentry, JFS_SBI(dip->i_sb)->nls_tab)))
                goto out1;

        /*
         * Either iAlloc() or txBegin() may block. Deadlock can occur if we
         * block there while holding dtree page, so we allocate the inode &
         * begin the transaction before we search the directory.
         */
        ip = ialloc(dip, mode);
        if (ip == NULL) {
                rc = -ENOSPC;
                goto out2;
        }

        tid = txBegin(dip->i_sb, 0);

        down(&JFS_IP(dip)->commit_sem);
        down(&JFS_IP(ip)->commit_sem);

        if ((rc = dtSearch(dip, &dname, &ino, &btstack, JFS_CREATE))) {
                jfs_err("jfs_create: dtSearch returned %d", rc);
                goto out3;
        }

        tblk = tid_to_tblock(tid);
        tblk->xflag |= COMMIT_CREATE;
        tblk->ip = ip;

        iplist[0] = dip;
        iplist[1] = ip;

        /*
         * initialize the child XAD tree root in-line in inode
         */
        xtInitRoot(tid, ip);

        /*
         * create entry in parent directory for child directory
         * (dtInsert() releases parent directory page)
         */
        ino = ip->i_ino;
        if ((rc = dtInsert(tid, dip, &dname, &ino, &btstack))) {
                jfs_err("jfs_create: dtInsert returned %d", rc);
                if (rc == -EIO)
                        txAbort(tid, 1);     /* Marks Filesystem dirty */
                else
                        txAbort(tid, 0);     /* Filesystem full */
                goto out3;
        }
```

Figure 5.8e continued: JFS Operations (Function) Source Code

```
ip->i_op = &jfs_file_inode_operations;
        ip->i_fop = &jfs_file_operations;
        ip->i_mapping->a_ops = &jfs_aops;

        insert_inode_hash(ip);
        mark_inode_dirty(ip);

        dip->i_ctime = dip->i_mtime = CURRENT_TIME;

        mark_inode_dirty(dip);

        rc = txCommit(tid, 2, &iplist[0], 0);
out3:
        txEnd(tid);
        up(&JFS_IP(dip)->commit_sem);
        up(&JFS_IP(ip)->commit_sem);
        if (rc) {
                ip->i_nlink = 0;
                iput(ip);
        } else
                d_instantiate(dentry, ip);
out2:
        free_UCSname(&dname);

out1:

        jfs_info("jfs_create: rc:%d", rc);
        return rc;
}


/*
 * NAME:    jfs_mkdir(dip, dentry, mode)
 *
 * FUNCTION:        create a child directory in the parent directory <dip>
 *                  with name = <from dentry> and mode = <mode>
 *
 * PARAMETER:       dip     - parent directory vnode
 *                  dentry  - dentry of child directory
 *                  mode    - create mode (rwxrwxrwx).
 *
 * RETURN:          Errors from subroutines
 *
 * note:
 * EACCESS: user needs search+write permission on the parent directory
 */
static int jfs_mkdir(struct inode *dip, struct dentry *dentry, int mode)
{
        int rc = 0;
        tid_t tid;                      /* transaction id */
        struct inode *ip = NULL;        /* child directory inode */
        ino_t ino;
        struct component_name dname;    /* child directory name */
        struct btstack btstack;
        struct inode *iplist[2];
        struct tblock *tblk;

        jfs_info("jfs_mkdir: dip:0x%p name:%s", dip, dentry->d_name.name);

        /* link count overflow on parent directory ? */
        if (dip->i_nlink == JFS_LINK_MAX) {
                rc = -EMLINK;
                goto out1;
        }

        /*
         * search parent directory for entry/freespace
         * (dtSearch() returns parent directory page pinned)
         */
        if ((rc = get_UCSname(&dname, dentry, JFS_SBI(dip->i_sb)->nls_tab)))
                goto out1;

        /*
         * Either iAlloc() or txBegin() may block. Deadlock can occur if we
         * block there while holding dtree page, so we allocate the inode &
         * begin the transaction before we search the directory.
         */
        ip = ialloc(dip, S_IFDIR | mode);
        if (ip == NULL) {
                rc = -ENOSPC;
                goto out2;
        }

        tid = txBegin(dip->i_sb, 0);

        down(&JFS_IP(dip)->commit_sem);
        down(&JFS_IP(ip)->commit_sem);

        if ((rc = dtSearch(dip, &dname, &ino, &btstack, JFS_CREATE))) {
                jfs_err("jfs_mkdir: dtSearch returned %d", rc);
                goto out3;
        }
```

Figure 5.8e continued: JFS Operations (Function) Source Code

```
tblk = tid_to_tblock(tid);
        tblk->xflag |= COMMIT_CREATE;
        tblk->ip = ip;

        iplist[0] = dip;
        iplist[1] = ip;

        /*
         * initialize the child directory in-line in inode
         */
        dtInitRoot(tid, ip, dip->i_ino);

        /*
         * create entry in parent directory for child directory
         * (dtInsert() releases parent directory page)
         */
        ino = ip->i_ino;
        if ((rc = dtInsert(tid, dip, &dname, &ino, &btstack))) {
                jfs_err("jfs_mkdir: dtInsert returned %d", rc);

                if (rc == -EIO)
                        txAbort(tid, 1);    /* Marks Filesystem dirty */
                else
                        txAbort(tid, 0);    /* Filesystem full */
                goto out3;
        }

        ip->i_nlink = 2;    /* for '.' */
        ip->i_op = &jfs_dir_inode_operations;
        ip->i_fop = &jfs_dir_operations;
        ip->i_mapping->a_ops = &jfs_aops;
        ip->i_mapping->gfp_mask = GFP_NOFS;

        insert_inode_hash(ip);
        mark_inode_dirty(ip);

        /* update parent directory inode */
        dip->i_nlink++;                 /* for '..' from child directory */
        dip->i_ctime = dip->i_mtime = CURRENT_TIME;
        mark_inode_dirty(dip);

        rc = txCommit(tid, 2, &iplist[0], 0);

out3:
        txEnd(tid);
        up(&JFS_IP(dip)->commit_sem);
        up(&JFS_IP(ip)->commit_sem);
        if (rc) {
                ip->i_nlink = 0;
                iput(ip);
        } else
                d_instantiate(dentry, ip);

out2:
        free_UCSname(&dname);

out1:

        jfs_info("jfs_mkdir: rc:%d", rc);
        return rc;
}
/*
 * NAME:  jfs_rmdir(dip, dentry)
 *
 * FUNCTION:      remove a link to child directory
 *
 * PARAMETER:     dip       - parent inode
 *                dentry    - child directory dentry
 *
 * RETURN:        -EINVAL   - if name is . or ..
 *                -EINVAL - if . or .. exist but are invalid.
 *                errors from subroutines
 *
 * note:
 * if other threads have the directory open when the last link
 * is removed, the "." and ".." entries, if present, are removed before
 * rmdir() returns and no new entries may be created in the directory,
 * but the directory is not removed until the last reference to
 * the directory is released (cf.unlink() of regular file).
 */
static int jfs_rmdir(struct inode *dip, struct dentry *dentry)
{
        int rc;
        tid_t tid;                      /* transaction id */
        struct inode *ip = dentry->d_inode;
        ino_t ino;
        struct component_name dname;
        struct inode *iplist[2];
        struct tblock *tblk;

        jfs_info("jfs_rmdir: dip:0x%p name:%s", dip, dentry->d_name.name);
```

Figure 5.8e continued: JFS Operations (Function) Source Code

```
/* directory must be empty to be removed */
        if (!dtEmpty(ip)) {
                rc = -ENOTEMPTY;
                goto out;
        }

        if ((rc = get_UCSname(&dname, dentry, JFS_SBI(dip->i_sb)->nls_tab))) {
                goto out;
        }

        tid = txBegin(dip->i_sb, 0);

        down(&JFS_IP(dip)->commit_sem);
        down(&JFS_IP(ip)->commit_sem);

        iplist[0] = dip;
        iplist[1] = ip;

        tblk = tid_to_tblock(tid);
        tblk->xflag |= COMMIT_DELETE;
        tblk->ip = ip;

        /*
         * delete the entry of target directory from parent directory
         */
        ino = ip->i_ino;
        if ((rc = dtDelete(tid, dip, &dname, &ino, JFS_REMOVE))) {
                jfs_err("jfs_rmdir: dtDelete returned %d", rc);
                if (rc == -EIO)
                        txAbort(tid, 1);
                txEnd(tid);
                up(&JFS_IP(dip)->commit_sem);
                up(&JFS_IP(ip)->commit_sem);

                goto out2;
        }

        /* update parent directory's link count corresponding
         * to ".." entry of the target directory deleted
         */
        dip->i_nlink--;
        dip->i_ctime = dip->i_mtime = CURRENT_TIME;
        mark_inode_dirty(dip);

        /*
         * OS/2 could have created EA and/or ACL
         */
        /* free EA from both persistent and working map */
        if (JFS_IP(ip)->ea.flag & DXD_EXTENT) {
                /* free EA pages */
                txEA(tid, ip, &JFS_IP(ip)->ea, NULL);
        }
        JFS_IP(ip)->ea.flag = 0;

        /* free ACL from both persistent and working map */
        if (JFS_IP(ip)->acl.flag & DXD_EXTENT) {
                /* free ACL pages */
                txEA(tid, ip, &JFS_IP(ip)->acl, NULL);
        }
        JFS_IP(ip)->acl.flag = 0;

        /* mark the target directory as deleted */
        ip->i_nlink = 0;
        mark_inode_dirty(ip);

        rc = txCommit(tid, 2, &iplist[0], 0);

        txEnd(tid);

        up(&JFS_IP(dip)->commit_sem);
        up(&JFS_IP(ip)->commit_sem);

        /*
         * Truncating the directory index table is not guaranteed. It
         * may need to be done iteratively
         */
        if (test_cflag(COMMIT_Stale, dip)) {
                if (dip->i_size > 1)
                        jfs_truncate_nolock(dip, 0);

                clear_cflag(COMMIT_Stale, dip);
        }
out2:
        free_UCSname(&dname);

out:
        jfs_info("jfs_rmdir: rc:%d", rc);
        return rc;
}
```

Figure 5.8e continued: JFS Operations (Function) Source Code

5.4 Remembering the Importance of File System Stability and Reliability

Finally, as with other types of embedded middleware, in order to insure the stability and the reliability of embedded systems developers should never assume that file systems come configured out-of-the-box for their own particular needs. Readers should remember to tune these parameters, then test and verify the file system according to the overall system's requirements. It is critical for middleware developers to tune parameters properly in order to insure that the file system supports the embedded design's frequency of I/O file system operations and the size of relative transactions.

For example, OS-related parameters in the Reliance file system when using this file system over vxWorks are shown in Table 5.2. So, when taking account memory usage and performance requirements of the device, an increase in simultaneous (multithreaded) read operations occurs when increasing the value of TFS_THREAD_LIMIT, but will also increase the latency of the serialized write operations. Reducing memory usage will occur when decreasing parameters

Table 5.2: Examples of Datalight's Reliance Tuning Parameters for vxWorks[14]

Reliance Parameter	Description
TFS_THREAD_WRITE_SIZEKB	Maximum amount of Kbytes that is written before allowing a context switch to higher priority threads access
TFS_THREAD_LIMIT	The number of threads allowed to operate inside the file system simultaneously
TFS_COORD_CACHE_ENTRIES	The number of 'coordinate' cache entries that is responsible for data related to frequently accessed files/directories
TFS_INDEX_CACHE_ENTRIES	The number of 'index' cache entries responsible for storing the location of metadata on the storage medium
TFS_CACHE_BUFFER_COUNT	The number of TFS_MAX_BLOCK_SIZE internal cache buffers
TFS_CACHE_WRITE_GATHER_KBSIZE	Enabling the writes of contiguous dirty buffers in cache as a single operation
TFS_ENABLE_DISCARD	Reports to a block device when sectors are no longer used
TFS_DISCARD_TABLE_SIZE	In bytes, the size of the discard table
TFS_DISCARD_TABLE_GROWTH	Enable/disable ability of discard table to dynamically grow in size
RELFS_DISCARD_SUPPORT_WRSTFFS	Enable/disable use of reliance with WindRiver's True Flash File System (TFFS)

such as TFS_COORD_CACHE_ENTRIES, TFS_INDEX_CACHE_ENTRIES and TFS_CACHE_BUFFER_COUNT; however, decreasing these values will also reduce performance. This means an improvement in performance will result when these parameters are increased, as long as there is enough of the right type of memory on the target boards.

The reliability of the embedded file system will also depend on the file system's internal design. As mentioned earlier in this chapter, many *(atomic) transactional* and *journaling* file systems employ some type of log management scheme as a means of increasing reliability by decreasing the chances that data will get corrupted or lost during file system transactions, or at least some type of data-recovery algorithm can be executed when necessary. Other embedded file systems (i.e., Datalight's Reliance) take reliability further within their internal design via the implementation of more complex schemes, such as utilizing transaction points or some similar mechanism which allows for the preservation of original data until file system transactions are 100% completed.

In short, Reliance (for example) continuously tracks used versus unused/free data blocks. This type of file system will then only utilize available storage space, and not overwrite any 'used' area on the medium. This is what insures that the state of this file system, prior to the start of any new transaction, remains safe on the storage media during the current processing of a current transaction. When the current transaction has completed without problems, then a *transaction point* is set. The Reliance file system then uses this transaction point to commit changes, and free up the data blocks that kept the original state and data safe. This file system scheme helps insure that if something goes wrong during a current file system transaction, the integrity of the original data is still preserved (see Figure 5.9).

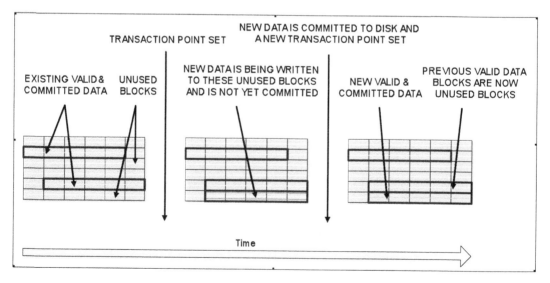

Figure 5.9: How Reliance File System Transaction Points Help Insure Reliability[14]

Finally, remember that tuning software parameters for components within your design will not be limited to the file system when this file system is utilized as 'middleware' within an embedded device. The reader needs to insure that 'overlying' application software components that utilize the file system are tuned properly for that particular file system implementation as well. Take, for instance, an FTP (file transfer protocol) server application that is configured internally to support some version of an embedded file system with certain stack requirements in order to support related tasks. The internal FTP server application code would need to be changed (i.e., size of the task stack increased) for the FTP server process to have additional stack space after being ported to a different file system to avoid a stack overflow, if using this other file system with the FTP server application requires more stack space to function without crashing when using the FTP server application residing on the embedded device.

For example, with a version of an FTP server application provided by WindRiver with vxWorks 6.5, the FTP server can be included when adding the component 'INCLUDE_ IPFTPS'. This FTP server application uses a stack size definition according to the value defined by IPCOM_PROC_STACK_DEFAULT, i.e.:

snippet from ipftps.c [15]

….

if (ipcom_proc_create(session->name, ipftps_session,

IPCOM_PROC_STACK_DEFAULT, &pid) !=

IPCOM_SUCCESS)

…..

This is the FTP server code that would be changed to give the server process more stack space, i.e.:

(snippet from ipftps.c [15])

….

If (ipcom_proc_create(session->name, ipftps_session,

IPCOM_PROC_STACK_LARGE, &pid) !=

IPCOM_SUCCESS)

….

If a stack overflow would occur when using a particular file system with the supplied definition of 'IPCOM_PROC_STACK_LARGE', for example, then modyfing this value to an even larger value in the corresponding header file (i.e., ipcom_pconfig.h) is necessary within the FTP server application.

5.5 Summary

As introduced in the various sections of this chapter, there are different file system design schemes that can be implemented in a particular file system. In order to understand a file system design, determine which file system design is the right choice for an embedded device, as well as understand the impact of a file system on a particular device, it is important to first understand the fundamentals of a file system. These fundamentals, introduced in this chapter, include what the purpose of a file system is, elements that commonly make up a file system, and real-world examples of some of the schemes implementing these elements. The reader can then apply these fundamentals to analyzing file system design features, such as:

- available API operations and/or an API that adheres to some type of industry standard interface
- maximum amount of memory that is needed by the file system
- non-blocking adherence for file systems implemented in real-time systems
- performance
- support of specific hardware and/or operating system

in order to determine if the file system design is the right one for a particular system, as well as the impact of the file system on the embedded device.

5.6 File System Problems

1. What is the purpose of a file system?
2. All file systems can only manage files located on the embedded system the file system resides on (True/False).
3. A file is:
   A. A set of data that has been grouped together and assigned a unique password
   B. A set of data that has been grouped together and assigned a unique name
   C. A set of names that has been grouped together and assigned a unique password
   D. None of the above.
4. What is a raw file? Give an example of a file system that supports raw files.
5. Outline the four-step model to understanding a file system design.
6. A file system implemented in the system software layer can exist as:
   A. Middleware that sits on top of the operating system layer
   B. Middleware that sits on top of other middleware components, for example a Java-based file system that resides on a Java Virtual Machine (JVM)
   C. Middleware that has been tightly integrated and provided with a particular operating system distribution

    D. None of the above

    E. All of the above.

7. One or more file systems can be implemented in an embedded system (True/False).

8. How do file systems view the hardware storage medium? Draw an example.

9. A file system can manage files on the following hardware:

    A. RAM

    B. CD

    C. Smart card

    D. Only B and C

    E. All of the above.

10. List and describe six types of file-system-specific device driver API functionality typically found in hardware storage medium device drivers.

11. What is the difference between an operating system character device and a block device?

12. A file system can require other underlying middleware components (True/False).

13. Draw and describe the layers of the General File System Model.

14. How do the design schemes of core elements of a file system impact performance?

15. Name and describe five examples of file system APIs.

5.7 End Notes

[1]   Microsoft Extensible Firmware Initiative FAT32 File System Specification. Version 1.03, December 6, 2000. Microsoft Corporation

[2]   http://redhat.brandfuelstores.com/

[3]   www.microsoft.com

[4]   http://shop.cxtreme.de

[5]   "Embedded Systems Architecture: A Comprehensive Guide for Engineers and Programmers". T. Noergaard. Elsevier 2005. p245.

[6]   http://www.westerndigital.com/en/products/Products.asp?DriveID=104

[7]   http://www.seagate.com/cda/products/discsales/marketing/detail/0,1081,771,00.html

[8]   http://www.babyusb.com/flashspecs2.htm

[9]   "Xscale Lite Datasheet" RLC Enterprises, Inc.

[10]   http://www.psism.com/pendrive.htm

[11]   'Corsair USB Flash Memory Datasheet'. Corsair.

[12]   http://www.linux-mtd.infradead.org/archive/

[13]   "vxWorks API Reference Guide: Device Drivers". Version 5.5

[14]   • Ditalight "FlashFx Pro API Guide"
      • source code
      • configuration files
      • Datalight FlashFX® Pro
      • "FlashFx Developers Guide for Wind River VxWorks", V3.10
      • "Reliance Developers Guide for Wind River VxWorks", V3.00

[15]   WindRiver sample code for FTP server application

Virtual Machines in Middleware

A powerful approach to understanding what a virtual machine (VM) is and how it works within an embedded system is by relating in theory to how an embedded operating system (OS) functions. Simply, a VM implemented as middleware software is a set of software libraries that provides an abstraction layer for software residing on top of the VM to be less dependent on hardware and underlying software. Like an OS, a VM can provide functionality that can perform everything from process management to memory management to IO system management depending on the specification it adheres to. What differentiates the inherent purpose of a VM in an embedded system versus that of an OS is introduced in the next section of this chapter, and is specifically related to the actual programming languages used for creating programs overlying a VM.

6.1 The First Step to Understanding a VM Implementation: The Basics to Programming Languages[1]

One of the main purposes of integrating a virtual machine (VM) is in relation to programming languages, thus this section will outline some programming language fundamentals. In embedded systems design, there is no single language that is the perfect solution for every system. In addition, many complex embedded systems software layers are inherently based on some combination of multiple languages. For example, within one embedded device the device driver layer may be composed of drivers written in assembly and C source code, the OS and middleware software implemented using C and C++, and different application layer

Table 6.1: General Evolution of Programming Languages[1]

	Language	**Details**
5th Generation	Natural languages	Programming languages similar to conversational languages typically used for AI (artificial intelligence) programming and design
4th Generation	Very high level (VHLL) and non-procedural languages	Very high level languages that are object-oriented, like C++, C#, and Java, scripting languages, such as Perl and HTML – as well as database query languages, like SQL for example
3rd Generation	High-order (HOL) and procedural languages, such as C and Pascal for example	High-level programming languages with more English-corresponding phrases. More portable than 2nd and 1st generation languages
2nd Generation	Assembly language	Hardware-dependent, representing machine code
1st Generation	Machine code	Hardware-dependent, binary zeros (0s) and ones (1s)

components implemented in C, C++, and embedded Java. So, let us start with the basics of programming languages for readers who are unfamiliar with the fundamentals, or would like a quick refresher.

The hardware components within an embedded system can only directly transmit, store, and execute *machine code*, a basic language consisting of ones and zeros. Machine code was used in earlier days to program computer systems, which made creating any complex application a long and tedious ordeal. In order to make programming more efficient, machine code was made visible to programmers through the creation of a hardware-specific set of instructions, where each instruction corresponded to one or more machine code operations. These hardware-specific sets of instructions were referred to as *assembly language*. Over time, other programming languages, such as C, C++, Java, etc., evolved with instruction sets that were (among other things) more hardware-independent. These are commonly referred to as *high-level* languages because they are semantically further away from machine code, they more resemble human languages, and are typically independent of the hardware. This is in contrast to a *low-level* language, such as assembly language, which more closely resembles machine code. Unlike high-level languages, low-level languages are hardware-dependent, meaning there is a unique instruction set for processors with different architectures. Table 6.1 outlines this evolution of programming languages.

Because machine code is the only language the hardware can directly execute, all other languages need some type of mechanism to generate the corresponding machine code. This mechanism usually includes one or some combination of *preprocessing*, *translation*, and *interpretation*. Depending on the language and as shown in Figure 6.1, these mechanisms

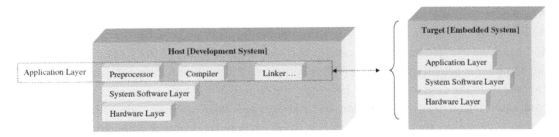

Figure 6.1: Programming Languages, Host, and Target[1]

exist on the programmer's ***host*** system, typically a non-embedded development system, such as a PC or Sparc station, or the ***target*** system (i.e., the embedded system being developed).

Preprocessing is an optional step that occurs before either the translation or interpretation of source code, and whose functionality is commonly implemented by a ***preprocessor***. The preprocessor's role is to organize and restructure the source code to make translation or interpretation of this code easier. As an example, in languages like C and C++, it is a pre-processor that allows the use of named code fragments, such as *macros*, that simplify code development by allowing the use of the macro's name in the code to replace fragments of code. The preprocessor then replaces the macro name with the contents of the macro during preprocessing. The preprocessor can exist as a separate entity, or can be integrated within the translation or interpretation unit.

Many languages convert source code, either directly or after having been preprocessed through use of a ***compiler***, a program that generates a particular target language – such as machine code and Java byte code – from the source language (see Figure 6.2).

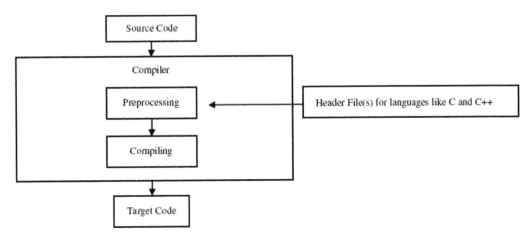

Figure 6.2: Compiling Native Code[1]

A compiler typically 'translates' all of the source code to some target code at one time. As is usually the case in embedded systems, compilers are located on the programmer's host machine and generate target code for hardware platforms that differ from the platform the compiler is actually running on. These compilers are commonly referred to as ***cross-compilers***. In the case of assembly language, the compiler is simply a specialized cross-compiler referred to as an ***assembler***, and it always generates machine code. Other high-level language compilers are commonly referred to by the language name plus the term 'compiler', such as 'Java compiler' and 'C compiler'. High-level language compilers vary widely in terms of what is generated. Some generate machine code, while others generate other high-level code, which then requires what is produced to be run through at least one more compiler or interpreter, as discussed later in this section. Other compilers generate assembly code, which then must be run through an assembler.

After all the compilation on the programmer's host machine is completed, the remaining target code file is commonly referred to as an ***object file***, and can contain anything from machine code to Java byte code (discussed later as an example in this chapter), depending on the programming language used. As shown in the C example in Figure 6.3, after linking this object file to any system libraries required, the object file, commonly referred to as an ***executable***, is then ready to be transferred to the target embedded system's memory.

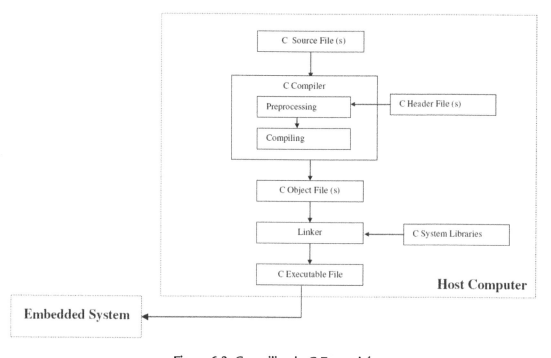

Figure 6.3: Compiling in C Example[1]

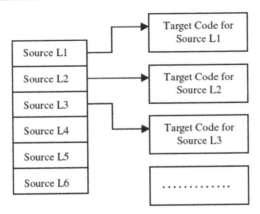

Figure 6.4: Interpretation of a Language[1]

6.1.1 Non-native Programming Languages that Impact the Middleware Architecture[1]

Where a compiler usually translates all of the given source code at one time, an *interpreter* generates (interprets) machine code one source code line at a time (see Figure 6.4).

One of the most common subclasses of interpreted programming languages is *scripting languages*, which include PERL, JavaScript, and HTML. Scripting languages are high-level programming languages with enhanced features, including:

- More platform independence than their compiled high-level language counterparts[2]
- Late binding, which is the resolution of data types on-the-fly (rather than at compile time) to allow for greater flexibility in their resolution[2]
- Importation and generation of source code at runtime, which is then executed immediately[2]
- Optimizations for efficient programming and rapid prototyping of certain types of applications, such as internet applications and graphical user interfaces (GUIs).[2]

With embedded platforms that support programs written in a scripting language, an additional component – an interpreter – must be included in the embedded system's architecture to allow for 'on-the-fly' processing of code. Note that while all scripting languages are interpreted, not all interpreted languages are scripting languages. For example, one popular embedded programming language that incorporates both compiling and interpreting machine code generation methods is *Java*. On the programmer's host machine, Java must go through a compilation procedure that generates Java byte code from Java source code (see Figure 6.5).

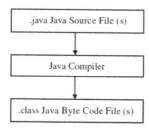

Figure 6.5: Embedded Java Compiling and Linking[1]

Java byte code is target code intended to be platform independent. In order for the Java byte code to run on an embedded system, one of the most commonly known types of virtual machines in embedded devices and used as the real-world example in this chapter, called a *Java Virtual Machine* (JVM), must reside on that system.

Real-world JVMs are currently implemented in an embedded system in one of three ways: in the hardware, as middleware in the system software layer, or in the application layer (see Figure 6.6). Within the scope of this chapter, it is when a virtual machine, like a JVM, is implemented as middleware that is addressed more specifically.

Scripting languages and Java aren't the only high-level languages that can automatically introduce an additional component as middleware within an embedded system. A real-world VM framework, called the *.NET Compact Framework* from Microsoft, allows applications written in almost any high-level programming language (such as C#, Visual Basic and Javascript) to run on any embedded device, independent of hardware or system software design.

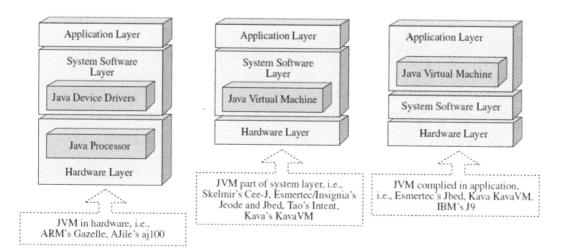

Figure 6.6: Embedded JVM[1]

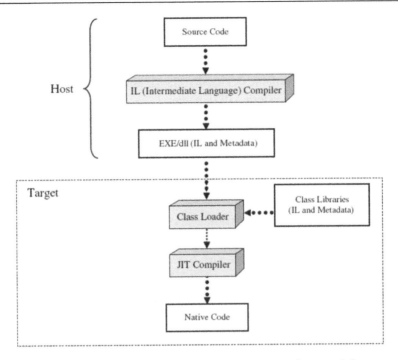

Figure 6.7: .NET Compact Framework Execution Model[1]

Applications that fall under the .NET Compact Framework must go through a compilation and linking procedure that generates a CPU-independent intermediate language file, called MSIL (Microsoft Intermediate Language), from the original source code file (see Figure 6.7). For a high-level language to be compatible with the .NET Compact Framework, it must adhere to Microsoft's **Common Language Specification**, a publicly available standard that anyone can use to create a compiler that is .NET compatible.

6.2 Understanding the Elements of a VM's Architecture[1]

After understanding the basics of programming languages, the key next steps for the reader in demystifying VM middleware include:

Step 2. Understand the APIs that are provided by a VM in support of its inherent purpose. In other words, know your standards relative to VMs that are specific to embedded devices (as first introduced in Chapter 3).

Step 3. Using the Embedded Systems Model, define and understand all required architecture components that underlie the virtual machine, including:

Step 3.1. Understanding the hardware (Chapter 2). If the reader comprehends the hardware, it is easier to understand why a VM implements functionality in a certain way relative to the hardware, as well as the hardware requirements of a particular VM implementation.

Step 3.2. Define and understand the specific underlying system software components, such as the available device drivers supporting the storage medium(s) and the operating system API (Chapter 2).

Step 4. Define the particular virtual machine or VM-framework architecture model, and then define and understand what type of functionality and data exists at each layer. This step will be addressed in the next few pages.

As mentioned at the start of this chapter, a virtual machine (VM) has many similarities in theory to the functionality provided by an embedded operating system (OS). This means a VM provides functionality that will perform everything from process management to memory management to I/O system management in addition to the translation of the higher-level language supported by the particular VM. Size, speed, and available out-of-the-box functionality are the technical characteristics of a VM that most impact an embedded system design, and essentially are the main differentiators of similar VMs provided by competing vendors. These characteristics are impacted by the internal design of three main subsystems within the VM, the:

- Loader
- Execution Engine
- API libraries.

As shown in Figure 6.8, for example, the .NET Compact Framework is made up of an execution engine referred to as a ***common language runtime*** (CLR) at the time this book was written, a class loader, and platform extension libraries. The CLR is made up of an execution engine that processes the intermediate MSIL code into machine code, and a garbage collector. The platform extension libraries are within the base class library (BCL), which provides additional functionality to applications (such as graphics, networking, and diagnostics). In order to run the intermediate MSIL file on an embedded system, the .NET Compact Framework must exist on that embedded system.

Another example is embedded JVMs implemented as middleware, which are also made up of a loader, execution engine, and Java API libraries (see Figure 6.9). While there are several embedded JVMs available on the market today, the primary differentiators between these JVMs are the JVM classes included with the JVM, and the execution engine that contains components needed to successfully process Java code.

6.2.1 The APIs

The APIs (application program interfaces) are application-independent libraries provided by the VM to, among other things, allow programmers to execute system functions, reuse

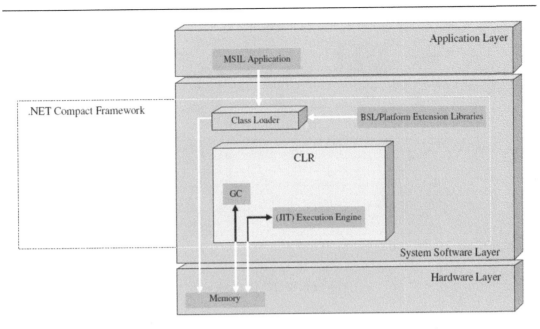

Figure 6.8: Internal .NET Compact Framework Components[1]

code, and more quickly create overlying software. Overlying applications that use the VM within the embedded device require the APIs, in addition to their own code, to successfully execute. The size, functionality, and constraints provided by these APIs differ according to the VM specification adhered to, but provided functionality can include memory management

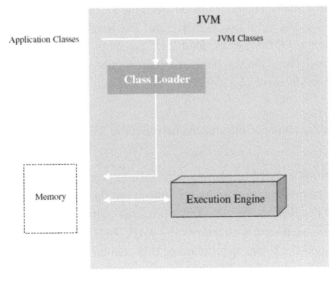

Figure 6.9: Internal JVM Components[1]

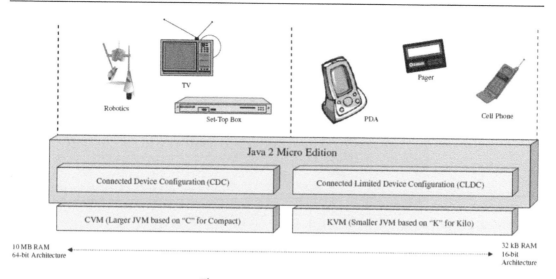

Figure 6.10: J2ME Devices[1]

features, graphics support, networking support, to name a few. In short, the type of applications in an embedded design is dependent on the APIs provided by the VM.

For example, different embedded Java standards with their corresponding APIs are intended for different families of embedded devices (see Figure 6.10). The type of applications in a Java-based design is dependent on the Java APIs provided by the JVM. The functionality provided by these APIs differs according to the Java specification adhered to, such as inclusion of the Real Time Core Specification from the J Consortium, Personal Java (pJava), Embedded Java, Java 2 Micro Edition (J2ME), and The Real Time Specification for Java from Sun Microsystems. Of these embedded Java standards, to date pJava and J2ME standards have typically been the standards implemented within larger embedded devices. PJava 1.1.8 was the predecessor of J2ME CDC that Sun Microsystems targeted to be replaced by J2ME.

Figure 6.11 shows an example of differences between the APIs of two different embedded Java standards.

There are later editions to 1.1.8 of pJava specifications from Sun, but as mentioned J2ME standards were intended to completely phase out the pJava standards in the embedded industry (by Sun) at the time this book was written. However, because the open source example used in this chapter is the *Kaffe JVM* implementation that is a clean room JVM based upon the pJava specification, this standard will be used as one of the examples to demonstrate functionality that is implemented via a JVM. Using this open source example, though based upon an older embedded Java standard, allows readers

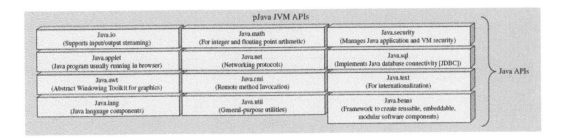

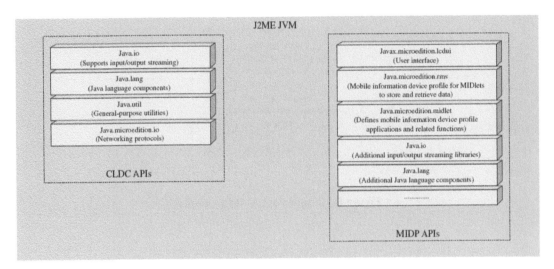

Figure 6.11: J2ME CLDC versus pJava APIs[1]

to have access to VM source code for hands-on purposes. The key is for the reader to use this open source example to get a clearer understanding of VM implementation from a systems-level perspective, regardless of whether the 'internal' functions used to implement one VM versus another differs from another because of the specification that VM adheres to (i.e., pJava versus J2ME, J2ME CDC versus J2ME CLDC, different versions of J2ME CLDC, and so on). The reader can use these examples as tools to understanding any VM implementation encountered, be it home-grown or purchased from a vendor.

To start, a high-level snapshot of the APIs provided by Sun's pJava standard are shown in Figure 6.12. In the case of a pJava JVM implemented in the system software layer, these libraries would be included (along with the JVM's loading and execution units) as middleware components.

Using specific networking APIs in the pJava specification as a more detailed example, shown in Figure 6.13 is the java.net package. The JVM provides an upper-transport layer API for

```
java.applet
java.awt
java.awt.datatransfer
java.awt.event
java.awt.image
java.beans
java.io
java.lang
java.lang.reflect
java.math
java.net
java.rmi
java.rmi.dgc
java.rmi.registry
java.rmi.server
java.security
java.security.acl
java.security.interfaces
java.sql
java.text
java.util
java.util.zip
```

Figure 6.12: pJava 1.1.8 API Example[3]

remote interprocess communication via the client–server model (where the client requests data, etc., from the server).

The APIs needed for client and servers are different, but the basis for establishing the network connection via Java is the socket (one at the client end and one at the server end). As shown in Figure 6.14, Java sockets use transport layer protocols of middleware networking components, such as TCP/IP discussed in the previous middleware example. Of the several different types of sockets (raw, sequenced, stream, datagram, etc.), the pJava JVM provides datagram sockets, in which data messages are read in their entirety at one time, and stream sockets, where data are processed as a continuous stream of characters. JVM datagram sockets rely on the UDP transport layer protocol, while stream sockets use the TCP transport layer protocol. pJava provides support for the client and server sockets, specifically one class for datagram sockets (called DatagramSocket, used for either client or server), and two classes for client stream sockets (Socket and MulticastSocket).

Interfaces
        ContentHandlerFactory
        FileNameMap
        SocketImplFactory
        URLStreamHandlerFactory
Classes
        ContentHandler
        DatagramPacket
        DatagramSocket
        DatagramSocketImpl
        HttpURLConnection
        InetAddress
        MulticastSocket
        ServerSocket
        Socket
        SocketImpl
        URL
        URLConnection
        URLEncoder
        URLStreamHandler
Exceptions
        BindException
        ConnectException
        MalformedURLException
        NoRouteToHostException
        ProtocolException
        SocketException
        UnknownHostException
        UnknownServiceException

Figure 6.13: java.net Package API Example[3]

A socket is created within a higher-layer application via one of the socket constructor calls, in the DatagramSocket class for a datagram socket, in the Socket class for a stream socket, or in the MulticastSocket class for a stream socket that will be multicast over a network (see Figure 6.15). As shown in the pseudocode example below of a Socket class constructor, within the pJava API, a stream socket is created, bound to a local port on the client device, and then connected to the address of the server.

In the J2ME set of standards, there are networking APIs provided by the packages within the CDC configuration and Foundation profile, as shown in Figure 6.18. In contrast to the pJava APIs shown in Figure 6.12, J2ME CDC APIs are a different set of libraries that would be included, along with the JVM's loading and execution units, as middleware components.

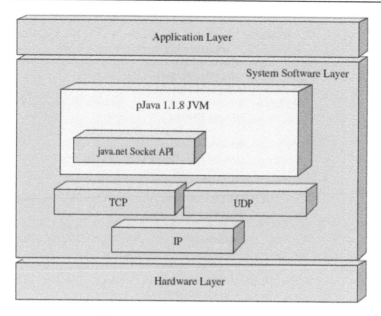

Figure 6.14: Sockets and a JVM[1]

As shown in Figure 6.16, the CDC provides support for the client sockets. Specifically, there is one class for datagram sockets (called DatagramSocket and used for either client or server) under CDC. The Foundation Profile, that sits on top of CDC, provides three classes for stream sockets, two for client sockets (Socket and MulticastSocket) and one for server sockets (ServerSocket). A socket is created within a higher-layer application via one of the socket constructor calls, in the DatagramSocket class for a client or server datagram socket, in the Socket class for a client stream socket, in the MulticastSocket class for a client stream socket that will be multicast over a network, or in the ServerSocket class for a server stream socket, for instance (see Figure 6.16). In short, along with the addition of a server (stream) socket API in J2ME, a device's middleware layer changes between pJava and J2ME CDC implementations in that the same sockets available in pJava are available in J2ME's network implementation, just in two different substandards under J2ME as shown in Figure 6.17.

The J2ME connected limited device configuration (CLDC, shown in Figure 6.18) and related profile standards are geared for smaller embedded systems by the Java community.

Continuing with networking as an example, the CLDC-based Java APIs provided by a CLDC-based JVM do not provide a .net package, as do the larger JVM implementations (see Figure 6.19).

Under the CLDC implementation, a generic connection is provided that abstracts networking, and the actual implementation is left up to the device designers. The Generic Connection

```
Socket(InetAddress address, boolean stream)
{
 X.create(stream);  //create stream socket
 X..bind(localAddress, localPort); //bind stream socket to port
 If problem ....
    X.close();//close socket
else
    X.connect(address, port); //connect to server
 }
```

Socket Class Constructor

Socket()
    Creates an unconnected socket, with the system-default type of SocketImpl.
Socket(InetAddress, int)
    Creates a stream socket and connects it to the specified port number at the specified IP address.
Socket(InetAddress, int, boolean)
    Creates a socket and connects it to the specified port number at the specified IP address.
    Deprecated.
Socket(InetAddress, int, InetAddress, int)
    Creates a socket and connects it to the specified remote address on the specified remote port.
Socket(SocketImpl)
    Creates an unconnected Socket with a user-specified SocketImpl.
Socket(String, int)
    Creates a stream socket and connects it to the specified port number on the named host.
Socket(String, int, boolean)
    Creates a stream socket and connects it to the specified port number on the named host. Deprecated.
Socket(String, int, InetAddress, int)
    Creates a socket and connects it to the specified remote host on the specified remote port.

MulticastSocket Class Constructors

MulticastSocket()
    Create a multicast socket.
MulticastSocket(int)
    Create a multicast socket and bind it to a specific port.

DatagramSocket Class Constructors

DatagramSocket()
    Constructs a datagram socket and binds it to any available port on the local host machine.
DatagramSocket(int)
    Constructs a datagram socket and binds it to the specified port on the local host machine.
DatagramSocket(int, InetAddress)
    Creates a datagram socket, bound to the specified local address.

Figure 6.15: Socket Constructors in Datagram, Multicast, and Socket Classes[3]

Framework (javax.microedition.io package) consists of one class and seven connection interfaces:

- Connection – closes the connection
- ContentConnection – provides metadata info
- DatagramConnection – create, send, and receive
- InputConnection – opens input connections
- OutputConnection – opens output connections
- StreamConnection – combines Input and Output
- Stream ConnectionNotifier – waits for connection.

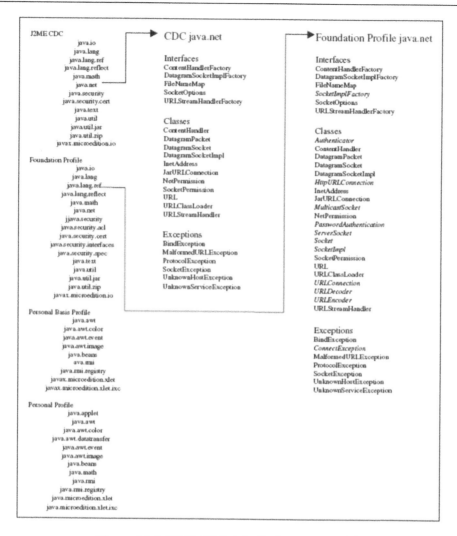

Figure 6.16: J2ME CDC 1.0a Package Example[4]

The Connection class contains one method (Connector.open) that supports the file, socket, comm, datagram and http protocols, as shown in Figure 6.20.

Another example is located within the Kaffe JVM open source example used in this chapter that contains its own implementation of a java.awt graphical library. AWT (abstract window toolkit) is a class library that allows for creating graphical user interfaces in Java. Figures 6.21a, b and c show a list of some of the java.awt libraries, as well as real-world source of one of the awt libraries being implemented.

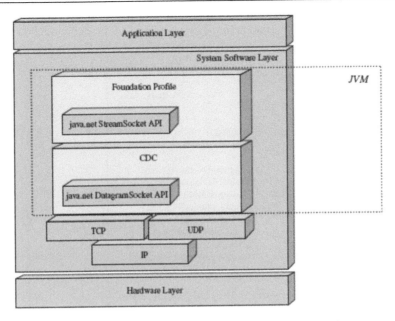

Figure 6.17: Sockets and a J2ME CDC-based JVM[1]

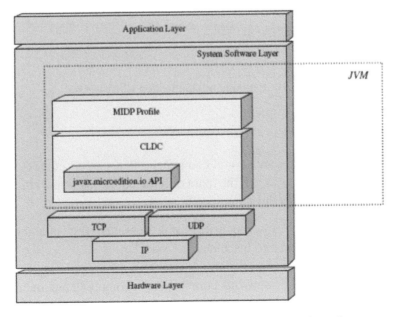

Figure 6.18: Sockets and a J2ME CLDC-based JVM[1]

```
J2ME CLDC 1.1
            java.io
            java.lang
            java.lang.ref
            java.util
        javax.microedition.io


J2ME MIDP 2.0
            java.lang
            java.util
        java.microedition.lcd.ui
        java.microedition.lcd.ui.game
        java.microedition.midlet
        java.microedition.rms
        java.microedition.io
        java.microedition.pki
        java.microedition.media
    java.microedition.media.control
```

Figure 6.19: J2ME CLDC APIs[4]

```
Http Communication :
        -Connection hc = Connector.open ("http:/www.wirelessdevnet.com");

Stream-based socket communication :
        -Connection sc = Connector.open ("socket://localhost:9000");

Datagram-based socket communication:
        -Connection dc = Connector.open ("datagram://:9000);

Serial port communication :
        -Connection cc = Connector.open ("comm:0;baudrate=9000");
```

Figure 6.20: Example of Connection Class in Use[1]

6.2.2 Execution Engine

Within an execution engine, there are several components that support process, memory, and I/O system management – however, the main differentiators that impact the design and performance of VMs that support the same specification are:

- The units within the VM that are responsible for process management and for translating what is generated on the host into machine code via:
  - interpretation
  - just-in-time (JIT), an algorithm that combines both compiling and interpreting
  - ahead-of-time compilation, such as dynamic adaptive compilers (DAC), ahead-of-time, way-ahead-of-time (WAT) algorithms to name a few.

A VM can implement one or more of these processing algorithms within its execution engine.

• The memory management scheme that includes a *garbage collector (GC)*, which is responsible for deallocating any memory no longer needed by the overlying application.

With interpretation in a JVM, shown in Figure 6.22 for example, every time the Java program is loaded to be executed, every byte code instruction is parsed and converted to native code, one byte code at a time, by the JVM's interpreter. Moreover, with interpretation, redundant portions of the code are reinterpreted every time they are run. Interpretation tends to have the lowest performance of the three algorithms, but it is typically the simplest algorithm to implement and to port to different types of hardware.

A JIT compiler (see Figure 6.23), on the other hand, interprets the program once, and then compiles and stores the native form of the byte code at runtime, thus allowing redundant code to be executed without having to reinterpret. The JIT algorithm performs better for redundant code, but it can have additional runtime overhead while converting the byte code into native code. Additional memory is also used for storing both the Java byte codes and the native compiled code. Variations on the JIT algorithm in real-world JVMs are also referred to as translators or dynamic adaptive compilation (DAC).

Kaffe java.awt.*

ActionEvt	EventQueue	MouseEvt
AdjustmentEvt	FlowLayout	NativeClipboard
AWTEvent	FocusEvt	NativeSelection
BarMenu	Font	OpaqueComponent
Button	FontMetrics	PaintEvt
Canvas	Frame	Panel
Checkbox	Graphics	PopupMenu
CheckboxGroup	GrpahicsLink	PopupWindow
CheckboxMenuItem	Image	....
Choice	ImageFrameLoader	
ClassAnalyzer	ImageLoader	
ClassProperties	ImageNativeProducer	
Componet	Insets	
ComponentEvt	ItemEvt	
Container	KeyEvt	
ContainerEvt	Label	
Cursor	List	
Defaults	MediaTracker	
DefKeyFilter	Menu	
Dialog	MenuBar	
Dimension	MenuComponent	
Event	MenuItem	
EventDispatchThread	MenuShortcut	

Figure 6.21a: Kaffe java.awt APIs[5]

Constructor Summary

`Checkbox()`
Creates a check box with an empty string for its label.

`Checkbox(String label)`
Creates a check box with the specified label.

`Checkbox(String label, boolean state)`
Creates a check box with the specified label and sets the specified state.

`Checkbox(String label, boolean state, CheckboxGroup group)`
Constructs a Checkbox with the specified label, set to the specified state, and in the specified check box group.

`Checkbox(String label, CheckboxGroup group, boolean state)`
Creates a check box with the specified label, in the specified check box group, and set to the specified state.

Method Summary

void	`addItemListener(ItemListener l)` Adds the specified item listener to receive item events from this check box.
void	`addNotify()` Creates the peer of the Checkbox.
AccessibleContext	`getAccessibleContext()` Gets the AccessibleContext associated with this Checkbox.
CheckboxGroup	`getCheckboxGroup()` Determines this check box's group.
ItemListener[]	`getItemListeners()` Returns an array of all the item listeners registered on this checkbox.
String	`getLabel()` Gets the label of this check box.
<T extends EventListener> T[]	`getListeners(Class<T> listenerType)` Returns an array of all the objects currently registered as *Foo*Listeners upon this Checkbox.
Object[]	`getSelectedObjects()` Returns an array (length 1) containing the checkbox label or null if the checkbox is not selected.
boolean	`getState()` Determines whether this check box is in the "on" or "off" state.
protected String	`paramString()` Returns a string representing the state of this Checkbox.
protected void	`processEvent(AWTEvent e)` Processes events on this check box.
protected void	`processItemEvent(ItemEvent e)` Processes item events occurring on this check box by dispatching them to any registered ItemListener objects.
void	`removeItemListener(ItemListener l)` Removes the specified item listener so that the item listener no longer receives item events from this check box.
void	`setCheckboxGroup(CheckboxGroup g)` Sets this check box's group to the specified check box group.
void	`setLabel(String label)` Sets this check box's label to be the string argument.
void	`setState(boolean state)` Sets the state of this check box to the specified state.

Figure 6.21b: java.awt Checkbox Class API[6]

```
/* << Checkbox.java >>
 *
 * This program is free software; you can redistribute it and/or modify
 * it under the terms of the GNU General Public License as published by
 * the Free Software Foundation; either version 2 of the License, or
 * (at your option) any later version.
 *
 * This program is distributed in the hope that it will be useful,
 * but WITHOUT ANY WARRANTY; without even the implied warranty of
 * MERCHANTABILITY or FITNESS FOR A PARTICULAR PURPOSE. See
 * the GNU General Public License for more details.
 *
 * You should have received a copy of the GNU General Public License
 * along with this program; if not, write to the Free Software
 * Foundation, Inc., 59 Temple Place, Suite 330, Boston, MA 02111-1307 USA
 */

package java.awt;

import java.awt.event.FocusEvent;
import java.awt.event.FocusListener;
import java.awt.event.ItemEvent;
import java.awt.event.ItemListener;
import java.awt.event.KeyEvent;
import java.awt.event.KeyListener;
import java.awt.event.MouseEvent;
import java.awt.event.MouseListener;

/**
 * class Checkbox -
 *
 * Copyright (c) 1998
 * Transvirtual Technologies, Inc. All rights reserved.
 *
 * See the file "license.terms" for information on usage and redistribution
 * of this file.
 */
public class Checkbox
 extends Component
 implements ItemSelectable, MouseListener, FocusListener, KeyListener
{
        private static final long serialVersionUID = 7270714317450821763L;
        CheckboxGroup group;
        int state;
        String label;
        ItemListener iListener;

        public static int counter;
        static int CHECKED = 1;
        static int HILIGHTED = 2;

public Checkbox () {
        this( "", false, null);
}

public Checkbox ( String label) {
        this( label, false, null);
}

public Checkbox ( String label, CheckboxGroup group, boolean state) {
        this( label, state, group);
}

public Checkbox ( String label, boolean state) {
        this( label, state, null);
}

public Checkbox ( String label, boolean state, CheckboxGroup group) {
        this.label = (label == null) ? "" : label;
        setCheckboxGroup( group);
        setState( state);
        setName("checkbox" + counter++);

        setForeground( Color.black);
        setFont( Defaults.TextFont);
        addMouseListener( this);
        addFocusListener( this);
        addKeyListener( this);
}

public synchronized void addItemListener ( ItemListener il) {
        iListener = AWTEventMulticaster.add( iListener, il);}
```

Figure 6.21c: Kaffe java.awt Checkbox Class Implemented[5]

```
void drawButton( Graphics g, int ext, int x0, int y0 ) {
        g.setColor( ((state & HILIGHTED) > 0) ? Defaults.BtnPointClr : Defaults.BtnClr);
        g.fill3DRect( x0, y0, ext, ext, true);

        if ( label.endsWith( " " ) )
                kaffePaintBorder( g);
        else {
                int d = BORDER_WIDTH;
                kaffePaintBorder( g, x0-d, y0-d, width-(x0+ext+d), height-(y0+ext+d) );
        }
}

void drawCheckMark( Graphics g, int ext, int x0, int y0 ) {
        g.setColor( Color.black);
        g.drawLine( x0+3, y0+4, x0+ext-5, y0+ext-4);
        g.drawLine( x0+2, y0+ext-4, x0+ext-5, y0+3);

        g.setColor( Color.white);
        g.drawLine( x0+3, y0+3, x0+ext-4, y0+ext-4);
        g.drawLine( x0+3, y0+ext-4, x0+ext-4, y0+3);
}

public void focusGained ( FocusEvent e) {
        state != HILIGHTED;
        repaint();
}

public void focusLost ( FocusEvent e) {
        state &= ~HILIGHTED;
        repaint();
}

public CheckboxGroup getCheckboxGroup () {
        return group;
}

ClassProperties getClassProperties () {
        return ClassAnalyzer.analyzeAll( getClass(), true);
}

public String getLabel () {
        return label;
}

public Object[] getSelectedObjects () {
        Object[] oa;
        if ( (state & CHECKED) > 0 ) {
                oa = new Object[1];
                oa[0] = this;
        }
        else
                oa = new Object[0];

        return oa;
}

public boolean getState () {
        return ((state & CHECKED) > 0);
}

public void keyPressed ( KeyEvent e) {
}

public void keyReleased ( KeyEvent e) {
}

public void keyTyped ( KeyEvent e) {
        char c = e.getKeyChar();

        switch ( c) {
                case ' ':
                case 0xA: //ENTER
                        if ( ((state & CHECKED) > 0) && (group != null) )
                                return;
                        setState ( (state & CHECKED) == 0 );
                        break;
        }
}

public void mouseClicked ( MouseEvent e) {
}

public void mouseEntered ( MouseEvent e) {
        state |= HILIGHTED;
        repaint();
}
```

Figure 6.21c continued: Kaffe java.awt Checkbox Class Implemented

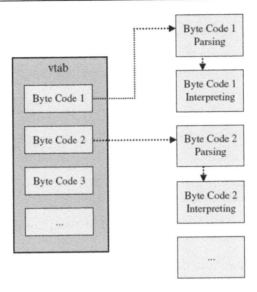

Figure 6.22: Interpretation[1]

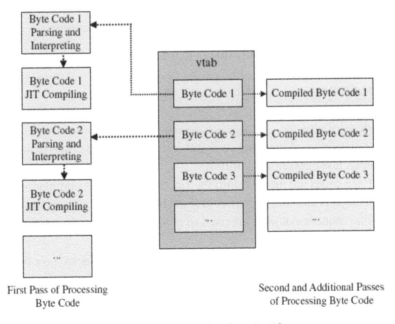

Figure 6.23: Just-in-Time (JIT)[1]

Finally, as shown in Figure 6.24, in WAT/AOT compiling all Java byte code is compiled into the native code at compile time, as with native languages, and no interpretation is done. This algorithm performs at least as well as the JIT for redundant code and better than a JIT for non-redundant code, but as with the JIT, there is additional runtime overhead when additional

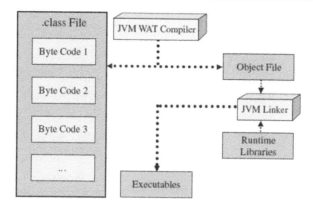

Figure 6.24: WAT (Way-Ahead-of-Time) Compiling[1]

Java classes dynamically downloaded at runtime have to be compiled and introduced to the system. WAT/AOT can also be a more complex algorithm to implement.

The Kaffe open source example used in this chapter contains a JIT (just-in-time) compiler called JIT3 (JIT version 3). The translate function shown in Figure 6.25 is the root of Kaffe's JIT3.[3] In general, the Kaffe JIT compiler performs three main functions:[7]

1. Byte code analysis. A *codeinfo* structure is generated by the 'verifyMethod' function that contains relevant data including:
    a. Stack requirements
    b. Local data usage
    c. Byte code attributes.
2. Instruction translation and machine code generation. Byte code translation is done at an individual block level generally as follows:
    a. Pass 1. Byte codes are mapped into intermediate functions and macros. A list of sequence objects containing master architecture-specific data are then generated.
    b. Pass 2. The sequence objects are used to generate the architecture-specific native instruction code.
3. Linking. The generated code is linked into the VM after all blocks have been processed. The native instruction code is then copied and linked.

6.2.2.1 Tasks versus Threads in Embedded VMs

As with operating systems, VMs manage and view other (overlying) software within the embedded system via some *process management* scheme. The complexity of a VM process management scheme will vary from VM to VM; however, in general the process management scheme is how a VM differentiates between an overlying program and the execution of that program. To a VM, a program is simply a *passive*, *static* sequence of instructions that could represent a system's hardware and software resources. The actual execution of a program

```
/* machine.c
 * Translate the Kaffe instruction set to the native one.
 *
 * Copyright (c) 1996-1999
 *              Transvirtual Technologies, Inc.  All rights reserved.
 *
 * Copyright (c) 2003, 2004
 *              Kaffe.org contributors. See ChangeLog for details. All rights reserved.
 *
 * Cross-language profiling changes contributed by
 * the Flux Research Group, Department of Computer Science,
 * University of Utah, http://www.cs.utah.edu/flux/
 *
 * See the file "license.terms" for information on usage and redistribution
 * of this file.
 */

....
/*
 * Translate a method into native code.
 *
 * Registers are allocated per basic block, using an LRU algorithm.
 * Contents of registers are spilled at the end of basic block,
 * depending on the edges in the CFG leaving the basic block:
 *
 * - If there is an edge from the basic block to an exception handler,
 * local variables are spilled on the stack
 *
 * - If there is only one non-exception edge, and the target basic
 * block is following the current block immediately, no spills are done
 *
 * - Otherwise, the local variables and the operand stack are spilled
 * onto the stack
 */
jboolean
translate(Method* xmeth, errorInfo* einfo)
{
...
        jint low;
        jint high;
        jvalue tmpl;
        int idx;
        SlotInfo* tmp;
        SlotInfo* tmp2;
        SlotInfo* mtable;

        bytecode* base;
        uint32 len;
        callInfo cinfo;
        fieldInfo finfo;
        Hjava_lang_Class* crinfo;
        codeinfo* mycodeInfo;

        nativeCodeInfo ncode;

        int64 tms = 0;
        int64 tme;

        static bool reinvoke = false;

        jboolean success = true;

        lockClass(xmeth->class);

        if (METHOD_TRANSLATED(xmeth)) {
                goto done3;
        }

        /* If this code block is native, then just set it up and return */
        if (methodIsNative(xmeth)) {
                void *func = native(xmeth, einfo);
                if (func != NULL) {
                        engine_create_wrapper(xmeth, func);
                        KAFFEJIT_TO_NATIVE(xmeth);
                } else {
                        success = false;
                }
                goto done3;
        }
```

Figure 6.25: Kaffe JIT 'Translate' Function[8]

```
/* Scan the code and determine the basic blocks */
success = analyzeMethod(xmeth, &mycodeInfo, einfo);
if (success == false) {
                /* It may happen that we already have translated it
                 * by implicit recursion in the verifier.
                 */
                if (METHOD_TRANSLATED(xmeth))
                 success = true;
                goto done3;
}
...
/* Only one in the translator at once. Must check the translation
 * hasn't been done by someone else once we get it.
 */
enterTranslator();

startTiming(&fulljit, "JIT translation");

if (Kaffe_JavaVMArgs.enableVerboseJIT) {
                tms = currentTime();
}
...
globalMethod = xmeth;

codeInfo = mycodeInfo;

/* Handle null calls specially */
if (METHOD_BYTECODE_LEN(xmeth) == 1 && METHOD_BYTECODE_CODE(xmeth)[0] == RETURN) {
                SET_METHOD_NATIVECODE(xmeth, (nativecode*)soft_null_call);
                goto done;
}

assert(reinvoke == false);
reinvoke = true;

maxLocal = xmeth->localsz;
maxStack = xmeth->stacksz;
maxArgs = sizeofSigMethod(xmeth, false);
if (maxArgs == -1) {
                goto done;
}
if (xmeth->accflags & ACC_STATIC) {
                isStatic = 1;
}
else {
                isStatic = 0;
                maxArgs += 1;
}


if (KaffeJIT3_setupExitWithOOM(einfo))
 {
 success = false;
 goto oom_error;
 }


....
base = (bytecode*)METHOD_BYTECODE_CODE(xmeth);
len = METHOD_BYTECODE_LEN(xmeth);
```

Figure 6.25 continued: Kaffe JIT 'Translate' Function

```
/*
 * Initialise the translator.
 */
initFakeCalls();

/* Do any machine dependent JIT initialization */
success = initInsnSequence(xmeth->localsz, xmeth->stacksz, einfo);
if (success == false) {
            goto done;
}

/***************************************/
/* Next reduce bytecode to native code */
/***************************************/

pc = 0;
start_function();
check_stack_limit();
if (Kaffe_JavaVMArgs.enableVerboseCall != 0) {
            softcall_trace(xmeth);
}
monitor_enter();
if (IS_STARTOFBASICBLOCK(0)) {
            end_basic_block();
            success = generateInsnSequence(einfo);
            if (success == false) {
                        goto done;
            }
            start_basic_block();
}

for (; pc < len; pc = npc) {

            assert(stackno <= maxStack+maxLocal);
            assert(stackno >= 0);

            npc = pc + insnLen[base[pc]];

/* Skip over the generation of any unreachable basic blocks */
if (IS_UNREACHABLE(pc)) {
while (npc < len && !IS_STARTOFBASICBLOCK(npc) & & !IS_STARTOFEXCEPTION(npc)) {
npc = npc + insnLen[base[npc]];
}
if (IS_STARTOFBASICBLOCK(npc)) {
end_basic_block();
start_basic_block();
stackno = STACKPOINTER(npc);
}
continue;
}

            /* Determine various exception conditions */
            checkCaughtExceptions(xmeth, pc);

            start_instruction();

            /* Note start of exception handling blocks */
            if (IS_STARTOFEXCEPTION(pc)) {
                        stackno = xmeth->localsz + xmeth->stacksz - 1;
                        start_exception_block();
            }

            switch (base[pc]) {
            default:
                        printf("Unknown bytecode %d\n", base[pc]);
                        leaveTranslator();
                        unlockClass(xmeth->class);
                        postException(einfo, JAVA_LANG(VerifyError));
            success = false;
                        break;
            #include "kaffe.def"
            }

            /* Note maximum number of temp slots used and reset it */
            if (tmpslot > maxTemp) {
                        maxTemp = tmpslot;
            }
            tmpslot = 0;
```

Figure 6.25 continued: Kaffe JIT 'Translate' Function

```
                        if (IS_STARTOFBASICBLOCK(npc)) {
                                end_basic_block();
                                success = generateInsnSequence(einfo);
                                if (success == false) {
                                        goto done;
                                }
                                start_basic_block();
                                stackno = STACKPOINTER(npc);
                        }
                }

        end_function();
        makeFakeCalls();

        assert(maxTemp < MAXTEMPS);

        if( finishInsnSequence(NULL, &ncode, einfo) )
        {
                installMethodCode(NULL, xmeth, &ncode);
        }
        else
        {
                success = false;
        }
        goto done;

oom_error:;
        KaffeJIT3_cleanupInsnSequence();

done:;

        KaffeJIT3_resetLabels();
        KaffeJIT3_resetConstants();
        tidyAnalyzeMethod(&codeInfo);

        reinvoke = false;

        globalMethod = NULL;


        if (Kaffe_JavaVMArgs.enableVerboseJIT) {
                tme = currentTime();
                jitStats.time += (int)(tme - tms);
                printf("<JIT: %s.%s%s time %dms (%dms) @ %p (%p)>\n",
                CLASS_CNAME(xmeth->class),
                xmeth->name->data, METHOD_SIGD(xmeth),
                (int)(tme - tms), jitStats.time,
                METHOD_NATIVECODE(xmeth), xmeth);
        }

        stopTiming(&fulljit);

        leaveTranslator();
done3:;

        unlockClass(xmeth->class);
        return (success);
}
...
```

Figure 6.25 continued: Kaffe JIT 'Translate' Function

is an *active*, *dynamic* event in which various properties change relative to time and the instruction being executed. A ***process*** (also commonly referred to as a ***task***) is created to encapsulate all the information that is involved in the executing of a program (i.e., stack, PC, the source code and data, etc.). This means that a program is only part of a task, as shown in Figure 6.26a.

Many embedded VMs also provide ***threads*** (*lightweight processes*) as an alternative means for encapsulating an instance of a program. Threads are created within the context of the OS task in which the VM is running, meaning all VM threads are bound to the VM task, and is a sequential execution stream within the task.

Unlike tasks, which have their own independent memory spaces that are inaccessible to other tasks, threads of a task share the same resources (working directories, files, I/O devices, global data, address space, program code, etc.), but have their own PCs, stack, and scheduling information (PC, SP, stack, registers, etc.) to allow for the instructions they are executing to be scheduled independently. Since threads are created within the context of the same task and can share the same memory space, they can allow for simpler communication and coordination relative to tasks. This is because a task can contain at least one thread executing one program in one address space, or can contain many threads executing different portions of one program in one address space (see Figure 6.26b), needing no intertask communication mechanisms. Also, in the case of shared resources, multiple threads are typically less expensive than creating multiple tasks to do the same work.

VMs must manage and synchronize tasks (or threads) that can exist simultaneously because, even when a VM allows multiple tasks (or threads) to coexist, one master processor on an embedded board can only execute one task or thread at any given time. As a result, multitasking embedded VMs must find some way of allocating each task a certain amount of time to use the master CPU, and switching the master processor between the various tasks. This is accomplished through task ***implementation***, ***scheduling***, ***synchronization***, and ***inter-task communication*** mechanisms.

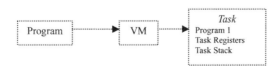

Figure 6.26a: VM Task

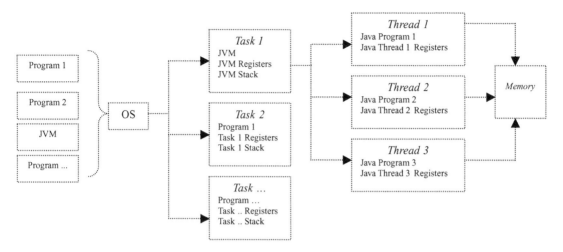

Figure 6.26b: VM Threads[1]

Jbed is a real-world example of a JVM that provides a *task*-based process management scheme that supports a multitasking environment. What this means is that multiple Java-based tasks are allowed to exist simultaneously, where each Jbed task remains independent of the others and does not affect any other Java task without the specific programming to do so (see Figure 6.27).

Jbed, for example, provides six different types of tasks that run alongside threads: *OneshotTimer Task* (which is a task that is run only once), *PeriodicTimer Task* (a task that is run after a particular set time interval), *HarmonicEvent Task* (a task that runs alongside a periodic timer task), *JoinEvent Task* (a task that is set to run when an associated task completes), *InterruptEvent Task* (a task that is run when a hardware interrupt occurs), and the *UserEvent Task* (a task that is explicitly triggered by another task). Task creation in Jbed

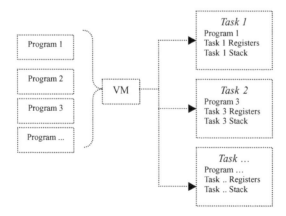

Figure 6.27: Multitasking in VMs

Task Creation Jbed Pseudocode

```
// Define a class that implements the Runnable interface for the software clock
public class ChildTask implements Runnable{

        //child task program Software Clock
        public void run () {
            integer seconds;

            while (softwareClock is RUNNING) {
                seconds = 0;
                while (seconds < 60) {
                    seconds = seconds+1;
                }
                ......
            }
    }
}

// parent task that enables software timer
void parentTask(void)
{
...
if sampleSoftware Clock NOT running {

        try{
            DURATION,
            ALLOWANCE,
            DEADLINE,
            OneshotTimer );
        }catch( AdmissionFailure error ){
```

Figure 6.28: Jbed Task Creation

is based upon a variation of the spawn model, called *spawn threading*. Spawn threading is spawning, but typically with less overhead and with tasks sharing the same memory space.

Figure 6.28 is a pseudocode example of task creation of a OneShot task, one of Jbed's six different types of tasks, in the Jbed RTOS where a parent task 'spawns' a child task software timer that runs only one time. The creation and initialization of the Task object is the Jbed (Java) equivalent of a task control block (TCB) which contains for that particular task data such as task ID, task state, task priority, error status, and CPU context information to name a few examples. The task object, along with all objects in Jbed, is located in Jbed's heap (in a JVM, there is typically only one heap for all objects). Each task in Jbed is also allocated its own stack to store primitive data types and object references.

Because Jbed is based upon the JVM model, a garbage collector (introduced in the next section of this chapter) is responsible for deleting a task and removing any unused code from memory once the task has stopped running. Jbed uses a non-blocking mark-and-sweep garbage collection algorithm which marks all objects still being used by the system and deletes (sweeps) all unmarked objects in memory.

In addition to creating and deleting tasks, a VM will typically provide the ability to *suspend* a task (meaning temporarily blocking a task from executing) and *resume* a task (meaning any blocking of the task's ability to execute is removed). These two additional functions are provided by the VM to support task *states*. A task's state is the activity (if any) that is going on with that task once it has been created, but has not been deleted.

Tasks are usually defined as being in one of three states:

- **Ready**: The process is ready to be executed at any time, but is waiting for permission to use the CPU.
- **Running**: The process has been given permission to use the CPU, and can execute.
- **Blocked** or **Waiting**: The process is waiting for some external event to occur before it can be 'ready' to 'run'.

Based upon these three states (Ready, Blocked, and Running), Jbed (for example) as a process state transition model is shown in Figure 6.29. *In Jbed, some states of tasks are related to the type of task, as shown in the table and state diagrams below. Jbed also uses separate queues to hold the task objects that are in the various states.*

The Kaffe open source JVM implements priority-preemptive-based 'jthreads' on top of OS native threads. Figure 6.30 shows a snapshot of Kaffe's thread creation and deletion scheme.

6.2.2.2 Embedded VMs and Scheduling

VM mechanisms, such as a *scheduler* within an embedded VM, are one of the main elements that give the illusion of a single processor simultaneously running multiple tasks or threads (see Figure 6.31). A scheduler is responsible for determining the order and the duration of tasks (or threads) to run on the CPU. The scheduler selects which tasks will be in what states (Ready, Running, or Blocked), as well as loading and saving the information for each task or thread.

There are many scheduling algorithms implemented in embedded VMs, and every design has its strengths and tradeoffs. The key factors that impact the effectiveness and performance of a scheduling algorithm include its *response time* (time for scheduler to make the context switch to a ready task and includes waiting time of task in ready queue), *turnaround time* (the time it takes for a process to complete running), *overhead* (the time and data needed to determine which tasks will run next), and *fairness* (what are the determining factors as to which processes get to run). A scheduler needs to balance utilizing the system's resources – keeping the CPU, I/O, as busy as possible – with task *throughput*, processing as many tasks as possible in a given amount of time. Especially in the case of fairness, the scheduler has to ensure that task *starvation*, where a task never gets to run, doesn't occur when trying to achieve a maximum task throughput.

State	Description
RUNNING	For all types of tasks, task is currently executing
READY	For all types of tasks, task in READY state
STOP	In Oneshot Tasks, task has completed execution
AWAIT TIME	For all types of tasks, task in BLOCKED state for a specific time period
AWAIT EVENT	In Interrupt and Joined tasks, BLOCKED while waiting for some event to occur

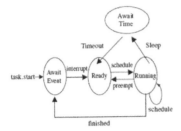

This state diagram shows some possible states for Interrupt tasks. Basically, an interrupt task is in an Await Event state until a hardware interrupt occurs – at which point the Jbed scheduler moves an Interrupt task into the Ready state to await its turn to run. At any time, the Joined Task can enter a timed waiting period.

State diagram for Jbed interrupt tasks

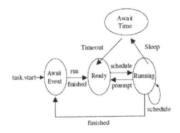

This state diagram shows some possible states for Joined tasks. Like the Interrupt task, the Joined task is in an Await Event state until an associated task has finished running – at which point the Jbed scheduler moves a Joined task into the Ready state to await its turn to run. At any time, the Joined Task can enter a timed waiting period.

State diagram for Jbed joined tasks

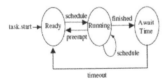

This state diagram shows some possible states for Periodic tasks. A Periodic task runs continuously at certain intervals and gets moved into the Await Time state after every run to await that interval before being put into the ready state.

State diagram for Jbed periodic tasks

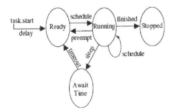

This state diagram shows some possible states for Oneshot tasks. A Oneshot task can either run once and then end (stop), or be blocked for a period of time before actually running.

State diagram for Jbed one shot tasks

Figure 6.29: Jbed Kernel and States[1]

One of the biggest differentiators between the scheduling algorithms implemented within embedded VMs is whether the algorithm guarantees its tasks will meet execution time deadlines. Thus, it is important to determine whether the embedded VM implements a scheduling algorithm that is ***non-preemptive*** or ***preemptive***. In ***preemptive scheduling***, the VM forces a context-switch on a task, whether or not a running task has completed executing or is cooperating with the context switch. Under *non-preemptive scheduling*, tasks (or threads) are given control of the master CPU until they have finished execution, regardless of the length of time or the importance of the other tasks that are waiting.

```
/*
 * Copyright (c) 1998 The University of Utah. All rights reserved.
 *
 * See the file "license.terms" for information on usage and
redistribution
 * of this file.
 *
 * Contributed by the Flux Research Group at the University of Utah.
 * Authors: Godmar Back, Leigh Stoller
 */


/*
 * This file implements jthreads on top of BeOS native threads and
 * was derived from oskit-pthreads/pjthread.c.
 *
 * Please address BeOS-related questions to alanlb@vt.edu.
 */
.....

/*
 * create a new jthread
 */
jthread_t
jthread_create(unsigned int pri, void (*func)(void *), int daemon,
 void *jlThread, size_t threadStackSize)
{
thread_id ntid;
jthread_t tid;

/*
 * Note that we create the thread in a joinable state, which is the
 * default. Our finalizer will join the threads, allowing the
 * thread system to free its resources.
 */

tid = allocator(sizeof(*tid));
assert(tid != 0);

            acquire_sem(threadLock);
tid->jlThread = jlThread;
tid->func = func;

tid->nextlive = liveThreads;
```

Figure 6.30: Kaffe JThread Creation and Deletion[8]

```
liveThreads = tid;
tid->status = THREAD_NEWBORN;

        ntid = spawn_thread(start_me_up, nameThread(jlThread),
                                    map_Java_priority(pri),
tid);
        tid->native_thread = ntid;

talive++;
if ((tid->daemon = daemon) != 0) {
tdaemon++;
}
        release_sem(threadLock);

        /* Check if we can safely save the per-thread info for
         * this thread. Yes, I know the per-thread stuff is lame,
         * but let's get this working first, shall we?
         */
        if (NULL == per_thread_info[ntid % MAX_THREADS].jtid) {
                resume_thread(ntid);
                return (tid);
        }
        else {
                kill_thread(ntid); /* stillborn */
                deallocator(tid);
                return NULL;

        }
}

/*
 * free a thread context
 */
void
jthread_destroy(jthread_t tid)
{
        status_t status;

        assert(tid);
        DBG(JTHREAD, dprintf("destroying %s\n", THREAD_NAME(tid));)

        atomic_and(&tid->stop_allowed, 0);
        wait_for_thread(tid->native_thread, &status);
        atomic_or(&tid->stop_allowed, 1);
        deallocator(tid);
}
```

Figure 6.30 continued: Kaffe JThread Creation and Deletion

Non-preemptive algorithms can be riskier to support since an assumption must be made that no one task will execute in an infinite loop, shutting out all other tasks from the master CPU. However, VMs that support non-preemptive algorithms don't force a context-switch before a task is ready, and the overhead of saving and restoration of accurate task information when switching between tasks that have not finished execution is only an issue if the non-preemptive scheduler implements a cooperative scheduling mechanism.

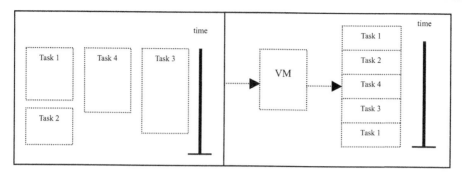

Figure 6.31: Interleaving Threads in VMs

As shown in Figure 6.32, Jbed contains an earliest deadline first (EDF)-based scheduler where the EDF/Clock Driven algorithm schedules priorities to processes according to three parameters: ***frequency*** (number of times process is run), ***deadline*** (when processes execution needs to be completed), and ***duration*** (time it takes to execute the process). While the EDF algorithm allows for timing constraints to be verified and enforced (basically guaranteed deadlines for all tasks), the difficulty is defining an exact duration for various processes. Usually, an average estimate is the best that can be done for each process.

Under the Jbed RTOS, all six types of tasks have the three variables 'duration', 'allowance', and 'deadline' when the task is created for the EDF scheduler to schedule all tasks (see Figure 6.33 for the method call).

The Kaffe open source JVM implements a *priority-preemptive*-based scheme on top of OS native threads, meaning jthreads are scheduled based upon their relative importance to each other and the system. Every jthread is assigned a priority, which acts as an indicator of orders

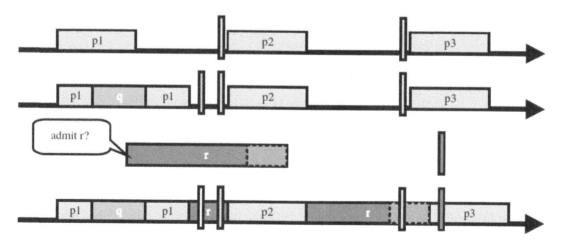

Figure 6.32: EDF Scheduling in Jbed

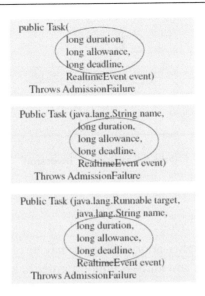

Figure 6.33: Jbed Method Call for Scheduling Task[1]

of precedence within the system. The jthreads with the highest priority always preempt lower-priority processes when they want to run, meaning a running task can be forced to block by the scheduler if a higher-priority jthread becomes ready to run. Figure 6.34 shows three jthreads (1, 2, 3 – where jthread 1 is the lowest priority and jthread 3 is the highest, and jthread 3 preempts jthread 2, and jthread 2 preempts jthread 1).

As with any VM with a priority-preemptive scheduling scheme, the challenges that need to be addressed by programmers include:

- JThread starvation, where a continuous stream of high-priority threads keeps lower-priority jthreads from ever running. Typically resolved by aging lower-priority jthreads (as these jthreads spend more time on queue, increase their priority levels).

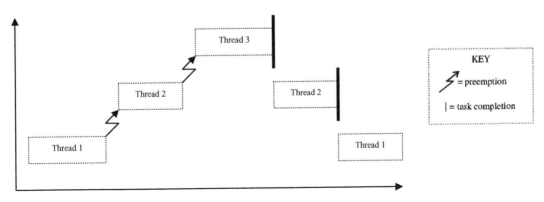

Figure 6.34: Kaffe's Priority-preemptive-based Scheduling

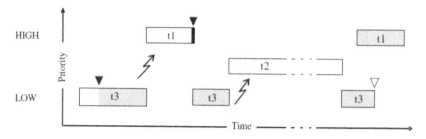

Figure 6.35: Priority Inversion[1]

- Priority inversion, where higher-priority jthreads may be blocked waiting for lower-priority jthreads to execute, and jthreads with priorities in between have a higher priority in running, thus both the lower-priority as well as higher-priority jthreads don't run (see Figure 6.35).
- How to determine the priorities of various threads. Typically, the more important the thread, the higher the priority it should be assigned. For jthreads that are equally important, one technique that can be used to assign jthread priorities is the Rate Monotonic Scheduling (RMS) scheme which is also commonly used with relative scheduling scenerios when using embedded OSs. Under RMS, jthreads are assigned a priority based upon how often they execute within the system. The premise behind this model is that, given a preemptive scheduler and a set of jthreads that are completely independent (no shared data or resources) and are run periodically (meaning run at regular time intervals), the more often a jthread is executed within this set, the higher its priority should be. The RMS Theorem says that if the above assumptions are met for a scheduler and a set of 'n' jthreads, all timing deadlines will be met if the inequality Σ Ei/Ti $\leq$ n(21/n − 1) is verified, where

> i = periodic jthread
> n = number of periodic jthreads
> Ti = the execution period of jthread i
> Ei = the worst-case execution time of jthread i
> Ei/Ti = the fraction of CPU time required to execute jthread i.

So, given two jthreads that have been prioritized according to their periods, where the shortest-period jthread has been assigned the highest priority, the 'n(21/n − 1)' portion of the inequality would equal approximately 0.828, meaning the CPU utilization of these jthreads should not exceed about 82.8% in order to meet all hard deadlines. For 100 jthreads that have been prioritized according to their periods, where the shorter period jthreads have been assigned the higher priorities, CPU utilization of these tasks should

---

To Benefit Most from a Fixed-Priority Preemptive Scheduling

Whether from either a pure OS or from an overlying VM perspective -- algorithms for assigning priorities to tasks or threads are typically classified as *fixed-priority* when threads/tasks are assigned priorities at design time, and do not change through the lifecycle of the thread or task, *dynamic-priority* when priorities are assigned to threads or tasks at run-time, or some combination of *both* algorithms. It must be determined by the reader on a VM to VM basis which scheduling schemes are supported. In the case of maximizing on the fixed-priority scheduling scheme, the keys to success include:

- to assign the priorities of threads or tasks according to their periods, so that the shorter the periods, the higher the priorities.

- to assign priorities using a fixed-priority algorithm (like the Rate Monotonic Algorithm, the basis of RMS) to assign fixed priorities to threads or tasks, and as a tool to quickly to determine if a set of threads or tasks is schedulable.

- to understand that in the case when the inequality of a fixed-priority algorithm, like RMS, is not met, an analysis of the specific thread or task set is required. RMS is a tool that allows for assuming that deadlines would be met in most cases if the total CPU utilization is below the limit ("most" cases meaning there are threads or tasks that are not schedulable via any fixed-priority scheme). It is possible for a set of threads or tasks to still be schedulable in spite of having a total CPU utilization above the limit given by the inequality. Thus, an analysis of each thread's or task's period and execution time needs to be done in order to determine if the set can meet required deadlines.

- to realize that a major constraint of fixed-priority scheduling is that it is not always possible to completely utilize the master CPU 100%. If the goal is 100% utilization of the CPU when using fixed priorities, then threads or tasks should be assigned harmonic periods. Meaning, a thread's period or task's period should be an exact multiple of all other threads or tasks with shorter periods.

- Based on the article "Introduction to Rate Monotonic Scheduling" by Michael Barr
Embedded Systems Programming, February 2002

Figure 6.36: Note on Scheduling

not exceed approximately 69.6% ($100 \times (21/100 - 1)$) in order to meet all deadlines. See Figure 6.36 for additional notes on this type of scheduling model.

6.2.2.3 VM Memory Management and the Garbage Collector[1]

A VM's memory heap space is shared by all the different overlying VM processes – so access, allocation, and deallocation of portions of the heap space need to be managed. In the case of VMs, a garbage collector (GC) is integrated within. Garbage collection discussed in this chapter isn't necessarily unique to any particular language. A garbage collector (GC) can be implemented within embedded devices in support of other languages that do not require VMs, such as C and C++.[8] Regardless, when creating a garbage collector to support any language, it becomes an integral component of an embedded system's architecture.

Applications written in a language such as Java or C# all utilize the same memory heap space of the VM and cannot allocate or deallocate memory in this heap or outside this heap that has been allocated for previous use (as can be done in native languages, such as using 'free' in the C language, though as mentioned above, a garbage collector can be implemented to support any language). In Java, for example, only the GC (garbage collector) can deallocate memory no longer in use by Java applications. GCs are provided as a safety mechanism for

Java programmers so they do not accidentally deallocate objects that are still in use. While there are several garbage collection schemes, the most common are based upon the copying, mark and sweep, and generational GC algorithms.

6.2.2.4 GC Memory Allocator[1]

Embedded VMs can implement a wide variety of schemes to manage the allocation of the memory heap, in combination with an underlying operating system's memory management scheme. With Kaffe, for example, the GC including a memory allocator for the JVM in addition to the underlying operating system's memory management scheme is utilized. When Kaffe's memory allocator is used to allocate memory (see Figure 6.37) from the JVMs heap space, its purpose is to simply determine if there is free memory to allocate – and if so, returning this memory for use.

6.2.2.5 Garbage Collection[1]

The copying garbage collection algorithm (shown in Figure 6.38) works by copying referenced objects to a different part of memory, and then freeing up the original memory space of unreferenced objects. This algorithm uses a larger memory area in order to work, and usually cannot be interrupted during the copy (it blocks the system). However, it does ensure that what memory is used is used efficiently by compacting objects in the new memory space.

```
/* gc-mem.c
 * The heap manager.
 *
 * Copyright (c) 1996, 1997
 *           Transvirtual Technologies, Inc.  All rights reserved.
 *
 * See the file "license.terms" for information on usage and redistribution
 * of this file.
 */


....

/**
 * Allocate a piece of memory.
 */
void*
gc_heap_malloc(size_t sz)
{
        size_t lnr;
        gc_freeobj* mem = NULL;
        gc_block** mptr;
        gc_block* blk;
        size_t nsz;
```

Figure 6.37: Kaffe's GC Memory Allocation Function[8]

```
        lockStaticMutex(&gc_heap_lock);
        startTiming(&heap_alloc_time, "gc_heap_malloc");


    if (KGC_SMALL_OBJECT(sz)) {

                /* Translate size to object free list */
                lnr = sztable[sz].list;
                nsz = freelist[lnr].sz;

                /* No available objects? Allocate some more */
                mptr = &freelist[lnr].list;
                if (*mptr != 0) {
                            blk = *mptr;
                            assert(blk->free != 0);
                }
                else {
                            blk = gc_small_block(nsz);
                            if (blk == 0) {
                                        goto out;
                            }
                            blk->next = *mptr;
                            *mptr = blk;

                }

                /* Unlink free one and return it */
                mem = blk->free;

                DBG(GCDIAG,
                 assert(gc_check_magic_marker(blk));
                 ASSERT_ONBLOCK(mem, blk);
                 if (mem->next) ASSERT_ONBLOCK(mem->next, blk));

                blk->free = mem->next;

                KGC_SET_STATE(blk, GCMEM2IDX(blk, mem), KGC_STATE_NORMAL);

                /* Once we use all the sub-blocks up, remove the whole block
                 * from the freelist.
                 */
                assert(blk->nr >= blk->avail);
                assert(blk->avail > 0);
                blk->avail--;
                if (blk->avail == 0) {
                            *mptr = blk->next;
                }
    }
    else {
                nsz = sz;
                blk = gc_large_block(nsz);
                if (blk == 0) {
                            goto out;
                }
                mem = GCBLOCK2FREE(blk, 0);
                KGC_SET_STATE(blk, 0, KGC_STATE_NORMAL);
                blk->avail--;
                assert(blk->avail == 0);
    }

    /* Clear memory */
    memset(mem, 0, nsz);

    assert(KGC_OBJECT_SIZE(mem) >= sz);

    out:
    stopTiming(&heap_alloc_time);
    unlockStaticMutex(&gc_heap_lock);

    return (mem);

}
```

Figure 6.37 continued: Kaffe's GC Memory Allocation Function

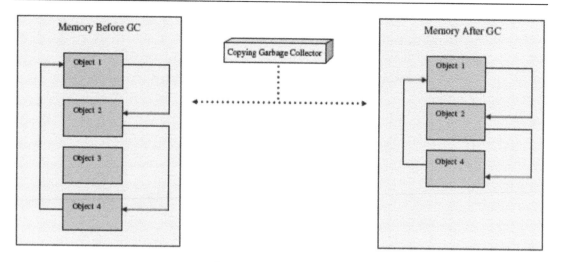

Figure 6.38: Copying GC[1]

The mark and sweep garbage collection algorithm (shown in Figure 6.39) works by 'marking' all objects that are used, and then 'sweeping' (deallocating) objects that are unmarked. This algorithm is usually non-blocking, meaning the system can interrupt the garbage collector to execute other functions when necessary. However, it doesn't compact memory the way a copying garbage collector does, leading to memory fragmentation, the existence of small, unusable holes where deallocated objects used to exist. With a mark and sweep garbage collector, an additional memory compacting algorithm can be implemented, making it a mark (sweep) and compact algorithm.

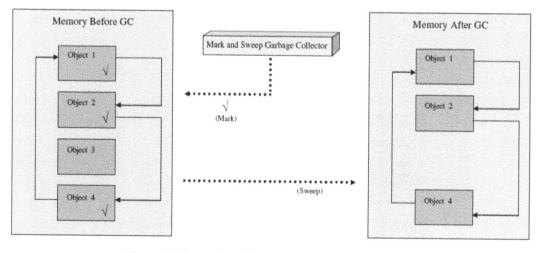

Figure 6.39: Mark and Sweep (No Compaction) GC[1]

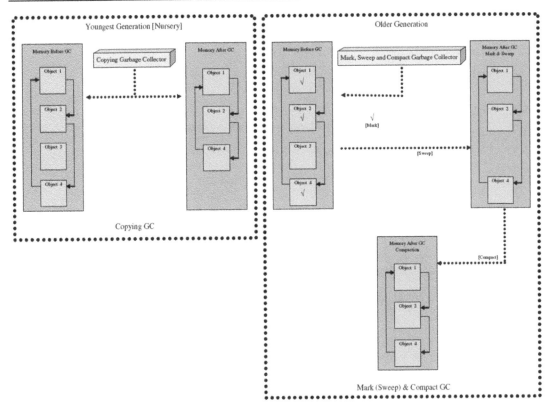

Figure 6.40: Generational GC[1]

Finally, the generational garbage collection algorithm (shown in Figure 6.40) separates objects into groups, called *generations*, according to when they were allocated in memory. This algorithm assumes that most objects that are allocated by a Java program are short-lived, thus copying or compacting the remaining objects with longer lifetimes is a waste of time. So, it is objects in the younger-generation group that are cleaned up more frequently than objects in the older-generation groups. Objects can also be moved from a younger-generation to an older-generation group. Different generational garbage collectors also may employ different algorithms to deallocate objects within each generational group, such as the copying algorithm or mark and sweep algorithms described previously.

The Kaffe open source example used in this chapter implements a version of a mark and sweep garbage collection algorithm. In short, the garbage collector (GC) within Kaffe will be invoked when the memory allocator determined more memory is required than free memory in the heap. The GC then schedules when the garbage collection will occur, and executes the collection (freeing of memory) accordingly. Figure 6.41 shows Kaffe's open source example of a mark and sweep GC algorithm for 'marking' data for collection.

```
/* gc-incremental.c
 * The garbage collector.
 * The name is misleading. GC is non-incremental at this point.
 *
 * Copyright (c) 1996, 1997
 *          Transvirtual Technologies, Inc.  All rights reserved.
 *
 * Copyright (c) 2003, 2004
 *          Kaffe.org contributors. See ChangeLog for details. All rights reserved.
 *
 * See the file "license.terms" for information on usage and redistribution
 * of this file.
 */

......

/*
 * Mark the memory given by an address if it really is an object.
 */
static void
gcMarkAddress(Collector* gcif UNUSED, void *gc_info UNUSED, const void* mem)
{
        gc_block* info;
        gc_unit* unit;

        /*
         * First we check to see if the memory 'mem' is in fact the
         * beginning of an object. If not we just return.
         */

        /* Get block info for this memory - if it exists */
        info = gc_mem2block(mem);
        unit = UTOUNIT(mem);
        if (gc_heap_isobject(info, unit)) {
                markObjectDontCheck(unit, info, GCMEM2IDX(info, unit));
        }
}

/*
 * Mark an object. Argument is assumed to point to a valid object,
 * and never, ever, be null.
 */
static void
gcMarkObject(Collector* gcif UNUSED, void *gc_info UNUSED, const void* objp)
{
gc_unit *unit = UTOUNIT(objp);
gc_block *info = gc_mem2block(unit);
DBG(GCDIAG, assert(gc_heap_isobject(info, unit)));
markObjectDontCheck(unit, info, GCMEM2IDX(info, unit));
}
```

Figure 6.41: Kaffe GC 'Mark' Functions[8]

6.2.3 VM Memory Management and the Loader

The loader is simply as its name implies. As shown in Figure 6.42a, it is responsible for acquiring and loading into memory all required code in order to execute the relative program

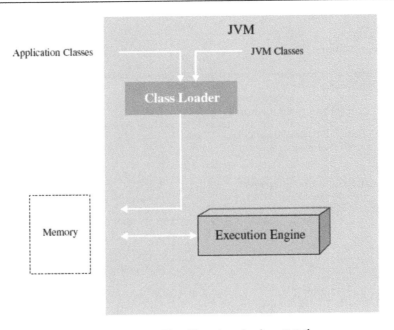

Figure 6.42a: The Class Loader in a JVM[1]

overlying the VM. In the case of a JVM like Kaffe, for example (see Figure 6.42b for open source snapshot), its internal Java class loader loads into memory all required Java classes required for the Java program to function.

```
struct Hjava_lang_Class*
java_lang_VMClassLoader_loadClass(Hjava_lang_String* jStr, jboolean
resolve)
{
        Hjava_lang_Class *clazz = NULL;
        errorInfo info;
        int error = 0;
Utf8Const *c;
char *name;
        int i;
        jboolean foundSlash = false;

        name = checkPtr(stringJava2C(jStr));
        for (i = strlen(name)-1; i >= 0; i--)
                if (name[i] == '/') {
                        foundSlash = true;
                        break;
                }
```

Figure 6.42b: Kaffe Class Loader Function[8]

```
            if (!foundSlash)
             classname2pathname(name, name);

            if (foundSlash || (!strncmp (name, "gnu/classpath/", 14) && strncmp (name, "gnu/classpath/tools/",
            strlen("gnu/classpath/tools/")))) {
                    struct Hjava_lang_Throwable *throwable;
                    throwable = (struct Hjava_lang_Throwable*)
                            execute_java_constructor(JAVA_LANG(ClassNotFoundException),  NULL,NULL,
                            "(Ljava/lang/String;)V", jStr);
                    throwException (throwable);
            }

            if( (c = utf8ConstFromString(name)) ){
                    clazz = loadClass(c, NULL, &info);
                    if( clazz ) {
                            if( processClass(clazz, resolve ? CSTATE_COMPLETE : CSTATE_PREPARED,
                            &info) == false )
                            {
                            error = 1;
                            }
                    }
                    else
{
                    error = 1;
                    }
                    utf8ConstRelease(c);
            }
            else {
                    postOutOfMemory(&info);
                    error = 1;
            }
            gc_free(name);
            if( error ) {
                    throwError(&info);
            }
            return( clazz );
}
```

Figure 6.42b continued: Kaffe Class Loader Function

6.3 A Quick Comment on Selecting Embedded VMs Relative to the Application Layer

Writing applications in a higher-level language that requires introducing an underlying VM in the middleware layer of an embedded system design, for better or worse, *will* require additional support relative to increased processing power and memory requirements. This is opposed to implementing the same applications in native C and/or assembly. So, as with integrating any type of middleware component, introducing a VM into an embedded system means planning for any additional hardware requirements and underlying system software by both the VM *and* the overlying applications that utilize the underlying VM middleware component. This is where understanding the fundamentals of the internal design of VMs, like the material presented in previous sections of this chapter, becomes critical to selecting the best design that meets your particular device's requirements.

For example, several factors, such as memory and performance, are impacted by the scheme a VM utilizes in processing the overlying application code. So, understanding the pros and cons of using a particular JVM that implements an interpretating byte-code scheme versus a just-in-time (JIT) compiler versus a way-ahead-of-time (WAT) compiler versus a dynamic adaptive compiler (DAC) is necessary. This means that, while using a particular JVM with a certain compilation scheme would introduce significant performance improvements, it may also introduce requirements for additional memory as well as introduce other limitations. For instance, pay close attention to the drawbacks to selecting a particular JVM that utilizes some type of ahead-of-time (AOT) or way-ahead-of-time (WAT) compilation which provides a big boost in performance when running on your hardware, but lacks the ability to process dynamically downloaded Java byte-code, whereas this dynamic download capability is provided by a competing JVM solution based on a slower, interpretating byte-code processing scheme. If on-the-field dynamic extensibility support is a non-negotiable requirement for the embedded system being designed, then it means needing to investigate further other options such as:

- selecting a competing JVM from another vendor that provides this dynamic-download capability out-of-the-box
- investigating the feasibility of deploying with a JVM based on a different byte-code processing scheme that runs a bit slower than the faster JVM solution that lacks dynamic download and extensibility support
- planning the resources, costs, and time to implement this required functionality within the scope of the project.

Another example would be when having to decide between a JIT implementation of a JVM versus going with the JIT-based .NET Compact Framework solution of comparable performance on your particular hardware and underlying system software. In addition to examining the available APIs provided by the JVM versus .NET Compact Framework embedded solutions for your application requirements, do not forget to consider the non-technical aspects of going with either particular solution as well. For example, this means taking into consideration when selecting between such alternative VM solutions, the availability of experienced programmers (i.e., Java versus C# programmers for instance). If there are no programmers available with the necessary skills for application development on that particular VM, factor in the costs and time involved in finding and hiring new resources, training current resources, and so on.

Finally do not forget that integrating the *right* VM in the *right* manner within the software stack which optimizes the performance of the solution is not enough to insure the design makes it to production successfully. To insure success taking an embedded design that introduces the complexity and stress to underlying components that incorporating an embedded VM produces, requires programmers to plan carefully how overlying applications will be written. This means it is *not* the most elegant nor the most brilliantly written application code that will insure the success of the design – but simply programmers that design applications in a manner that properly utilizes the underlying VM's powerful strengths and avoids its weaknesses. A Java application, for example, that is written as a masterpiece by even the cleverest programming guru will not be worth much, if when it runs on the device it was intended for this application is so slow and/or consumes so much of the embedded system's resources that the device simply cannot be shipped!

In short, the key to selecting which embedded VMs best match the requirements of your design, and successfully taking this design to production within schedule and costs, includes:

- determining if the VM has been ported to your target hardware's master CPU's architecture in the first place. If not, it means determining how much time, cost, and resources would be required to port the particular VM to your target hardware and underlying system software stack
- calculating additional processing power and memory requirements to support the VM solution and overlying applications
- specifying what additional type of support and/or porting is needed by the VM relative to underlying embedded OS and/or other middleware system software
- investigating the stability and reliability of the VM implementation on real hardware and underlying system software
- planning around the availability of experienced developers
- evaluating development and debugging tool support
- checking up on the reputation of vendors
- insuring access to solid technical support for the VM implementation for developers
- writing the overlying applications properly.

6.4 Summary

This chapter introduced embedded VMs, and their function within an embedded device. A section on programming languages and the higher-level languages that introduce the requirement of a VM within an embedded system was included in this chapter. The major components that make up most embedded VMs were discussed, such as an execution engine, the garbage collector, and loader to name a few. More detailed discussions of process management, memory management, and I/O system management relative to VMs and their architectural components were also addressed in this chapter. Embedded Java virtual machines (JVMs) and the .NET Compact Framework were utilized as real-world examples to demonstrate concepts.

The next chapter in this section introduces database concepts, as related to embedded systems middleware.

6.5 Problems

1. What is a VM? What are the main components that make up a VM's architecture?
2.
   A. In order to run Java, what is required on the target?
   B. How can the JVM be implemented in an embedded system?
3. Which standards below are embedded Java standards?
   A. pJava – Personal Java
   B. RTSC – Real Time Core Specification
   C. HTML – Hypertext Markup Language
   D. A and B only
   E. A and C only.
4. What are the main differences between all embedded JVMs?
5. Name and describe three of the most common byte processing schemes.
6.
   A. What is the purpose of a GC?
   B. Name and describe two common GC schemes.
7.
   A. Name three qualities that Java and scripting languages have in common.
   B. Name two ways that they differ.
8.
   A. What is the .NET Compact Framework?
   B. How is it similar to Java?
   C. How is it different?
9. The .NET compact framework is implemented in the device driver layer of the Embedded Systems Model (True/False).
10.
    A. Name three embedded JVM standards that can be implemented in middleware.
    B. What are the differences between the APIs of these standards?
    C. List two real-world JVMs that support each of the standards.
11. VMs do not support process management (True/False).
12. Define and describe two types of scheduling schemes in VMs.
13. How does a VM typically perform memory management? Name and describe at least two components that VMs can contain to perform memory management.

6.6 End Notes

[1] 'Embedded Systems Architecture'. Noergaard. 2005 and http://msdn.microsoft.com/en-us/library/w6ah6cw1 .aspx

[2] Personal Java 1.1.8 API documentation, java.sun.com

[3] 'I/Opener', Morin and Brown, Sun Expert Magazine, 1998.

[4] Java 2 Micro Edition 1.0 API Documentation, java.sun.com

[5] 'Boehm-Demers-Weiser conservative garbage collector: A garbage collector for C and C++', Hans Boehm, http://www.hpl.hp.com/personal/Hans_Boehm/gc/

[6] Kaffe Open Source Code Libraries.

[7] pJava 1.1.8 and CLDC Documentation from Sun Microsystems.

[8] Kaffe.jit3 FAQ.

[9] http://download.java.net/jdk7/docs/api/java/awt/Checkbox.html

CHAPTER 7

An Introduction to the Fundamentals of Database Systems

Chapter Points

- Introduces fundamental database concepts
- Discusses different database models and relevance to database middleware
- Shows examples of real-world embedded database middleware

7.1 What is a Database System?

Like a file system, a database management system (DBMS), also commonly referred to as simply a database system, is another scheme that can be used to reliably and efficiently manage data within an embedded system. A database system can be accessible and directly utilized by the embedded system's user, by other middleware software, by applications in the system to manage data for the application, or some combination of the above. Database systems are commonly used instead of file systems within a design when using a file system instead of a DBMS would result in a great deal of redundancy of the 'same' data in 'different' files. So, when using a file system introduces the challenge of insuring that redundant data within the system need to be constantly updated to insure consistency – then a database as an alternative option is commonly considered. A database is also considered, for example, when managing access to the same data within a file system requires additional overhead when working to insure reliable and secure access to more than one overlying software component and/or user to that data, without corrupting that data in the process.

Keep in mind, a particular database design may not 100% eliminate redundant data. In fact, a database based upon for example the relational model may introduce some redundant data. A database can be used to ensure that the redundant data remain consistent. For example, an IP address for a given device can be changed everywhere that IP address is used via an efficient look up (indexes) scheme. Remember, a database is not intended to be a direct "alternative" to a file system, and in some DBMS designs is most often implemented on top of the file system. It is simply an approach commonly used instead of direct manipulation of files within a file system.

Demystifying Embedded Systems Middleware. DOI: 10.1016/B978-0-7506-8455-2.00007-8

At the highest level, a database system is made up of two major components: (1) the database(s) and (2) the overlying middleware and/or application software used to manage the access to the database(s). Within the database system, a database manages data by allowing for:

- the organization, storage, and management of interrelated data
- querying of data via a query language
- the generation of reports based on data analysis
- data integrity, redundancy, and security.

Thus, in contrast to the wide variety of data that is typically stored in a file system, in the case of data stored in a database system, simply put the data are interrelated. As with file systems, data within a database system are not limited to the data belonging to users, other middleware, and/or applications utilizing the database system. This is because an underlying infrastructure must be in place to store the data, manipulate these data, insure the integrity of the data, and provide secure access to these data.

As with file systems, depending on the database the storage medium can be volatile RAM, and/or non-volatile memory such as: Flash, CD, floppy disk, and hard disk to name a few. Keep in mind that the database itself and the data it manages may or may not reside on the same device. This means, as shown in Figure 7.1, the data the database manages can be located on some type of hardware storage medium located on the embedded system board or located on some other storage medium accessible to the embedded system (i.e., over a network, on a CD, etc.)

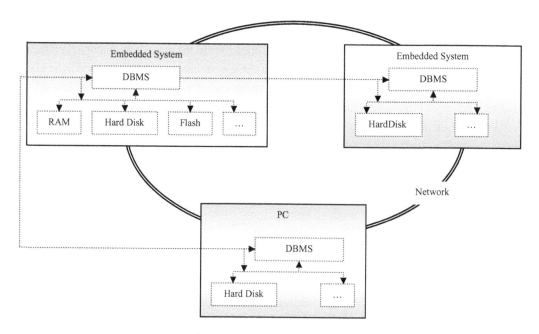

Figure 7.1: Database Access

Ultimately, managing data within the database is accomplished by utilizing **metadata** stored within the database system's *data dictionary* region. Metadata is all the additional components that the database middleware uses to maintain the context, or state, of the system, for example run-time structures describing active connections, and other "metadata" components that are specific to the architecture of that particular database. The database's data dictionary is simply a region which contains information that describes, for example:

- the type and attributes of data being stored within the database
- the structure and location of the data within the database
- the type(s) of object(s) storing the data
- database features and constraints, such as triggers and referential integrity
- details to manage database users, such as permissions and accounts details.

To be useful in the embedded device, a database system must then have a reliable and efficient 'data modeling' scheme to create the components that store data, process data, and locate the data these metadata describes on the embedded device's storage medium(s). The data model drives how the fundamental database subsystems are designed internally, and ultimately how the user/application data will be managed. There are several types of data models used in real-world database designs on the market today. However, the most common schemes implemented within database systems on embedded devices are based upon a record-based model, an object-based model, or some hybrid combination of both.

7.2 Record-based versus Object-oriented Database Models

Important note

Within the scope of this text, the relational algebra that is an important foundation to understanding languages like SQL and relational databases in general is kept at a minimum since this book is intended to be an introduction to database fundamentals.
However, it is useful and necessary to review relational algebra mathematical fundamentals if the intent of the reader is to do 'more' than just selecting/using a database for a particular design – but planning to do the hardcore design and programming of a relational database code.

A record-based database system structures data as *records* within the database, and then relates records to one another via the data contained within the record. Depending on the internal database design, these records can be fixed-length or variable-length. While there are several types of record-based database models, one of the most common is the **relational** database model – where records are grouped and organized into more complex *tables* (note: tables are not more complex than records; they are simply groupings of like records). Each table within the relational database model has a unique name. Each table then represents a unique set of relationships, where the data contained within each row represents a relation.

The types of columns that make up the tables within a relational database are the *attributes* of the data within that table. In Figure 7.2, attributes include 'CDId', 'CDName', 'Genre', 'Price', and 'NumberInStock' for example. When defining a table and its corresponding attributes, domains for these attributes are specified that define the allowed type of data. For example, the domain for 'CDId' may be defined to be unique integers assigned to independent compact disks (CDs), whereas the domain for 'CDName' may be defined as a set of CD names of an alphanumeric string of some 'n' maximum length. Thus, tables within a relational database can then be related to other tables via the shared attributes (keys) within a table, such as the example shown in Figure 7.2.

Overlying middleware software, application software, and/or a user directly communicates with a database system via some type of programming language (see Figure 7.3a) and via database system APIs. Basically, every database system has some type of DML (data-manipulation language) and/or Data Definition Language (DDL) to allow communication. The DML, as its name implies, is what allows for the *manipulation* of the data within a database – meaning the reading, writing, and deleting of these data within the database. DDLs are used to specify a set of definitions that define the underlying database scheme itself. So, to function within the embedded device, the database system uses the DML and DDL to translate and understand all that is required of it. Everything from managing the structure of the database to actually querying the data contained within is done via communicating through the DML, the DDL, or a language that acts as some combination of both a DML and DDL.

An example of a common real-world language utilized in many database systems, especially dominant in the relational database sphere, is based on a common industry standard called *SQL* (structured query language). SQL is a type of computer database language, meaning a language used to create, maintain and control a database. In reality, SQL is much more than a query language; it has DML, DDL and DCL (data control language) elements within it. For example, the DML includes INSERTIUPDATE/DELETE statements in addition to SELECT

- *Tables represent data and their relationships*
- *Each table is a matrix of record rows and columns that contain some unique set of data.*
- *Fields in the table contain data that relate one table to another*

CDId	CDName	Genre	Price	NumberInStock
1	Taking the Long Way	Country	$21.99	5
2	Home	Country	$19.99	2
3	I'm Not Dead	Rock	$15.49	9
4	Up!	Country/Pop	$19.99	0
5	B'Day	R&B/Soul	$19.99	1

CDId	ArtistId	Song
1	2	Not Ready to Make Nice
2	2	Traveling Solder
3	3	The One that Got Away
4	4	Up

ArtistId	Artist Name
1	Beyonce
2	Dixie Chicks
3	Pink
4	Shania Twain

Figure 7.2: Tables

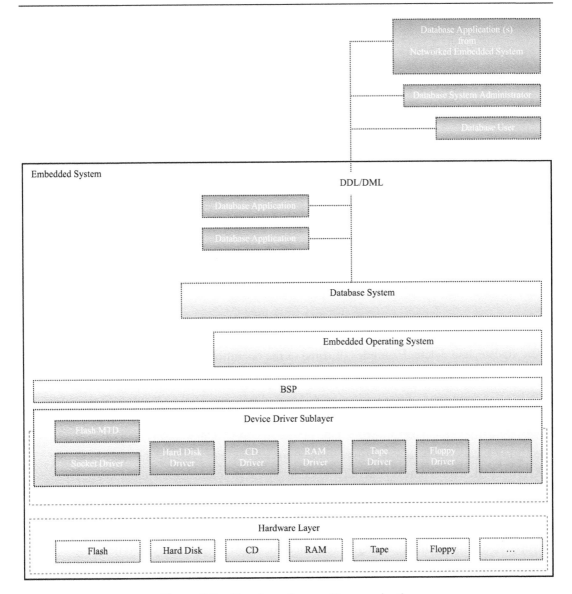

Figure 7.3a: Database System Communication

statements for querying. The Perst database system used as a real-world example in this chapter utilizes a procedural query language based on a derivation of the SQL standard, called JSQL (see Figure 7.3b).

In general, database query languages are considered either non-procedural (where only the specific data within the database are specificed) or procedural (where both the data and the *program logic* to perform on the data can be specified). Procedural refers to the presence of logic statements like if-then-else and do-while. Operations are selection, projection, join,

"JSQL is a subset of the SQL language. JSQL uses a notation that is more geared to objects and object-oriented programming, and can be used to select object instances according to the selection condition. Table rows are considered as object instances and the table - as a class of these objects. So the result of each query execution is a set of objects of one class. The main differences between JSQL and standard SQL are:

1.   There are no joins of several tables and nested sub queries. A Query always returns a set of objects from one table.

2.   The standard Java types are used for atomic table columns.

3.   There are no null values, only null references.

4.   Arrays can be used as record components. A special **exists** quantor is provided for locating elements in arrays.

5.   User methods can be defined for table records (objects) as well as for record components.

6.   References between objects are supported, including user-defined reference resolvers.

7.   As long as the query language is deeply integrated with the Java language, the case sensitive mode

Figure 7.3b: JSQL[1]

insert, update, and delete. Examples of some of the operations that act as foundations for procedural query languages are shown in Table 7.1.

SQL itself is composed of a combination of both a DML and DDL. Meaning, SQL is used for everything from defining and deleting relations *to* executing commands for modifyng the database (deleting data, inserting data, etc.) *to* insuring data integrity and security via specifying access rights *to* managing overall transactions. For creating the table in Figure 7.XX, the SQL expression is generally based upon the structure 'create table x $(A_1, D_1, A_2, D_2, A_3, D_3, ... A_n, D_n$, {integrity-contsraint$_i$}, ...)' where 'x' is the name of the table, A_i define the attributes of the table, and D_i are defining the domains of these attributes. Integrity-constraint$_i$ is how to insure that changes made to the database do not result in some type of corruption. So, for example, an SQL expression for creating CDTable could be:

    create table CDTable (CDId integer not null)
    CDName char(30)
    Genre char(10)
    Price float
    NumberInStock integer
    check (Genre in ('Country', 'Rock', 'Country/Pop',
    'R&B/Soul', 'Opera', 'Classical'))

For extracting data, generally, SQL expressions are made up of three parts:

1.   *select*, as described in Table 7.1 for the 'select' operation relative to attributes to be copied (*select* $A_1, A_2, A_3, ... A_n$ from ... -- A_i is an attribute).

Table 7.1: Examples of Procedural Query Language Operations

Operations	Descriptions
Assignment	Using a temporary relation variable to write a relational expression (allowing for modification of the database, itself) ($\leftarrow$) for deletion, insertion, and updating for example
Cartesian Product	Returns a relation (table of rows) representing each possible pairing of rows from the original tables specified within the Cartesian product ($\times$)
Division	Querying for all rows that contain some specified subset of attributes ($\div$)
Natural Join	Combines into one operation the Cartesian product and selection operations ($\bowtie$)
Project	Selects columns (attributes) from specified tables that satisfy the supplied arguments
Rename	Allows for renaming of relations (table of rows) that come from the same table due to another operation on that table
Select	Selects rows from specified tables that satisfy the supplied argument requirements
Set Difference	Results in finding the rows in a specified table that does not exist in other tables (-)
Set Intersection	Returns a relation (table of rows) that contains rows that are in all specified tables that meet argument requirements ($\cap$)
Union	Allows the union of specified tables, that have an equal number of attributes with identical domains ($\cup$)

2. *from*, the Cartesian product that lists relations to be used (select $A_1, A_2, A_3, \ldots A_n$ *from* $r_1, r_2, r_3, \ldots r_n$ where …. -- r_i is a relation).
3. *where,* the selection predicate (select $A_1, A_2, A_3, \ldots A_n$ from $r_1, r_2, r_3, \ldots r_n$ *where* P – P is the predicate).

So, for example, given the table in Figure 7.2, to use SQL to find the names of CDs (CDName) that cost less than \$20 the SQL expression could look as follows:

```
select CDName
from CDTable
where Price < 20
```

Table 7.2: Example of SQL Query and Table

CDId	CDName	Genre	Price	NumberInStock
1	Taking the Long Way	Country	$21.99	5
2	Home	Country	$19.99	2
3	I'm Not Dead	Rock	$15.49	9
4	Up!	Country/Pop	$19.99	0
5	B'Day	R&B/Soul	$19.99	1

and would return all the CDNames listed in Table 7.2 with a price less than $20 (rows two through five).

For modifying the database, SQL expressions are generally made up of:

1. *(type of database modification), i.e., 'delete from', 'insert into', 'update'.*
2. *where,* the selection predicate (select $A_1,A_2,A_3, \ldots A_n$ from $r_1,r_2,r_3, \ldots r_n$ *where* P – P is the predicate).

So, an SQL example could be updating a row into the CDTable (Table 7.3) with the following SQL expression that would increase the number of CDs in stock for one of the listed CDs:

> *update* CDName set NumberInStock=NumberInStock + 5
> *where* CDId = 4

Real-world Advice

Is a database that supports the SQL API the right choice for an embedded design? Whether or not a database system that supports SQL is the right fit for a particular embedded system's requirements will depend on how deterministic the database access needs to be. SQL is interpreted at run-time, with the actual execution plan determined by the database system's SQL optimizer. This makes it difficult for embedded programmers to understand what a database system is doing when processing SQL statements as opposed to languages embedded programmers are more familiar with, such as C or C++. So, this increases the likelihood execution plans leading to unexpected or even inferior performance. However, the ability in SQL to express complex queries that would otherwise require laborious (and potentially error-prone) C/C++ programming may outweigh the performance its determinism disadvantages. In short, any non-deterministic behavior and additional overhead disadvantages of utilizing a database that supports an SQL API relative to other types of databases needs to be weighed against the advantage of its simplicity in use relative to supporting complex queries. Based on the article 'COTS Databases for Embedded Systems' by Steve Graves.

Record-based 'hierarchical' database systems can also implement *trees* that use pointers to define relations between the different records (see Figure 7.4). Another example is a record-based 'network' database model, where records are related via links into arbitrary *graphs*.

There are several types of object-based models in database system design, from object-oriented to entity/relationship to semantic. However, all of these object-oriented models are in

Table 7.3: Example of SQL Query and Updating Table

CDId	CDName	Genre	Price	NumberInStock
1	Taking the Long Way	Country	$21.99	5
2	Home	Country	$19.99	2
3	I'm Not Dead	Rock	$15.49	9
4	Up!	Country/Pop	$19.99	5
5	B'Day	R&B/Soul	$19.99	1

∞   *Trees represent data and their relationships*

∞   *Each record node in the tree contains some unique set of data.*

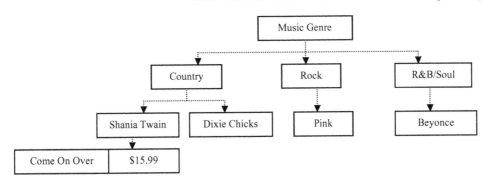

Figure 7.4: Trees

general based on object-oriented programming fundamentals where all components within a database system are considered *objects*. Within this type of database, all objects encapsulate *state* and associated *behavior* data. Another example is a record-based "network" database model, where records are related via links (pointers) into arbitrary *graphs*." An object state is simply some set of parameters that defines the attributes of that object. The behavior of objects is defined via their methods (functions) that operate on the object's state data.

The various types of relationships between these objects are defined via their *classes* as shown in Figure 7.5. A class is simply a way to group objects that share identical states (attributes) and behaviors (methods). Basically, objects are created via the relative instantiation of a class. Via classes then, more complex relationships between data are supported, such as *inheritance* for example, which allows new sets of objects (classes) to be derived from a current class. Objects, and inherently their classes in which they are instantiated from, are the basis in which database queries are made. In some database systems these queries are implemented via an existing DML that is expanded to provide object-oriented support, whereas other database solutions affect queries through an application programming interface (API) that is used within an objectoriented programming language such as Java or C#."

Databases can also be implemented with a design that is some hybrid combination of both object-based and record-based schemes – the most common type being ***object-relational*** databases. Overlying requirements on these types of hybrid databases typically include having powerful querying capabilities, being able to manage complex data (i.e., CAD or multimedia data), and decent performance on handling a large number of database accesses. This means that these hybrid models exist to support requirements that would utilize the best of both worlds.

For example, relational database models support simple data types and the use of the 'safer' querying languages (like SQL) that provide better database protection. Object-oriented models provide the support of more complex data types and offer more flexibility via the use

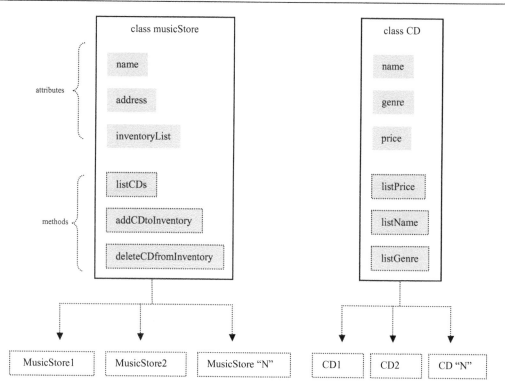

Figure 7.5: Classes

of conventional programming languages (e.g. C/C++ or Java). The hybrid object-relational databases provide support for *both* the simpler *and* complex data types *in synergy with* data querying features typically only found in relational databases in addition to object-oriented data modeling capabilities. The open source Perst database example, used in this chapter, is a real-world example based on a hybrid object-relational approach.

7.3 Why Care About The Different Database Models?

It is important for a middleware developer to understand the different database models, since these different database models created were done to meet different requirements. The model a particular database design adheres to determines how that database logically organizes data, defines the constraints on the data, and the inter-relationships supported. Here it is important to understand a database scheme at the logical level in order to understand how data are represented and managed. This means understanding whether it is via some set of tables within a relational (record-based) type of database versus within an object-oriented database's set of defined classes and instantiated objects. Furthermore, understanding the type of data structures and relative operations used to manage within a database design is key, for example, to predicting the type of performance to expect of the database given the underlying software and hardware components.

The database system itself then implements the data model, via the syntax provided by its language. There is not one database solution that fits all embedded systems' needs. So developers need to understand the pros and cons of each design to insure the database approach is one that maximizes the strengths of its internal design. For example, some database models best support functions that include monitoring inventory and/or managing lists of sales customers, such as the relational database model, for example. This type of database is the approach of choice when data management requirements support that the database is not expected to make major changes too often, and standard operations on data are all that will typically be required (i.e., create table, update table, and so on). Given these standard operations, transactions on data are then expected to be atomic and of shorter duration. The relational database model is also functioning at its best when the data are typically similar in size and structure, allowing for these data to be managed via smaller, fixed-length records.

Other types of database models, such as the object-oriented approach, better support needs relative to complex object graphs such as found in social networking, audio/video multimedia requirements and engineering functions such as CAD (computer-aided design). Object-oriented databases support the management of more complex objects with more freedom to support varying types of data. Databases based on this model also provide better support for non-atomic, asynchronous transactions. Another strength to object-oriented-based database models is considered to be the ability to manage an object (and associated data) with less risk of impacting and corrupting other database components. This is because of the underlying messaging scheme inherent in this approach, where an object's interface (and associated data) can only be accessed and manipulated via some set of messages the object will have defined as acceptable for processing.

In short, the goal of any database design is to successfully manage data without unnecessary redundancy, as well as to insure the integrity of these data, and manage them efficiently. If, given the specific requirements, a particular database design utilized in a real-world system results in:

- data corruption and/or loss
- unnecessary data redundancy
- inability of the database to manage a particular type of data
- unacceptable degradation in system performance

it is time for the developer to investigate a different approach.

7.4 The Fundamentals of Database Design: The First Steps

The first steps to understanding an embedded database design are as follows:

Step 1. As with any other middleware component – understand what the purpose of the database is within the system and how it achieves this purpose. Then, simply keep this in mind regardless of how complex a particular database implementation is. As introduced at the start of this chapter, the purpose of a database is to manage data stored on some type of storage medium located within the embedded device and/or some remotely accessible storage medium, and modern database designs can achieve this in a few different ways.

Step 2. Understand the APIs that are provided by a DBMS and the associated database in support of a database's inherent purpose. These APIs can, of course, differ from database to database, but in general include some set such as the open source example shown in Figures 7.6a and 7.6b.

Step 3. Using the Embedded Systems Model, define and understand all required architecture components that a database requires, specifically:

Step 3.1. Know your database-specific standards, as discussed in Chapter 3.

Step 3.2. Understand the hardware (see Chapter 2). If the reader comprehends the hardware, it is easier to understand why a particular database implements functionality in a certain way relative to the storage medium, as well as the hardware requirements of a particular database implementation.

Step 3.3. Define and understand the specific underlying system software components, such as the available device drivers supporting the storage medium(s) and the operating system API (see Chapter 2).

Step 4. Define the database architecture models on the market today, based on an understanding of the generic database models, and then define and understand what type of functionality and data exist at each layer. This includes database-specific data, such as data structures and the functions included at each layer.

Method Summary

boolean	addRecord(java.lang.Class table, IPersistent record) Add new record to the specified table.
boolean	addRecord(IPersistent record) Add new record to the table.
void	beginTransaction() Begin transaction
void	commitTransaction() Commit transaction
boolean	createIndex(java.lang.Class table, java.lang.String key, boole Add new index to the table.
boolean	createTable(java.lang.Class table) Create table for the specified class.
boolean	deleteRecord(java.lang.Class table, IPersistent record) Delete record from the specified table.
boolean	deleteRecord(IPersistent record) Delete record from the table.
boolean	dropIndex(java.lang.Class table, java.lang.String key) Drop index for the specified table and key.
boolean	dropTable(java.lang.Class table) Drop table associated with this class.
boolean	excludeFromIndex(java.lang.Class table, IPersistent record, ja Exclude record from specified index.
boolean	[illegible]

```
/* << database.java >>
 * This program is free software;
you can redistribute it and/or
modify
 * it under the terms of the GNU
General Public License as published
by
 * the Free Software Foundation;
either version 2 of the License, or
 * (at your option) any later
version.
 *
 * This program is distributed in
the hope that it will be useful,
 * but WITHOUT ANY WARRANTY;
without even the implied warranty
of
 * MERCHANTABILITY or FITNESS FOR A
PARTICULAR PURPOSE. See
 * the GNU General Public License
```

Figure 7.6a: Perst API Example[1]

Step 3. Using the Embedded Systems Model, define and understand all required architecture components that a database requires, specifically:

Step 3.1. Know your database-specific standards, as discussed in Chapter 3.

Step 3.2. Understand the hardware (see Chapter 2). If the reader comprehends the hardware, it is easier to understand why a particular database implements functionality in a certain way relative to the storage medium, as well as the hardware requirements of a particular database implementation.

Step 3.3. Define and understand the specific underlying system software components, such as the available device drivers supporting the storage medium(s) and the operating system API (see Chapter 2).

Step 4. Define the database architecture models on the market today, based on an understanding of the generic database models, and then define and understand what type of functionality and data exist at each layer. This includes database-specific data, such as data structures and the functions included at each layer.

Figure 7.6b: Open Source Perst API Source Code Examples[1]

7.5 Real-world Database System Model

When an application or user initiates communication, then an embedded database system contains a number of components to process this incoming communication. What these components are and how these components are designed essentially determine what underlying system software and hardware requirements need to be met in order to utilize them successfully within a design. So, to start, it is recommended the reader begin to familiarize themselves with these components. Figure 7.7 shows a general database systems model made up of some combination of a

- transaction manager
- query compiler
- execution engine
- resource manager
- storage and buffer manager.

Important note

Remember, real-world database systems may have different names than what is listed above for the various components, may have split the functions of these components into additional elements, and/or merged the functionality of various components into other database system subsystems. The key here is to use the subsystems within the model and examples as a reference in understanding the fundamentals of any database system design.

An incoming query can impact the data within the database system, as well as trigger actions that impact the structure of the database itself. Database systems typically group incoming queries, as well as other database system actions in general, into independent, atomic tasks

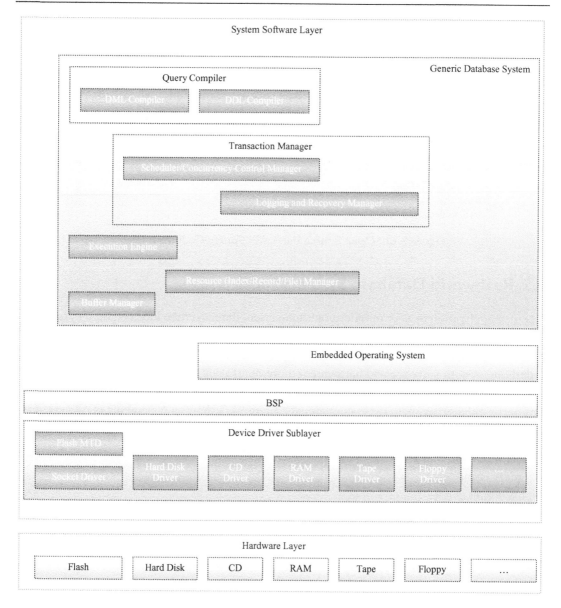

Figure 7.7: General Database System Model

called *transactions*. In order to manage transactions, a **transaction manager** resides within
the system to support:

1. Scheduling, which manages multiple concurrent, independent transactional database
    system tasks. Depending on the database system, an underlying operating system and/or
    virtual machine's scheduler is utilized or an independent scheduler may be implemented

by the database system's designers themselves. Scheduling within a Java virtual machine (JVM), utilized by the open source Java-based Perst example used in the chapter, was discussed further in Chapter 6.

2. Logging and Recovery, which is responsible for insuring that the database can be recovered from mid-transactional failures via utilizing logs kept on the transactions and being able to rollback to a non-corrupted version of the database system.

For database systems based on (or implementing) SQL, the query coming into the database system is first received and translated by some type of **query compiler**. This query compiler is responsible for translating DML (data-manipulation language) and/or Data Definition Language (DDL) incoming queries for processing. After translation, the query compiler transmits the result, commonly referred to as the query plan, to the execution engine for further processing. Some sample code translating JSQL in Perst is shown in Figure 7.8a.

Upon receiving the query plan from the query compiler, the **execution engine** actually processes the actions within the plan to manage the data request. The execution engine communicates and transmits requests to a resource manager that manages the indices, records, files, and/or objects (depending on the database design) relative to the data being processed. In the case of the open source Java-based Perst example used in the chapter, the Java virtual machine's execution engine is mainly utilized and is discussed further in Chapter 6.

```
/* << QueryImpl.java >>
 * This program is free software; you can redistribute it and/or modify
 * it under the terms of the GNU General Public License as published by
 * the Free Software Foundation; either version 2 of the License, or
 * (at your option) any later version.
 *
 * This program is distributed in the hope that it will be useful,
 * but WITHOUT ANY WARRANTY; without even the implied warranty of
 * MERCHANTABILITY or FITNESS FOR A PARTICULAR PURPOSE. See
 * the GNU General Public License for more details.
 *
 * You should have received a copy of the GNU General Public License
 * along with this program; if not, write to the Free Software
 * Foundation, Inc., 59 Temple Place, Suite 330, Boston, MA 02111-1307 USA
 */
....
public class QueryImpl implements Query {
public Iterator select(Class cls, Iterator iterator, String query) throws CompileError
{
this.query = query;
buf = query.toCharArray();
str = new char[buf.length];
this.cls = cls;

compile();
return execute(iterator);
}
```

Figure 7.8a: Open Source Perst Query Translation Source Code Example[1]

```
public Iterator execute(Iterator iterator)
{
Iterator result = applyIndex(tree);
if (result == null) {
if (storage.listener != null) {
storage.listener.sequentialSearchPerformed(query);
}
result = new FilterIterator(this, iterator, tree);
}
if (order != null) {
ArrayList list = new ArrayList();
while (result.hasNext()) {
list.add(result.next());
}
sort(list);
return list.iterator();
}
return result;
}

private void sort(ArrayList selection) {
int i, j, k, n;
OrderNode order = this.order;
Object top;

if (selection.size() == 0) {
return;
}
for (OrderNode ord = order; ord != null; ord = ord.next) {
if (ord.fieldName != null) {
ord.resolveName(selection.get(0).getClass());
}
}
for (n = selection.size(), i = n/2, j = i; i >= 1; i--) {
k = i;
top = selection.get(k-1);
do {
if (k*2 == n ||
order.compare(selection.get(k*2-1), selection.get(k*2)) > 0)
{
if (order.compare(top, selection.get(k*2-1)) >= 0) {
break;
}
selection.set(k-1, selection.get(k*2-1));
k = k*2;
} else {
if (order.compare(top, selection.get(k*2)) >= 0) {
break;
}
selection.set(k-1, selection.get(k*2));
k = k*2+1;
}
} while (k <= j);
selection.set(k-1, top);
}
for (i = n; i >= 2; i--) {
top = selection.get(i-1);
selection.set(i-1, selection.get(0));
selection.set(0, top);
for (k = 1, j = (i-1)/2; k <= j;) {
if (k*2 == i-1 ||
order.compare(selection.get(k*2-1), selection.get(k*2)) > 0)
{
if (order.compare(top, selection.get(k*2-1)) >= 0) {
break;
}
selection.set(k-1, selection.get(k*2-1));
k = k*2;
} else {
if (order.compare(top, selection.get(k*2)) >= 0) {
break;
}
selection.set(k-1, selection.get(k*2));
k = k*2+1;
}
}
selection.set(k-1, top);
}
```

Figure 7.8a continued: Open Source Perst Query Translation Source Code Example

```
....
final void compile() {
pos = 0;
vars = 0;
tree = checkType(Node.tpBool, disjunction());
OrderNode last = null;
order = null;
if (lex == tknEof) {
return;
}
if (lex != tknOrder) {
throw new CompileError("ORDER BY expected", pos);
}
int tkn;
int p = pos;
if (scan() != tknBy) {
throw new CompileError("BY expected after ORDER", p);
}

do {
p = pos;
if (scan() != tknIdent) {
throw new CompileError("field name expected", p);
}
OrderNode node;
Field f = ClassDescriptor.locateField(cls, ident);
if (f == null) {
Method m = lookupMethod(cls, ident, defaultProfile);
if (m == null) {
if (!cls.equals(Object.class)) {
throw new CompileError("No field '"+ident+"' in class "+
cls.getName(), p);
}
node = new OrderNode(ident);
} else {
node = new OrderNode(m);
}
} else {
node = new OrderNode(ClassDescriptor.getTypeCode(f.getType()), f);
}
if (last != null) {
last.next = node;
} else {
order = node;
}
last = node;
p = pos;
tkn = scan();
if (tkn == tknDesc) {
node.ascent = false;
tkn = scan();
} else if (tkn == tknAsc) {
tkn = scan();
}
} while (tkn == tknComma);
if (tkn != tknEof) {
throw new CompileError("',' expected", p);
}
}
}
}
.....
```

Figure 7.8a continued: Open Source Perst Query Translation Source Code Example

7.5.1 Resource Manager

A *resource manager* is responsible for keeping track of the data within the data structures of the database, to allow for efficient retrieval of data from storage via the buffer and storage manager. Relative to the buffer and storage management (introduced in the next section), while some database designs will utilize their own scheme for managing data directly on the hardware, in other database systems the storage and buffer manager is actually the file system residing on the embedded device, and it is the file system APIs that are utilized by the overlying database system layers.

This is important because, for example, a relational database that utilizes an underlying file system will do so by mapping its internal records into files sequentially or some other method such as some *indexing* or *hashing* approach. In this case, how this type of database manages its records within these files given the underlying hardware, and the file internal design system itself, will impact how the database performs. Specifically, it is relative to *overhead*, meaning computing how much additional compile *and* runtime memory is required for the particular scheme to execute efficiently, as well as how much time it takes to locate and access these records – then add, delete, or modify data within.

There are several indexing and hashing algorithms that can be implemented into a database design to insure efficiency and avoid overhead when searching for data. Indexing schemes involve traversing some type of index 'structure' to insert, delete, and modify data. Hashing schemes involve the use of a function to calculate the data's address in memory directly.

In general, indexing schemes are based upon individual *indices* being assigned to data – records and/or objects depending on the type of database (relational, object-oriented, object-relational hybrid, for example). The indices are essentially the fundamental components used within indexing resource management schemes to organize and track data. For example, a B+-tree index is a multilevel index in the form of a tree that is made up of different types of *nodes*, specifically some combination of root, non-leaf, and leaf nodes. As shown in Figure 7.8b, a B+-tree node is typically made up of *key values* (K_1, K_2, … K_{n-1}) and *pointers* (P_1, P_2, … P_n). Key values within a node are the one or more sorted attributes used to search for another node within the B+-tree or the data itself. Non-leaf node pointers are references to the child nodes with the relative search key values less than (on the left branch) or greater than (on the right branch). The number of pointers within a 'non-leaf' node are between 'n/2' and 'n', thus having between 'n/2' and 'n' number of child nodes. Except for the last pointer within a leaf node (P_n), leaf node pointers P_1 …P_{n-1} reference the data with the relative search key value. The last pointer (P_n) within a leaf node is used to link to another leaf node.

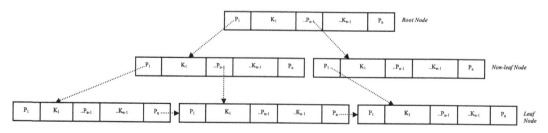

Figure 7.8b: B+-Tree High Level Diagram

So, traversing a B+-tree for a particular query retrieving all data with the search key value 'x', for example, requires traversing the tree from the root to the desired leaf node. This means starting at the root node of the tree and comparing to search keys K_1, K_2, … K_n to search key 'x'. The closest root node search key value that is greater than 'x' is whose pointer is used to traverse to the next level of the tree. This means, if $K_2 < $ 'x' $< K_3$, than it is the non-leaf node the pointer P_3 is pointing to that is traversed to. Within the non-leaf nodes, the comparison of 'x' to search key values within the node continues until arriving at a leaf node that contains the desired search key values.

```
/* << BTree.java >>
 * This program is free software; you can redistribute it and/or modify
 * it under the terms of the GNU General Public License as published by
 * the Free Software Foundation; either version 2 of the License, or
 * (at your option) any later version.
 *
 * This program is distributed in the hope that it will be useful,
 * but WITHOUT ANY WARRANTY; without even the implied warranty of
 * MERCHANTABILITY or FITNESS FOR A PARTICULAR PURPOSE. See
 * the GNU General Public License for more details.
 *
 */
....

package org.garret.perst.impl;
import org.garret.perst.*;
import java.util.*;
import java.lang.reflect.Array;

class Btree extends PersistentResource implements Index {
  int root;
  int height;
  int type;
  int nElems;
  boolean unique;


        .....


  class BtreeSelectionIterator implements PersistentIterator {
  BtreeSelectionIterator(Key from, Key till, int order) {
  this.from = from;
  this.till = till;
  this.order = order;
  reset();
  }


  ...

  public Object next() {
  if (!hasNext()) {
  throw new NoSuchElementException();
  }
  StorageImpl db = (StorageImpl)getStorage();
  int pos = posStack[sp-1];
  currPos = pos;
  currPage = pageStack[sp-1];
  Page pg = db.getPage(currPage);
  Object curr = getCurrent(pg, pos);
  if (db.concurrentIterator) {
  currKey = getCurrentKey(pg, pos);
  }
  gotoNextItem(pg, pos);
  return curr;
  }
```

Figure 7.8c: Open Source Perst B-Tree Source Code Example[1]

```
public int nextOid() {
if (!hasNext()) {
throw new NoSuchElementException();
}
StorageImpl db = (StorageImpl)getStorage();
int pos = posStack[sp-1];
currPos = pos;
currPage = pageStack[sp-1];
Page pg = db.getPage(currPage);
int oid = getReference(pg, pos);
if (db.concurrentIterator) {
currKey = getCurrentKey(pg, pos);
}
gotoNextItem(pg, pos);
return oid;
}

private int getReference(Page pg, int pos) {
return (type == ClassDescriptor.tpString || type == ClassDescriptor.tpArrayOfByte)
? BtreePage.getKeyStrOid(pg, pos)
: BtreePage.getReference(pg, BtreePage.maxItems-1-pos);
}


protected Object getCurrent(Page pg, int pos) {
StorageImpl db = (StorageImpl)getStorage();
return db.lookupObject(getReference(pg, pos), null);
}

protected final void gotoNextItem(Page pg, int pos)
{
StorageImpl db = (StorageImpl)getStorage();
if (type == ClassDescriptor.tpString) {
if (order == ASCENT_ORDER) {
if (++pos == end) {
while (--sp != 0) {
db.pool.unfix(pg);
pos = posStack[sp-1];
pg = db.getPage(pageStack[sp-1]);
if (++pos <= BtreePage.getnItems(pg)) {
posStack[sp-1] = pos;
do {
int pageId = BtreePage.getKeyStrOid(pg, pos);
db.pool.unfix(pg);
pg = db.getPage(pageId);
end = BtreePage.getnItems(pg);
pageStack[sp] = pageId;
posStack[sp] = pos = 0;
} while (++sp < pageStack.length);
break;
}
}
} else {
posStack[sp-1] = pos;
if (sp != 0 && till != null && -BtreePage.compareStr(till, pg, pos) >= till.inclusion) {
sp = 0;
}
} else { // descent order
if (--pos < 0) {
while (--sp != 0) {
db.pool.unfix(pg);
pos = posStack[sp-1];
pg = db.getPage(pageStack[sp-1]);
if (--pos >= 0) {
posStack[sp-1] = pos;
do {
int pageId = BtreePage.getKeyStrOid(pg, pos);
db.pool.unfix(pg);
pg = db.getPage(pageId);
pageStack[sp] = pageId;
posStack[sp] = pos = BtreePage.getnItems(pg);
} while (++sp < pageStack.length);
posStack[sp-1] = --pos;
break;
}
}
} else {
posStack[sp-1] = pos;
}
if (sp != 0 && from != null && BtreePage.compareStr(from, pg, pos) >= from.inclusion) {
sp = 0;
}
}
} else if (type == ClassDescriptor.tpArrayOfByte) {
if (order == ASCENT_ORDER) {
if (++pos == end) {
while (--sp != 0) {
db.pool.unfix(pg);
pos = posStack[sp-1];
pg = db.getPage(pageStack[sp-1]);
...
...
```

Figure 7.8c continued: Open Source Perst B-Tree Source Code Example

The Perst open source example, shown in Figure 7.8c, is based on a *multilevel indexing-based B-tree* implementation and a partial snapshot of the Perst B+-tree traversing scheme is shown below.

7.5.2 Buffer and Storage Management

Storage and buffer management is the liaison to underlying system software and manages retrieval and transmission of data to and from the user and the supported storage mediums, including RAM and whatever non-volatile memory is supported by the database. This means it is responsible for managing the requests and the allocation of buffer space in volatile and non-volatile memory. Because access to non-volatile memory is typically much slower than accessing data in volatile memory, the storage and buffer manager for a particular database is based on a scheme that attempts to minimize the number of accesses to non-volatile memory. However, because there is only a limited amount of 'faster' volatile memory available to the database, some type of data swapping and replacement scheme must be implemented. The most common types of data swapping and replacement schemes implemented in different database designs are similar to schemes used in underlying operating systems, such as:

- *Optimal*, utilizes a future reference time to swap out data that won't be used in the near future
- *Least recently used* (LRU), data that are used the least recently are swapped out
- *FIFO* (first in, first out), swaps out data that are the oldest, regardless of how often those data are accessed by the database. FIFO is a simpler algorithm than LRU, but typcially is much less efficient
- *Not recently used* (NRU), data that are not used within a certain time period are swapped out
- *Second chance*, a more-complex FIFO scheme that uses a reference bit that sets to '1' when data access occurs. So, if this bit is '0', then associated data are swapped out.

The storage and buffer manager is also what is responsible for managing data integrity within the database, in the cases of synchronizing more than one application/user that must access the database concurrently or recovering system problems, for example. Therefore, some type of scheme that manages the blocking and unblocking of data writes, as well as the write-through of data from volatile memory (i.e., cache, DRAM, etc.) to non-volatile memory (i.e., Flash, Hard Disk, etc.), falls under this database subsystem.

7.6 Utilizing Embedded Databases in Real-world Designs and the Application Layer

Embedded targets constrained by limited memory and processing power typically shy away from the use of a database system to manage data. So, the key is investigating how well the embedded database solution integrates the overlying applications and data management code to allow for better performance, including decreasing the amount of required memory and CPU cycles to process and manage data. When an embedded device can support the costs of introducing a faster master CPU, more memory, and so on, then utilizing an embedded database within the architecture is feasible.

In general, utilizing a database over other types of methods to manage data on an embedded device boils down to the desire for:

- increasing reliability
- improved data management efficiency
- insuring data integrity
- higher availability and operational continuity
- scalability
- predictability and determinism for real-time requirements
- decreasing overlying application development time.

Because the most time-consuming processing relative to a database involves the management of data relative to the non-volatile storage device (be it Flash, Hard Disk, etc.) it is important to understand the importance of having enough cache or even volatile main memory on the target if the team selects an IMDS (in-memory database system) to allow for better performance when managing data, for example. It is also important to understand the database write-through scheme that insures all changes made in volatile memory are saved properly to the non-volatile storage device in the cases of a system failure and power disruptions. This means understanding a particular database system's scheme for managing redundant data as well as managing the transactions and logging that allows for the ability to insure consistent data and even recover data if a problem occurs with the device.

As with other types of middleware, selecting which embedded database supports the system requirements means insuring the database implementation supports the underlying platform. Figure 7.9 shows a sample snapshot of a datasheet of a real-world embedded database, called eXtremeDB. This datasheet outlines some underlying platform and development tool support information, as well as the type of complex data types that can be supported by eXtremeDB. In the case of the version of eXtremeDB referred to in Figure 7.9, the embedded operating systems that this embedded database has been ported to support include various flavors of vxWorks, Integrity, QNX, and Nucleus embedded OSs to name a few.

7.7 Summary

There are several different database design schemes that can be implemented in a particular database system. In order to understand a database system design, determine which database design is the right choice for an embedded device, as well as understand the impact of a database on a particular device – it is important to first understand the fundamental

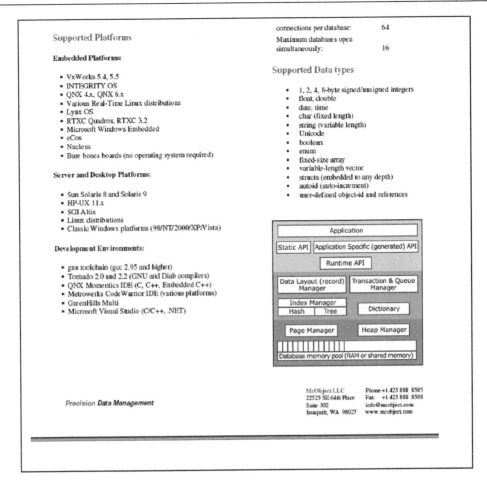

Figure 7.9: MCObject extremeDB Datasheet[3]

components that make up a database system. These fundamentals, introduced in this chapter, included what the purpose of a database is, elements that commonly make up a database, and a real-world example of an object-oriented database system scheme implementing some of these elements. The reader can then apply these fundamentals to analyzing database design features, such as:

- available API operations
- maximum amount of memory that is needed by the database system
- performance
- support of specific hardware, operating system, or underlying middleware

in order to determine if the database system design is the right one for a particular system, as well as the impact of the database system on the embedded device. This chapter has introduced the basic concepts of database systems design.

The next section of this book will compare different types of middleware, including discussing how to determine which middleware is the best-fit for particular requirements, as well as the hardware and system software requirements when using particular middleware components such as a database system implementation.

7.8 Problems

1. What is the purpose of a database system?
2. All database systems can only manage files located on the embedded system the file system resides on (True/False).
3. What does DBMS stand for? What is the difference between a database system and a DBMS?
4. A database system can utilize a file system within its design (True/False).
5. Outline the four-step model to understanding a database system design.
6. A database system implemented in the system software layer can exist as:
   A. Middleware that sits on top of the operating system layer
   B. Middleware that sits on top of other middleware components, for example a Java-based file system that resides on a Java Virtual Machine (JVM)
   C. Middleware that has been tightly integrated and provided with a particular operating system distribution
   D. None of the above
   E. All of the above.
7. One or more database systems can be implemented in an embedded system (True/False).
8. Name and describe three data modeling schemes.
9. A database system can manage files on the following hardware:
   A. RAM
   B. CD
   C. Smart card
   D. Only B and C
   E. All of the above.
10. How does an application communicate with a database system?
11. A database system will never require other underlying middleware components (True/False).
12. Draw and describe the layers of the General Database System Model.
13. Name and describe five examples of database system APIs.

7.9 End Notes

[1] Perst API Guide, User's Guide, and Open Source.
[2] http://www.mcobject.com/standardedition.shtml#Shared%20Memory%20Databases
[3] MCObject Datasheet, 2010.
[4] 'COTS Databases for Embedded Systems' Steve Graves.

CHAPTER 8

Putting It All Together
Complex Messaging, Communication, and Security

Chapter Points

- Identifies the main types of complex messaging, communication, and security middleware
- Defines each of the different types of middleware
- Outlines the pros and cons of utilizing one model over another

---

As application requirements increase in complex, distributed embedded systems these requirements usually impose additional software prerequisites in underlying layers to support these desirable applications within the device itself. Overlaying complex networking and communication middleware on top of core middleware is increasingly becoming a popular approach in embedded systems design to support these additional requirements. There are several different types of complex networking and communication middleware that build on the core middleware discussed in the previous chapters. In general, the more complex type of middleware that is introduced in this chapter falls under some combination of the following:

- Message-oriented and Distributed Messaging, i.e.,
    - Message Queues
    - Message-oriented Middleware (MOM)
    - Java Messaging Service (JMS)
    - Message Brokers
    - Simple Object Access Protocol (SOAP)
- Distributed Transaction, i.e.,
    - Remote Procedure Call (RPC)
    - Remote Method Invocation (RMI)
    - Distributed Component Object Model (DCOM)
    - Distributed Computing Environment (DCE)

Demystifying Embedded Systems Middleware. DOI: 10.1016/B978-0-7506-8455-2.00008-X

- Transaction Processing (TP), i.e.,
  - Java Beans (TP) Monitor
- Object Request Brokers, i.e.,
  - Common Object Request Broker Object (CORBA)
  - Data Access Object (DAO) Frameworks
- Authentication and Security, i.e.,
  - Java Authentication and Authorization Support (JAAS)
- Integration Brokers.

8.1 Message-oriented Middleware and Distributed Transaction

Message-oriented middleware (MOM) is software that provides *message* passing capabilities between overlying middleware and/or application software within an embedded system (Figure 8.1). MOM is typically used when some type of point-to-point and/or autonomous publish-subscribe messaging scheme is optimal. The APIs and functionality provided by MOM software allow for simplifying the ability to design overlying software components, because the APIs abstract out underlying networking protocol details and other underlying system components for developers. For overlying middleware and application software that adhere to the supplied APIs, this type of middleware is what allows for the interoperability of the overlying software components that communicate via these messages.

When MOM software also provides message-passing capabilities between overlying software of independent devices connected across a network, MOM middleware is further classified as *distributed* messaging middleware. In this case, the APIs and functionality provided by this type of more complex MOM software also aid in the portability of overlying middleware and application software to other devices with vastly different underlying system software and hardware components.

MOM middleware is typically based on some hybrid combination of client–server and the peer-to-peer architecture model of message-passing communication. This means an MOM server

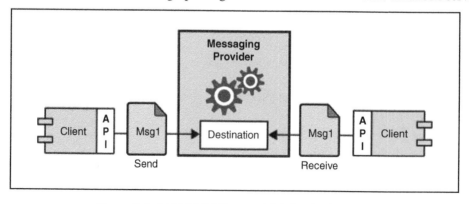

Figure 8.1: MOM Middleware High-level Diagram[1]

has control in managing the MOM clients. MOM clients can pass messages peer-to-peer, not requiring the centralized MOM server to manage and control all communication. Depending on the implementation, MOM servers can manage MOM client messages either concurrently (where more than one MOM client request can be handled in parallel) and/or iteratively (a.k.a. one-at-a-time). How MOM servers and clients manage communication is ultimately dependent on whether the messaging scheme is based upon a synchronous message-passing model, an asynchronous message-passing model, or a model based upon some combination of both.

In general, synchronous message passing is based upon a request–response type of handshaking scheme. This type of MOM communication requires that an MOM receiver exists in some form at the time the MOM sender transmits a message, in order to eventually unblock the waiting sender with a response message. For example, an MOM client that transmits a message to another MOM client and/or MOM server will block waiting for some type of response message from the receiver. The advantage of synchronous MOM messaging communication is that it is simpler and straightforward. On the flip-side, if there are hard real-time scheduling requirements for the device, it is risky to use MOM software that blocks the system waiting for a response message.

Disadvantages with an underlying synchronous scheme are highlighted in an embedded device with requirements to support multiple, complex interactions and nested calls between MOM senders and receivers. There is an increase in connection overhead in relation to this scheme due to the system resources required to manage the sessions. If some type of connection pool on an as-needed basis is not implemented, then managing these resources becomes very expensive and limits to the number of connections allowed are introduced.

In an asynchronous message-passing scheme the transmitter can send a message at will independent of the availability of the receiver. The receiver can process received messages when available as well. This means it is a form of non-blocking, connectionless communication in which the transmitter and/or receiver does not have to wait for the response from the other to continue to perform other tasks.

Typically, MOMs transmit messages under some combination of a:

- *broadcasting* messaging scenario
- *multicast* messaging scheme
- message *queuing* scheme.

Message-oriented and -distributed messaging middleware that is based on queuing, a.k.a. *message queuing middleware* (MQM), implements message queues that can transmit, receive, store, and forward messages (Figure 8.2). MQM is typically utilized within embedded devices in which performance is a challenge, as well as within devices that do not have a constant and/or stable networking access. Given the utilization of message queues, MQM addressed handshaking and performance goals by allowing the embedded device to process messages according to available system resources, as well as independent of the networking connections.

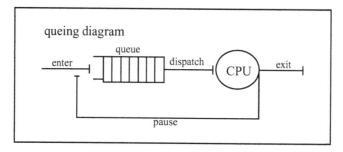

Figure 8.2: Message Queues[2]

What differentiates MQMs are the types of queues supporting the messaging model, as well as the specific attributes of these message queues. These attributes include everything from size to naming convention to security access such as public versus private, permanent versus temporary, journaled versus non-transactional (no back-up copies) to name a few.

While there are many types of message queue model schemes available, one of the most common is based on some type of the FIFO (First In, First Out) queue model. Under FIFO, a queue stores ready messages (messages ready to be processed). Messages are added to the queue at the end of the queue, and are retrieved to be processed from the start of the queue. In the FIFO queue, all messages are treated equally regardless of their importance or receiver (Figure 8.3). Variations on an MQM based upon the FIFO queuing scheme include queues in which messages in the queue are processed in the order in which the smaller-sized messages are processed first, and/or messages are processed according to their importance (priority) in the queue for example.

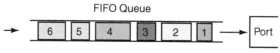

Figure 8.3: FIFO Queue[3]

8.1.1 Building on MOM: Transactional Middleware and RPC

One of the most common types of distributed transaction middleware is the RPC (remote procedure call). The RFC 5531 'RPC: Remote Procedure Call Protocol Specification Version 2'[5] is a common industry standard which defines an RPC model for implementation where a thread of control logically winds through a caller and receiver task. As shown in Figure 8.4, RPC middleware simply allows synchronous communication across remote systems, where the caller on one embedded device can invoke a native language-based routine residing on a remote system in a manner similar to invoking a local procedure. In general, RPC implements a scheme in which:

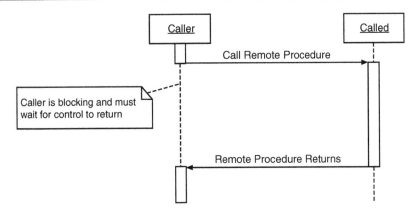

Figure 8.4: RPC Synchronous Communication[4]

- a caller first sends a message to the receiver task on a remote system
- the caller blocks, waiting for a reply message from the remote system
- after the reply message is received by the caller, the caller's execution is resumed.

RPC is built upon some type of underlying core networking middleware such as TCP and/or UDP depending on the type of RPC scheme. RPC is also fundamentally overlying some type of MOM foundation (see Figure 8.5). RPC also acts as a basis for other types of distributed transaction middleware found within a variety of computer systems, such as *Remote Method*

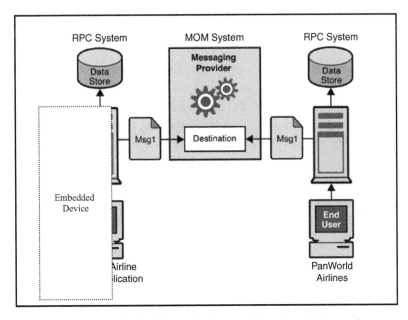

Figure 8.5: RPC-MOM Middleware High-level Diagram[1]

Invocation (a.k.a RMI; a variation of RPC but originating in the Java space), *Distributed Component Object Model* (DCOM), and *Distributed Computing Environment* (DCE).

8.1.2 Building on RPC: Object Request Brokers

An Object Request Broker (ORB) provides a layer to allow for creating an ***individual*** overlying middleware and/or application component that resides as multiple objects, on the same device and/or across more than one device. ORBs are an approach to allow for software interoperability, since they allow for integration within one individual application or middleware component – even if the integrated software came from vastly different vendors with different APIs. As shown in Figure 8.6a, an ORB acts as the foundation to the Common Object Request Broker Architecture (CORBA), and is based on industry standards from the Object Management Group (OMG).

A similar philosophy behind using ORBs lies behind the the popularity of using DAO (Data Access Object) design patterns in embedded systems designs. DAO originated in the Java space (Figure 8.6b), and has been used as a basis for DAO frameworks in real-world designs for abstracting, encapsulating, and managing accesses to various heterogeneous underlying resources in the form of objects.

An ORB handles any translation and transformation (marshalling) of data between overlying heterogeneous objects to allow for this intercommunication. Each object within the individual overlying component to an ORB integrates an ORB interface. It is the ORB interface that allows the objects that make up the overlying application and/or middleware software to communicate and provide remote invocation access to functionality.

An overlying ORB object becomes accessible to other overlying ORB objects for remote invocations over a network. Thus, depending on the implementation, an ORB is built upon the Internet Inter-ORB Protocol (IIOP) and other underlying core networking middleware for this support across networked devices. Also, depending on the ORB, more complex middleware such as RPC components can also act as a foundation.

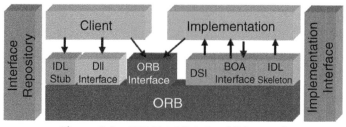

Figure 8.6a: CORBA High-level Diagram[6]

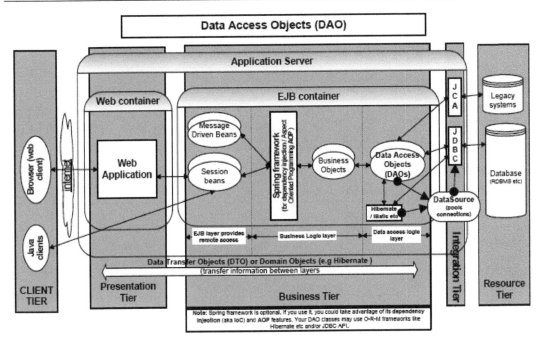

Figure 8.6b: J2EE DAO High-level Diagram[7]

When an ORB, such as within CORBA, manages the routing requests and responses between a client and a distributed object, an IDL (Interface Description Language) is used to describe the transmitted data. As defined by the OMG, IDL interface definitions are stored in some type of interface repository that the ORB then utilizes for tracking and managing communication with objects. The ORB then can also activate and deactivate objects by request, and can provide the types of services such as those shown in Table 8.1.

8.2 Authentication and Security Middleware

Authentication and security middleware is software that is used as a foundation for implementing security schemes for overlying middleware and application software. This type of middleware is required in the case of using RPC middleware, for example, in which without some type of authentication and security middleware component, data are transported in an insecure manner with the routine call.

In general, authentication and security middleware typically provides at the very least some type of code security features. Middleware that helps to insure code security, validation, and verification can be implemented within an embedded device independently, and/or can be based upon core middleware components such as JVM or .NET components that with their very implementation and their respective higher-level languages contain this type of support.

Table 8.1: Examples of CORBA Services

ORB Service Type	Description
Concurrency control	Managing data locks in support of multitasking environment
Event	Objects specify what notifications for events are of interest
Externalization	Manages data transmission and translation between ORB objects and some format of a data stream
Licensing	Manages objects that require active licenses for usage from a vendor
Live cycle	Definitions for creation, deletion, copying, and moving objects
Naming	Searching for objects by name
Persistence	Storage/retrieval of ORB data from non-volatile memory via a file system and/or database
Properties	Manages ORB object description details
Query	Manages database queries for ORB objects
Transaction	Manages transactions and insuring data integrity
Security	Managing authentication and authorization issues relative to data and ORB objects

This can include everything from insuring valid type operations are performed, i.e., array bounds checking, type checks and conversions, to checking for stack integrity (i.e., overflow) and memory safety. For example, an embedded JVM and associated byte processing scheme will include support class loading verification and security, as well as garbage collection and memory management.

When it then comes to securing the actual data managed within an embedded device, *cryptography* algorithms are one of the most reliable implementations for insuring security via a middleware layer. Cryptography schemes utilize some combination of encryption keys, obfuscator tools, digital signatures, and/or certificates to name a few. This allows the sender to perform some type of encryption on the data before transmission, to help insure that 'only' the 'intended' receiver can decrypt the data. Again, here core middleware with an embedded JVM implementation for example can be used as a foundation to include some set of Java-based APIs for cryptography support for:

- algorithms such as AES, DSA, DES, SHA, PKCS#5, and RC4 to name a few
- asymmetric vs. symmetric ciphers
- digital signatures

- key generators and factories
- message authentication codes
- message digests.

Access control is policy-based and provides support to insure only code that is *allowed* to execute on the embedded device is permitted. In this case, specific policies associated with a particular overlying software component are used to check and to enforce the access control scheme to provide protection. Evidence-based CAS (Code Access Security) enforces the check on code for everything from its origin to searching for 'dangerous' code within the software before permitting execution on the device. Again, if utilizing an embedded JVM scheme, then this can be built upon the pre-existing class loading implementation within the JVM.

Finally, authentication and authorization middleware simply provides functionality to determine whether an overlying component is what it claims to be. For example, this can include a scheme for verifying logins and passwords. After authentication, results are passed on to an authorization scheme that actually executes what is necessary to allow access to the device's resources.

8.3 Integration Brokers

The implementation of an **integration broker** in an embedded system is typically due to the necessity of integrating vastly different types of overlying middleware and/or applications that must be able to process each other's data. This overlying software can reside within the same device, or across networks within other devices. Figure 8.7 shows an example of such an ecosystem. Integration brokers allow applications and other middleware to exchange different formats of data, by managing the translation and transmission of these data. This means overlying software is not required to concern itself with any communication requirements of the software receiving the data. To achieve this, integration brokers provide some set of functionality that supports:

- Auditing and Monitoring
- Connectivity
- Policy Management
- Scalability
- Security and Authentication
- Stability
- Transactional Integrity
- Workflow Management.

Integration brokers inherently support an interoperability interface and communication scheme that is an *alternative* to point-to-point with a design in which point-to-point communication would result in too many connections to be managed and maintained

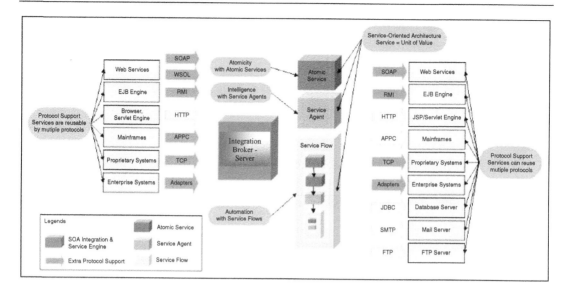

Figure 8.7: Example of Integration Broker Ecosystem[8]

efficiently. This means the number of connections would decrease when overlying software utilizes the broker for intercommunication. In this case, dependencies between overlying software that communicate via the broker are non-existent, leaving only the dependency of this overlying middleware/application on the definition of the integration broker's interoperability interface.

An integration broker is fundamentally built upon other types of middleware, such as ORBs, RPC, TP monitors, or MOM. Understanding what an integration broker's middleware foundation is is important because, for instance, an MOM-based implementation will require overlying software that use the broker to communicate via messages. An integration broker based on some type of MOM implementation (via an integrated message broker) could then also include support for functionality ranging from message routing to message queuing and translation, whereas an RPC and/or ORB-type of RMI base requires overlying applications and/or middleware to trigger communication via procedure (RPC)/method (RMI) calls made, for example.

An integration broker is not only made up of some type of underlying communication broker, be it an MOM, RPC, and so on. At the highest level, as shown in Figure 8.8, an integration broker is also composed of components that handle the event *listening* and *generation* that resides upon some type of core networking middleware. For example, an integration broker's 'TCP listener' component that utilizes underlying TCP sockets, whereas a 'file listener' component utilizes an underlying file system. An integration broker's *transformer* component handles any translation of data required as these data pass through the broker on their way

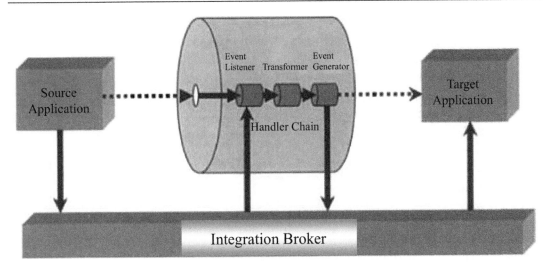

Figure 8.8: High-level Integration Broker Diagram[9]

to the destination. This for example would include an 'FTP adaptor' subcomponent that supports FTP or an 'HTTP adaptor' subcomponent that supports HTTP.

8.4 Summary

For embedded devices that have enough memory and processing power, overlaying complex networking and communication middleware on top of core middleware is increasingly becoming a popular approach in embedded systems design to support additional requirements. What was introduced in this chapter included:

- Message-oriented and -distributed Messaging Middleware
- Distributed Transaction and Transaction Processing Middleware
- Object Request Brokers
- Authentication and Security Middleware
- Integration Brokers.

This chapter pulled it all together for the reader relative to demystifying these types of complex networking and communication middleware, and how they build upon the core middleware discussed in the previous chapters. The next and final chapter of this book concludes with a holistic view of demystifying designing an embedded system with middleware.

8.5 Problems

1. What are the three types of complex messaging and communication middleware?
2. RPC middleware is based upon MOM (True/False).

3. What do MOM and MQM stand for? What is the difference between MOM and MQM?

4. RPC is based upon an asynchronous communication model (True/False).

5. Outline the main components that make up an integration broker.

6. RPC does not require underlying core networking in an embedded system (True/False).

7. What is cryptography?

8. What is the difference between authentication and authorization?

9. An integration broker cannot be based upon a message broker (True/False).

10. What does FIFO stand for? How is it used in MOM middleware?

8.6 End Notes

[1] 'Sun Java System Message Queue 4.1 Technical Overview'. http://docs.sun.com/app/docs/doc/819-7759/aeraq?a=view

[2] Embedded Systems Architecture, Noergaard, 2005. Elsevier.

[3] http://www.opalsoft.net/qos/DS-22.htm

[4] 'Middleware for Communications'. Qusay Mahmoud. P. 2.

[5] RFC 5531. http://tools.ietf.org/html/rfc5531

[6] 'Distributed Computing'. Oliver Mueller.

[7] 'Java/J2EE Job Interview Companion'. K. Arulkumaran & A. Sivayini.

[8] http://www.s-integrator.org/

[9] 'Business Services Orchestration: the hypertier of information technology'. Sadiq & Racca. P. 264.

CHAPTER 9

The Holistic View to Demystifying Middleware

Chapter Points

- Putting it all together with the holistic approach of demystifying middleware
- Outline the importance of more then just middleware technology to insure success
- About selecting the 'best-fit' middleware software for the design

---

This book has taken the holistic systems approach to demystifying building an embedded system with middleware components. The reason for the approach used in this book corresponds to why systems engineering has been going strong for over 50 years. Because one of the most powerful methods of insuring the success of a software engineering team is to accept and address the reality that the successful engineering of a product with the added complexity of middleware software will be impacted by more than the pure middleware technology alone.

As introduced in Chapter 1, successfully completing complex embedded designs, such as one that incorporates middleware software within schedule and costs, means having the wisdom to recognize that it takes **Rule #1 – more than technology**. For better or worse, successfully building an embedded system with middleware requires more than just the technology alone. It means understanding and planning for both the technical as well as the non-technical aspects of the project, be they social, political, legal, and/or financial influences in nature. Developers that recognize this from day one are most likely to reach production, within deadlines, costs, as well as with the least amount of stress. The key is for the team to identify, understand, and engage these different project influences from the start and throughout the life cycle of the project.

Next requires **Rule #2 – discipline in following development processes and best practices**. There are various best practices team members need to adhere to in order to avoid costly mistakes. These best practices can include everything from following programming language-specific guidelines to doing code inspections to having a hardcore testing strategy, for

Demystifying Embedded Systems Middleware. DOI: 10.1016/B978-0-7506-8455-2.00009-1

example. These best practices can be incorporated into any development team's agreed-upon process model. Team members *not* following healthy, disciplined processes and development practices are one of the most costly mistakes made in complex embedded systems projects that incorporate middleware software. So, the key questions to start asking are – *does* the team follow a common software development process and *how well* is it *really* working for the project? Simply put, if the team is *not* delivering high-quality code, within budget and on time – something is wrong!

In the industry, there are several different process models used today, under various 'names'. Newer software process schemes and improvements to current process models are also being introduced constantly. In general, most of the approaches used in embedded design teams are typically based upon one or some hybrid combination of the following schemes:

- **Big-bang**, designing with essentially no planning or processes in place before and during the development of a system.
- **Code-and-fix**, in which product requirements are defined but no formal processes are in place before the start of development.
- **XP** (extreme programming) and **TDD** (test driven development), development driven by re-engineering and ad-hoc testing of code over and over until the team gets it right, or the project runs out of money and/or time.
- **Waterfall**, where the process for developing a system occurs in steps, and where results of one step flow into the next step.
- **Hybrid Spiral**, in which the system is developed in steps, and throughout the various steps feedback is obtained and incorporated back into the process.
- **Hybrid Iterative Models**, such as **RUP** (Rational Unified Process), which is a framework that allows for adapting different processes for different stages of project.
- **Scrum**, another framework for adapting different processes for different stages of project, as well as team members with various roles. Incorporates shorter-term, more stringent deadlines and continuous communication between team members.

So, how can a team objectively evaluate how well a software development process model is working for them? Start by doing practical and efficient assessments by first outlining the software development goals the team wishes to achieve. Then, document what challenges team members are facing, as well as what existing processes team members are following. This means objectively investigating and documenting in some fashion software development:

- one-shot as well as reoccurring activities
- functional roles of team members at various stages of a project
- measuring and metering software development efforts that capture objective snapshots of what is working versus what is crippling development efforts
- project management, release management, and configuration management efforts

- testing and verification efforts
- infrastructure and training used to get programmers up and running.

Then, follow through with defining improvements to these existing processes that all developers on the team must adhere to. This means looking into the different possibilities of what the development team is ready to implement, in terms of more disciplined, tougher measures relative to software development efforts. There are standard industry approaches, such as via CMMI (Capability Maturity Model Integration), that a team can use to introduce improvements and increase discipline in order to save money, save time, as well as improve the quality of the software.

In short, what is recommended is **Rule #3 – teamwork**. Get together with fellow team members to discuss the various process models, and determine via consensus together what is the best 'fit' for your particular team. This means there is not yet 'one' particular process that has been invented that is the right approach for 'all' teams in the industry, or even 'all' projects for one particular team. In fact, most likely what works for the team is some hybrid combination of a few models, and this model will need to be tuned according to the types of team members and how they function best, the project's goals, and system requirements.

Then all team members, from junior to the most senior technical members of the team, as well as leadership, align together to come to an agreed *consensus* for a process model that will achieve the business results (**Rule #4 – alignment behind leadership**). Each team member then understands the big picture, the part each plays in it, and commits to the *discipline* to follow through. If, along the way, it is discovered the process isn't optimally working as expected, team members get together again, openly and respectfully discuss the challenges and frustrations together in a constructive manner, then immediately tune and adjust the process, with team members each doing their part to improve software development efforts. Finally, do not forget **Rule #5 – strong ethics and integrity among each and every team member**, to continue moving forward as agreed upon *together* towards success.

9.1 Does using Middleware in your Embedded System Design Actually 'Make Sense'?

Software, by its inherent nature, is what makes any embedded system 'configurable', 'portable', and so on. This could mean swapping in an embedded Linux operating system versus vxWorks, interchanging device drivers in the BSP – as well as adding more middleware, whether it is some sort of home-grown C++ framework, a J2ME JVM, .NET Compact Framework, or using a database instead of the file system alone, for example. So, it is critical to have the wisdom to remember not to let one particular middleware component get in the way of the success of the 'whole' software design by buying into any 'ultimate and only one solution' illusions being sold by anyone about any particular type of middleware. In short, one of the main keys to success in taking an embedded system to production is by *not overcomplicating the design*!

Remember, the ultimate goal is not to build a particular '.NETCE' or a 'DAO' embedded device. This is because including particular middleware components, in themselves, does not insure scalability, reliability, configurability, profitability and more importantly will not insure success. The key to understanding what is out there and if/when the reader should use it in an embedded design is to keep in mind the purpose of that middleware to begin with – and whether using it actually makes sense given the project scope, technology limitations, schedule, costs, and available resources. So, when determining whether to design middleware into the architecture of an embedded system, there are a few basic guidelines shown in Table 9.1.

Whether you are straight out of school or have been in the industry for the last 10–15+ years or so, it is imperative to ask the right questions for 'yourself'. So, *always* do your research, *never* be afraid to ask the hard questions out loud with team members, and then make up your own mind about what answers are discovered. *Never* be afraid to use your 'own two eyes' to help the team determine better ways to design the software in order to help insure the success of your team and organization. Remember, using a particular middleware solution is simply *a means to an end* – and could be a very good one under the right circumstances. However, if one approach in the engineering creative process does not work, do not hesitate to investigate another as soon as possible to insure a win–win!

9.2 Buy an Off-the-shelf Middleware Solution or Do-it-yourself?

When determining whether the team should do-it-yourselves versus buying commercial, it requires asking:

1. What are the *risks* to creating the middleware component ourselves in-house versus purchasing from a vendor?
2. What are the *costs* to creating the middleware component ourselves in-house versus purchasing from a vendor?
3. What are the *tradeoffs* to creating the middleware component ourselves in-house versus purchasing from a vendor?
4. Does the team have the *expertise* in-house, money, time, and resources to create and support this middleware component?
5. Is there a *commercially available* version of this middleware component? If so, *why* can it not be used in the design?

When investigating off-the-shelf embedded middleware solutions, one of the most commonly effective ways of selecting between commercially available components is to build a *matrix*

Table 9.1: Guidelines to Asking the Right Questions

	What to Think About
When is using middleware *probably not* a good idea?	If rather than simplifying the architecture and development of the device, the middleware is more complicated than the actual requirements of the device itself – and would actually introduce a greater degree of cost, time, risk, and stress into the project and team, relative to the benefits
	If the reader is constrained by the hardware *in any way* (i.e., it's too slow, not enough RAM, etc.). Basic rule of thumb in developing software for more limited embedded hardware boards in terms of processing power and memory is the less software, fewer abstraction layers, etc. the better
	When the project team is mainly made up of *non-experienced* and/or *non-embedded* programmers who do *not* comprehend the importance of understanding the difference between developing software for an embedded target with more limited processing power and RAM (for example) versus developing code for a PC that has GHz of processing power and gigabytes memory
When might it be a good idea to *replace* a current middleware design and investigate a *new* middleware *approach*?	Hardware support is there, but the project is made up of a bunch of people who have been trying to rewrite middleware code several times over an extended period and who still cannot get it to work stably enough to ship on the embedded hardware
	Better technology has emerged on the market
	New or changes to project requirements that render decisions relative to the original middleware design irrelevant
When could a current middleware solution be *too risky* to support?	If the middleware does not work reliably after years and years of re-engineering, and there is no interest or effort in rethinking, improvement, or change in how things are done
	The middleware is not capable of doing what was promised/intended, and is being kept hidden as the project deadlines are continuously delayed
	Substandard products are delivered to the confusion and anger of application developers, customers, upper management, and so on, which essentially renders it an actual liability risk within the design
	When there are expensive project cost overruns, the loss of highly-qualified team members leaving stressful team environment, and/or huge software liability risks introduced to the whole organization if the product is deployed
	Counter-productive agendas on the project have hijacked the middleware design, and will not allow for any discussions, questions, are selfish and/or bullying with their own ideas to the exclusion and detriment of the rest of the design and team
	If the middleware is being bypassed and/or not being used for the purposes of how it was intended by other developers, i.e., for portability, performance, stability, etc.

(continued)

Table 9.1: Guidelines to Asking the Right Questions   (*Continued*)

	What to Think About
When would using a particular middleware solution in an embedded design *make sense?*	Hardware and underlying system software will support it, or can be modified to support it
	Required according to some industry standard, customer requirements, etc.
	Skilled, support team resources to implement the design are there
	If it helps insure the success of the project by helping simplify designing the device
	Allows for the dynamic configurability, portability, maintainability, etc. of the device to make maintaining software for different product variations better

Table 9.2: Example of Matrix

	Requirement 1	Requirement 2	Requirement 3	Requirement '...'	Requirement 'N'
Middleware Product 'Vendor A'	NO	NOT YET (in 6 months)	YES (Features ...)	...	...
Middleware Product 'Vendor B'	YES (Features ...)	NO	NOT YET (in 3 months)	...	...
Middleware Product 'Vendor C'	NOT YET (Next Year)	YES (Features ...)	YES (Features ...)	...	...
Middleware Product 'Vendor ...'	....	....	...	...	...

(see Table 9.2) of required features for each component. The matrix is then filled with the vendor's product information that fulfills the particular requirements. When considering a third-party commercial solution for a particular middleware component, the types of questions to be asked (that should be addressed in the matrices) include:

- What is the vendor's reputation? When can the vendor deliver what you need? Has the product ever been deployed in another commercially available embedded system successfully?
- What will it cost to use this middleware component? What is the cost of technical support from this vendor? Can you buy the source code or is it object code only?

- How is this middleware software product tested? How reliable is the vendor's software in real-world stress conditions? Can you get some type of test plan and report to review? Can you re-run the vendor's tests for verification on your system before your own product deploys?
- What are the specific requirements for the vendor's middleware component in terms of underlying hardware and software? How compatible is this vendor's middleware product with your hardware, programming language(s), tools, etc.? Will it require special debugging tools?
- Have you actually seen this middleware component running in a real system and on real hardware? Or have you just seen the sales pitch and some sexy marketing 'documentation'?

In short, target the off-the-shelf middleware component that has been stably ported and supporting the hardware and underlying software. Make sure developers can design and debug with this third-party commercial middleware component, because without the proper tool support it will be a nightmare for the team. Finally, if considering more than one off-the-shelf component, then create more than one matrix representing these different vendors' components that can then be cross-referenced, and distributed among team members for review. This helps to insure that all requirements have been addressed, and are ultimately met.

9.2.1 More on Keys to Success with Developing 'Do-it-yourself' Middleware Software...

Once again – do *not* overcomplicate the design! In other words, start with a skeleton for the middleware solution, and then hang code off of this skeleton within different phases of the project. This approach allows for the sacrificing of less essential features in the first (and any future) release, allowing for the team to ship a high-quality middleware software solution within the embedded system, on time.

Partitioning the middleware design can be done in many ways including keeping function sizes within a certain number of LOC (lines of code), by features, top-down decomposition, via underlying software (such as action via tasks/threads with an operating system), and/or utilizing additional underlying hardware. Partitioning the middleware solution into smaller modules with fewer dependencies between modules results in:

- an increased likelihood that the finished middleware design will meet requirements
- middleware will be completed within time and budget
- application developers know how to plan around a solid and reliable system underlying middleware solution at every phase of the project
- fewer complex bugs introduced in the middleware software to delay release.

In short, less is more! Maintain control of the middleware design by closely managing the requirements and features.

Next, do not underestimate the impact of the programming language(s) on the embedded design. Carefully consider what *programming language* to implement the in-house embedded middleware component in. This is because there is not yet one programming language that is perfect for every embedded system. Objectively weigh the pros and cons of going with programming language options. The real questions to ask include:

1. How mature and stable is the language specification, the compiler, debugger, and so on for your particular target hardware? For example, going with a 'native' programming language when implementing a particular middleware solution does mean losing the power that comes associated with going object-oriented, such as encapsulation, modularity, polymorphism, and inheritance to name a few. However, it will not matter how many C++ 'wicked-smart' developers are on the team, if the C++ compiler available from the vendor for the embedded target hardware is extremely unstable, immature, and/or buggy.
2. Is there a requirement to support a particular language or language standard? For example, who needs to brush up on embedded Java to implement an MHP (multimedia home platform)-compliant STB (set-top box).
3. How complex is the programming language to debug, test, and maintain in-house?

Bad, buggy code can be written in *any* language. So, once the programming language(s) has been agreed upon by the team, then embedded developers need to follow the best practices for that particular programming language. Meaning there are general best practices that are independent of programming language, such as not using magic numbers hard-coded in the source code, not manually editing automatically generated code, capturing/handling all exceptions/errors, and initializing all defined variables for example.

Of course, the cheapest way to debug is to not insert any bugs in the first place. This means slowing down and using best practice programming language techniques for that language. Developers need to *not* be pressured to make rushed source code without investigating properly, thinking about the changes (a lot), having time and the discipline to spec out in writing the changes, and then doing it right. As shown with the examples in Table 9.3, it is important to understand and follow the best practices that exist which are unique to that particular programming language.

Table 9.3: Examples of Programming Language Best Practices[2]

Programming Language	Example	Best Practice
C	Functions	Check that the parameters they are passed are workable
		Calls provide easily identifiable points of checking the state of the system allowing early detection of memory corruption or other unexpected states
		Return values from function calls are checked for expected values

Table 9.3: Examples of Programming Language Best Practices   *Continued*

Programming Language	Example	Best Practice
	Rules	Non-obvious code that relies on rules of precedence inherent in a programming language is avoided
		Expressions that mix operators from the set >, >=, <, <=, ==, !=, with operators from the set <<, >>, ^, &, |, &&, || are fully parenthesized and do not rely on inherent precedence rules
	Switch statements	Explicitly list all known cases and have default cases that warn of unhandled cases
	Conditional statements	Explicitly stated in code that conforms to this standard
		Non-Boolean values are not tested as Boolean values and Boolean values are not tested as non-Boolean values
		Value ranges are used when checking real number values to protect against error introduced due to loss of precision
	GOTO	Goto statements are avoided because of difficulty in validating usage and debugging
	…	…
C++	Global static variables	Global static objects should be avoided
		If used, an instantiator class which counts references to an object should be implemented
	Errors and exceptions	Should be specified in the signature of the class method declaration and definition
		Create your own exception hierarchies reflecting the domain and define relevant exception classes derived from the standard exception class
		Use ASSERT()'s or comparable debugging macros liberally to trap potential programming errors
		Validate all parameters passed to any public, protected, or even private function
		Handle all potentially invalid parameters or environmental conditions in a graceful, consistent, and documented manner
	…	…
Java or C#	Classes	Do not make any instance or class variable public, make them private
	Methods	Methods should not have more than five arguments. If more than five, use a structure to pass the data
…	…	…

Once the team is aligned on the middleware design, then *do not* skimp on the tools. One of the most common mistakes middleware developers make is not using the appropriate design and debug tools for implementing complex software solutions such as middleware software within the embedded design. For instance, editors, compilers, linkers, and debuggers are essential and non-negotiable, meaning it is impossible to do efficient development without these tools. Another example for embedded middleware development is developers who include *software design patterns* as an integral part of their toolbox for object-oriented design and development. There are several, different types of software design patterns that have been published with characteristics that are intended to address different types of design requirements from the object-oriented point of view.

Specific characteristics of different software design patterns can be used by developers as models for helping to determine possible design implementations that address their specific requirements, such as additional encapsulation versus a need for greater flexibility to name a few. In the case of flexibility, for instance, where underlying hardware, such as the underlying storage medium, remains unchanged, then, for example, the 'factory method' software design pattern may be a feasible approach. On the flip side, with requirements that include supporting an underlying storage medium that will change, then, for example, embracing the 'abstract factory' software design pattern may be considered as an alternative.

As with any other model, remember not to blindly use any software design pattern within an embedded systems middleware project. This means developers can start by questioning why a particular pattern was used in a particular software component, and what were the resulting pros and cons of the approach. For inherited source code, developers can start to look for software design patterns within components to help with the reverse-engineering process in understanding the code. Remember, a software design pattern elegantly and brilliantly implemented for some type of Java Enterprise server-based system may *not* be the best approach for a J2ME (Java 2 Micro Edition)-based embedded systems solution or some C++ rework of that same design pattern. The software team needs to also investigate and think through the *type* of system(s) a particular software design pattern has successfully been implemented for (underlying hardware and system software differences).

Additional examples of commonly used types of tools by embedded systems developers are shown in Figure 9.1. In short, embedded developers need a solid software toolbox to help insure success when building an embedded system that has the added complexity of middleware components. So, this means asking key questions, such as:

- Will the tool help write better source code, faster?
- Who is actually using what tool?
- Why and how is the tool being used?

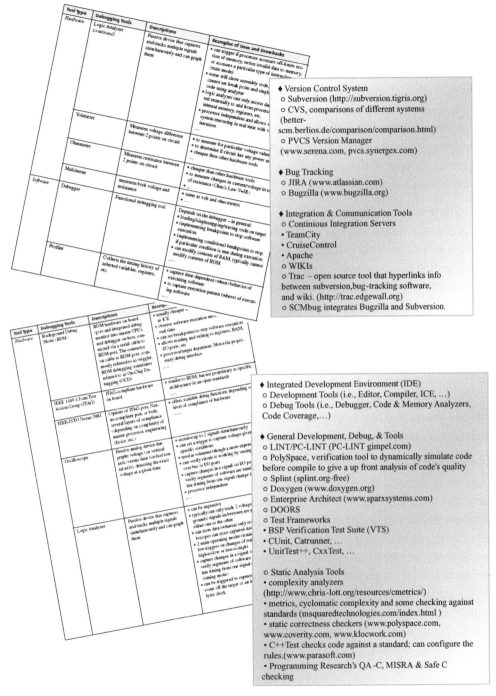

Figure 9.1: Examples of Development and Debugging Tools[3]

9.2.2 Always Ask, Is the Code 'Good Enough'? – Systems Integration, Testing and Verification from Day 1

With the complexity that middleware adds to an embedded design, it is important to have an integration, verification, and test strategy from day one, meaning, as soon as embedded hardware is available with device drivers to bring up the board, developers not verifying and even unit testing these available software components guarantees headaches later. Make the quality of your source code visible along with feature feedback from the start, by executing a disciplined test strategy as soon as you have any software that you plan to ship the board with running on the system. Verify and test everything, including the prototype that would be used as the final design's foundation.

Plan the different *types* of testing from the start. It does not matter 'how', i.e., whether it's via individual responsible engineers that are assigned the role or with a formal test group – as long as it gets done. This includes everything from unit testing to integration testing to regression testing to stress and acceptance testing. This also includes planning for not only test-to-pass scenarios, but more importantly the test-to-fail scenarios. Figure out what the limits of the design are, to insure that the end system deployed can be expected to function even under unexpected stress conditions in the field. This allows the team to determine early on if the hardware is faulty or the foundational source code is buggy. Fix the hardware and/or source code defects as they are found, and do not defer. Having an unstable system with unreliable hardware and/or bad code is worse than having 'no' system. Track defects and measure their rate in specific components as they are found. This is in order to monitor these defects, and insure that highly problematic components are replaced or rewritten – since they become more expensive to debug than to replace.

A common mistake within embedded design teams is *not* code inspecting and/or testing proactively and adequately. It does not matter if this software is home-grown or a BSP, operating system, and/or networking middleware from an external vendor, for example. Do not assume that because a particular off-the-shelf software component has come from an expensive external vendor, it is bug-free and production-ready. More importantly, do not assume that any software that comes out of the box is tuned to your own embedded system's requirements until team members see it running and have verified it with their own eyes. In fact, many embedded software vendors deliver their software components with additional configuration files that are accessible to their customers, because the expectation is that their customers will tune their software to meet the requirements of their particular embedded device.

To be the most effective, code inspections should be incorporated into the test strategy from the start, and these code inspections need to do more than look for 'pretty' code. Insure language 'best practices' have been followed for particular language and actually look for bugs. It is cheaper and faster to do stringent code inspections *before* testing, and as soon as source code compiles.[1] A team's code inspection process should include some type of checklist (see Table 9.4) of what is being checked for and where the results are being *documented*.

Insure that the right 'type' of team members are doing the actual code inspections in order to be the most effective and most efficient. For example, insuring developers with knowledge

Table 9.4: Example of Code Inspection Checklist for 'C' Source

Parameter/Function Name	Number of Errors		Error Type	
	Major	Minor		
			Code does not meet firmware standards	
				Header Block
				Naming Consistency
				Comments
				Code Layout and Elements
			Recommended Coding Practices	
				Auto-generated code not manually edited
				Don't use magic numbers hard coded in the source code, i.e., place constant numerical values directly into the source
				Avoid using global variables
				Initialize all defined variables
				Function size and complexity unreasonable
				Unclear expression of ideas in the code
				Poor encapsulation
				Data types do not match
				Poor logic – won't function as needed
				Exceptions and error conditions not caught (e.g., return codes from malloc())?
				Switch statement without a default case (if only a subset of the possible conditions used)?
				Incorrect syntax – such as proper use of ==, =, &&, &, etc.
				Non-reentrant code in dangerous places
...	...	...	Other ...	

and understanding of the 'hardware' or even the actual hardware engineers to code inspect device drivers that manage various hardware components on the target. In general, an effective code inspection team targets including team members that support the following type roles:[2]

1. *Moderator*, who is responsible for managing the code inspection process and meeting(s)
2. *Reader* (not the developer that created the source code being inspected), reads out loud the source code and relative specifications for the operational investigation
3. *Recorder*, fills in code inspection checklist report, and documents any agreed-upon open items
4. *Author* (the developer of the source code being inspected), helps explain source to code inspection team, to discuss errors found, and future rework that needs to be done.

9.3 Conclusion – See the Pattern Yet?

It is powerful for the team to start with getting the *full* systems picture via defining the high-level architectural profile of the embedded system from the top at the application layer and then work your way down. This is why, for instance, application examples were presented in previous middleware chapters – because the middleware selected needs to fulfill the needs of your system's applications. So, at the application layer, start with describing the applications, how they will function within the system, interrelationships, and external interfaces. Then, from the application layer outline what underlying system software and hardware functionality these applications would explicitly and implicitly require.

Today in the embedded market, there is not 'one' middleware solution that supports all requirements for all systems. Furthermore, many complex embedded designs will require more than one middleware component to meet various middleware requirements. Thus, developers must investigate how individual middleware software will be successfully integrated in order to insure compatibility during the system's operation. Basically weigh the pros and cons of utilizing a completely integrated middleware stack from various vendors that could result in a successful system implementation against what risk is introduced, additional costs, and potential schedule risk. Compare this to implementing and integrating various components of the complete stack independently by the team.

For example, larger embedded operating system vendors (such as WindRiver) typically supply integrated middleware packages with different components that support their respective embedded OS over various hardware solutions. There are also companies that supply various middleware packages with integrated core and/or more complex middleware components that can be ported by customers.

Remember, the idea is to keep things as simple as possible. So, make sure that you have a solid understanding of core middleware components (Figure 9.2) and how to select between

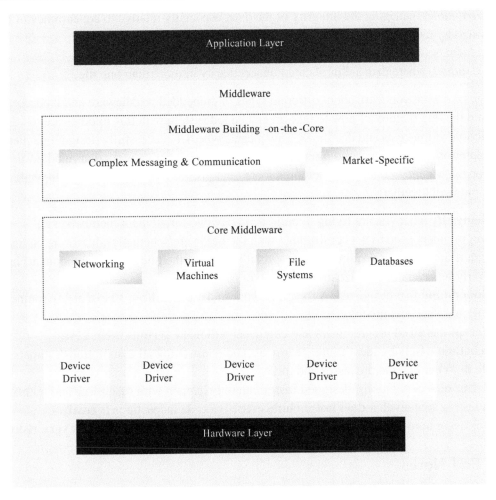

Figure 9.2: General Types of Middleware

these. For example, how to select between a using a file system versus a database within the embedded design, to manage data? While, on one side, a database system may come with functionality lacking in a file system the team must also investigate the underlying requirements of a database versus a file system. If the hardware is not powerful enough to support a more fully featured database system the best approach may be utilizing a file system and insuring applications are written to meet specific data management requirements. A database would be considered a better fit than just a file system for the design, for example, if given the type of data being managed and how applications will utilize these data results in data managed in a file system becoming:

- *inconsistent*, where data changes are not updated in all files properly
- *insecure*, managing access of multiple applications to specific files that application is permitted to have access to

- *corruptible*, managing the integrity of the data, especially relative to concurrency of data access by multiple applications accessing the file simultaneously
- *isolated*, searching through files for particular data with a specific format
- *redundant*, where data are duplicated unnecessarily in more than one file.

Finally, take what you learn about different types of embedded middleware and incorporate that into the 'big picture' of how to take an embedded system with this middleware to production within costs and deadlines. This means, whether successfully designing a piece of middleware software or building a more complex system with a particular middleware component, accepting that to win requires more than just the middleware technology alone. In short, put it all together:

- **Demystify** what you are trying to build from day one for all team members, i.e., programmers need to get comfortable with the hardware/schematics, hardware engineers understand software requirements (especially those of the middleware components and overlying layers that utilize them), and so on.
- **Understand** that design teams forced to work under **unhealthy, stressful** environments will make **serious mistakes!**
- **Accept** that **tired** programmers and engineers will make **serious mistakes!**
- **Schedule wisely** by identifying clear goals and questioning all assumptions/estimates.
- **Do not cheat** on the processes and tools.
- **Better** quality solutions, designed **faster** can only happen with **discipline** and **teamwork.**
- A testing strategy that **does not include** extensive code inspections is **costly.**
- Releasing **inadequate** and/or **untested** code on the embedded hardware is **very risky!**

9.4 End Notes

[1] 'A Boss's Quick-Start to Firmware Engineering'. Jack Ganssle.
[2] 'A Guide to Code Inspections'. Jack Ganssle. Code Inspection Process, Wind River Services.
[3] Embedded Systems Architecture, Noergaard, 2005. Elsevier.

Abbreviations and Acronyms

A

A2A	Application-to-Application
AC	Alternating Current
ACK	Acknowledge
ACL	Access Control List
A/D	Analog-to-Digital
ADC	Analog-to-Digital Converter
ALU	Arithmetic Logic Unit
AM	Amplitude Modulation
AMI	Application Messaging Interface
amp	Ampere
ANSI	American National Standards Institute
AOT	Ahead-of-Time
API	Application Programming Interface
APPC	Advanced Program to Program Communication
ARIB-BML	Association of Radio Industries and Business of Japan
AS	Address Strobe
ASCII	American Standard Code for Information Interchange
ASIC	Application Specific Integrated Circuit
ASP	Application Service Provider or Active Server Pages
ATM	Asynchronous Transfer Mode, Automated Teller Machine
ATMI	Application to Transaction Model Interface
ATSC	Advanced Television Standards Committee
ATVEF	Advanced Television Enhancement Forum

B

BDM	Background Debug Mode
BER	Bit Error Rate
BIOS	Basic Input/Output System
BML	Broadcast Markup Language

BMP	Bean Managed Persistence
BOA	Basic Object Adaptor
BOM	Bill of Materials
bps	Bits per Second
BSP	Board Support Package
BSS	'Block Started by Symbol', 'Block Storage Segment', 'Blank Storage Space', ...
BTP	Business Transaction Protocol

C

CAD	Computer Aided Design
CAE	Common Application Environment
CAN	Controller Area Network
CAS	Column Address Select
CASE	Computer Aided Software Engineering
CBIC	Cell-Based IC or Cell-Based ASIC
CBQ	Class Based Queuing
CCF	Common Connector Framework
CCI	Common Client Interface
CCM	Common CORBA Model
CDC	Connected Device Configuration
CDN	Content Delivery Network
CEA	Consumer Electronics Association
CEN	European Committee for Standardization
CGI	Common Gateway Interface
CIDL	Component Implementation Definition Language
CISC	Complex Instruction Set Computer
CLDC	Connected Limited Device Configuration
CLI	Call Level Interface
CLR	Common Language Runtime
CLS	Common Language Specification
CMI	Common Messaging Interface
CMOS	Complementary Metal Oxide Silicon
CMP	Container Managed Persistence
COFF	Common Object File Format
COM	Component Object Model
COPS	Common Open Policy Service
CORBA	Common Object Request Broker Architecture
CPI	Container Provided Interface
CPLD	Complex Programmable Logic Device
CPU	Central Processing Unit

CRT	Cathode Ray Tube
CTG	CICS Transaction Gateway
CTS	Clear-to-Send

D

DAC	Digital-to-Analog Converter
DAD	Document Access Definition
DAG	Data Address Generator
DASE	Digital TV Applications Software Environment
DAVIC	Digital Audio Visual Council
dB	Decibel
DC	Direct Current
D-Cache	Data Cache
DCE	Data Communications Equipment or Distributed Computing Environment
DCOM	Distributed Component Object Model
DDL	Data Definition Language
Demux	Demultiplexor
DHCP	Dynamic Host Configuration Protocol
DII	Dynamic Invocation Interface
DIMM	Dual Inline Memory Module
DIP	Dual Inline Package
DLL	Dynamic Link Library
DMA	Direct Memory Access
DNS	Domain Name Server, Domain Name System, Domain Name Service
DOM	Document Object Model
DPRAM	Dual Port RAM
DRAM	Dynamic Random Access Memory
DRDA	Distributed Relational Database Architecture
DSL	Digital Subscriber Line
DSP	Digital Signal Processor
DTD	Data Type Definition
DTE	Data Terminal Equipment
DTP	Distributed Transaction Processing
DTVIA	Digital Television Industrial Alliance of China
DVB	Digital Video Broadcasting

E

ECI	External Call Interface
EDA	Electronic Design Automation

EDF	Earliest Deadline First
EDI	Electronic Data Interchange
EDO	RAM Extended Data Out Random Access Memory
EEMBC	Embedded Microprocessor Benchmarking Consortium
EEPROM	Electrically Erasable Programmable Read Only Memory
EIA	Electronic Industries Alliance
ELF	Extensible Linker Format
EMI	Electromagnetic Interference
EPROM	Erasable Programmable Read Only Memory
ESD	Electrostatic Discharge
EU	European Union

F

FAT	File Allocation Table
FCFS	First Come First Served
FDA	Food and Drug Administration – USA
FDMA	Frequency Division Multiple Access
FET	Field Effect Transistor
FIFO	First In First Out
FFS	Flash File System
FM	Frequency Modulation
FPGA	Field Programmable Gate Array
FPU	Floating Point Unit
FSM	Finite State Machine
FTP	File Transfer Protocol

G

GB	Gigabyte
Gbit	Gigabit
GCC	GNU C Compiler
GDB	GNU Debugger
GHz	Gigahertz
GND	Ground
GPS	Global Positioning System
GUI	Graphical User Interface

H

HAVi	Home Audio/Video Interoperability
HDL	Hardware Description Language
HL7	Health Level Seven

HLDA	Hold Acknowledge
HLL	High-Level Language
HTML	HyperText Markup Language
HTTP	HyperText Transport Protocol
Hz	Hertz

I

IC	Integrated Circuit
I_2C	Inter Integrated Circuit Bus
I-Cache	Instruction Cache
ICE	In-Circuit Emulator
ICMP	Internet Control Message Protocol
IDE	Integrated Development Environment
IEC	International Engineering Consortium
IEEE	Institute of Electrical and Electronics Engineers
IETF	Internet Engineering Task Force
IGMP	Internet Group Management Protocol
INT	Interrupt
I/O	Input/Output
IP	Internet Protocol
IPC	Interprocess Communication
IR	Infrared
IRQ	Interrupt ReQuest
ISA	Instruction Set Architecture
ISA	Bus Industry Standard Architecture Bus
ISO	International Standards Organization
ISP	In-System Programming
ISR	Interrupt Service Routine
ISS	Instruction Set Simulator
ITU	International Telecommunication Union

J

JIT	Just-In-Time
J2ME	Java 2 MicroEdition
JTAG	Joint Test Access Group
JVM	Java Virtual Machine

K

kB	Kilobyte
kbit	Kilobit

kbps	Kilobits per second
kHz	Kilohertz
KVM	K Virtual Machine

L

LA	Logic Analyzer
LAN	Local Area Network
LCD	Liquid Crystal Display
LED	Light Emitting Diode
LIFO	Last In First Out
LSb	Least Significant Bit
LSB	Least Significant Byte
LSI	Large Scale Integration

M

mΩ	Milliohm
MΩ	Megaohm
MAN	Metropolitan Area Network
MCU	Microcontroller
MHP	Multimedia Home Platform
MIDP	Mobile Information Device Profile
MIPS	Millions of Instructions per Second, Microprocessor without Interlocked Pipeline Stages
MMU	Memory Management Unit
MOSFET	Metal Oxide Silicon Field Effect Transistor
MPSD	Modular Port Scan Device
MPU	Microprocessor
MSb	Most Significant Bit
MSB	Most Significant Byte
MSI	Medium Scale Integration
MTU	Maximum Transfer Unit
MUTEX	Mutual Exclusion

N

NAK	NotAcKnowledged
NAT	Network Address Translation
NCCLS	National Committee for Clinical Laboratory Standards
NFS	Network File System
NIST	National Institute of Standards and Technology

NMI	Non-Maskable Interrupt
nsec	Nanosecond
NTSC	National Television Standards Committee
NVRAM	Non-Volatile Random Access Memory

O

OCAP	Open Cable Application Forum
OCD	On Chip Debugging
OEM	Original Equipment Manufacturer
OO	Object Oriented
OOP	Object-Oriented Programming
OS	Operating System
OSGi	Open Systems Gateway Initiative
OSI	Open Systems Interconnection
OTP	One Time Programmable

P

PAL	Programmable Array Logic, Phase Alternating Line
PAN	Personal Area Network
PC	Personal Computer
PCB	Printed Circuit Board
PCI	Peripheral Component Interconnect
PCP	Priority Ceiling Protocol
PDA	Personal Data Assistant
PDU	Protocol Data Unit
PE	Presentation Engine, Processing Element
PID	Proportional Integral Derivative
PIO	Parallel Input/Output
PIP	Priority Inheritance Protocol, Picture-In-Picture
PLC	Programmable Logic Controller, Program Location Counter
PLD	Programmable Logic Device
PLL	Phase Locked Loop
POSIX	Portable Operating System Interface X
POTS	Plain Old Telephone Service
PPC	PowerPC
PPM	Parts Per Million
PPP	Point-to-Point Portocol
PROM	Programmable Read Only Memory
PSK	Phase Shift Keying

PSTN	Public Switched Telephone Network
PTE	Process Table Entry
PWM	Pulse Width Modulation

Q
| QA | Quality Assurance |

R
RAM	Random Access Memory
RARP	Reverse Address Resolution Protocol
RAS	Row Address Select
RF	Radio Frequency
RFC	Request For Comments
RFI	Radio Frequency Interference
RISC	Reduced Instruction Set Computer
RMA	Rate Monotonic Algorithm
RMS	Root Mean Square
ROM	Read Only Memory
RPM	Revolutions Per Minute
RPU	Reconfigurable Processing Unit
RTC	Real Time Clock
RTOS	Real Time Operating System
RTS	Request To Send
RTSJ	Real Time Specification for Java
R/W	Read/Write

S
SBC	Single Board Computer
SCC	Serial Communications Controller
SECAM	Système Electronique pour Couleur avec Mémoire
SEI	Software Engineering Institute
SIMM	Single Inline Memory Module
SIO	Serial Input/Output
SLD	Source Level Debugger
SLIP	Serial Line Internet Protocol
SMPTE	Society of Motion Picture and Television Engineers
SMT	Surface Mount
SNAP	Scalable Node Address Protocol
SNR	Signal-to-Noise Ratio

SoC	System-on-Chip
SOIC	Small Outline Integrated Circuit
SPDT	Single Pole Double Throw
SPI	Serial Peripheral Interface
SPST	Single Pole Single Throw
SRAM	Static Random Access Memory
SSB	Single Sideband Modulation
SSI	Small Scale Integration

T

TC	Technical Committee
TCB	Task Control Block
TCP	Transmission Control Protocol
TDM	Time Division Multiplexing
TDMA	Time Division Multiple Access
TFTP	Trivial File Transfer Protocol
TLB	Translation Lookaside Buffer
TTL	Transistor–Transistor Logic

U

UART	Universal Asynchronous Receiver/Transmitter
UDM	Universal Design Methodology
UDP	User Datagram Protocol
ULSI	Ultra Large Scale Integration
UML	Universal Modeling Language
UPS	Uninterruptible Power Supply
USA	United States of America
USART	Universal Synchronous–Asynchronous Receiver–Transmitter
USB	Universal Serial Bus
UTP	Untwisted Pair

V

VHDL	Very High Speed Integrated Circuit Hardware Design Language
VLIW	Very Long Instruction Word
VLSI	Very Large Scale Integration
VME	VersaModule Eurocard
VoIP	Voice Over Internet Protocol
VPN	Virtual Private Network

W

WAN	Wide Area Network
WAT	Way-Ahead-of-Time
WDT	Watchdog Timer
WLAN	Wireless Local Area Network
WML	Wireless Markup Language
WOM	Write Only Memory

X

XCVR	Transceiver
XHTML	eXtensible HyperText Markup Language
XML	eXtensible Markup Language

Embedded Systems Glossary

A

Absolute Memory Address The physical address of a specific memory cell.

Accumulator A special processor register used in arithmetic and logical operations to store an operand used in the operation, as well as the results of the operation.

Acknowledge (ACK) A signal used in bus and network 'handshaking' protocols as an acknowledgment of data reception from another component on the bus (on an embedded board for bus handshaking) or from another embedded system via some networking transmission medium (for network handshaking).

Active High Where a logic value of '1' is a higher voltage than a logic value of '0' in a circuit.

Active Low Where a logic value of '0' is a higher voltage than a logic value of '1' in a circuit.

Actuator A device used for converting electrical signals into physical actions, commonly found in flow-control valves, motors, pumps, switches, relays and meters.

Adder A hardware component that can be found in a processor's CPU that adds two numbers.

Address Bus An address bus carries the addresses (of a memory location, or of particular status/control registers) between board components. An address bus can connect processors to memory, as well as processors to each other.

Ahead-of-Time Compiler (AOT) See *Way-Ahead-of-Time Compiler.*

Alternating Current (AC) An electric current whose voltage source changes polarity of its terminals over time, causing the current to change direction with every polarity change.

Ammeter A measurement device that measures the electrical current in a circuit.

Ampere The standard unit for measuring electrical current, defined as the charge per unit time (meaning the number of coulombs that pass a particular point per second).

Amplifier A device that magnifies a signal. There are many types of amplifiers (log, linear, differential, etc.), all differing according to how they modify the input signal.

Amplitude A signal's size. For an AC signal it can be measured via the high point of an AC wave from the equilibrium point (center) to the wave's highest peak or by performing the RMS (root mean square) mathematical scheme – which is by 1) finding the square of the waveform function, 2) averaging the value of the result of step (1) over time, and 3) taking the square root of the results of step (2). For a DC signal, it is its voltage level.

Amplitude Modulation (AM) The transmission of data signals via modifying (modulating) the amplitude of a waveform to reflect the data (i.e., a '1' bit being a wave of some amplitude, and a '0' bit being a wave with a different amplitude).

Analog Data signals represented as a continuous stream of values.

Analog-to-Digital Converter (A/D Converter) A device that converts analog signals to digital signals.

AND Gate A gate whose output is 1 when both inputs are 1.

Anion A negative ion, meaning an atom that gains electrons.

Anode (1) The negatively charged pole (terminal) of a voltage source. (2) The positively charged electrode of a device (i.e., diode), which accepts electrons (allowing a current to flow through the device).

Antenna A transducer made up of conductive material (wires, metal rod, etc.) used to transmit and receive wireless signals (radio waves, IR, etc.).

Antialiased Fonts Fonts in which a pixel color is the average of the colors of surrounding pixels. It is a commonly used technique in digital televisions for evening (smoothing) displayed graphical data.

Application Layer The layer within various models (OSI, TCP/IP, Embedded Systems Model, etc.), which contains the application software of an embedded device.

Application Programming Interface (API) A set of subroutine calls that provide an interface to some type of component (usually software) within an embedded device (OS APIs, Java APIs, MHP APIs, etc.).

Application Server Middleware Middleware that allows access to legacy software via a browser.

Application-Specific Integrated Circuit (ASIC) An application-specific ISA-based IC that is customized for a particular type of embedded system or in support of a particular application within an embedded system. There are mainly full-custom, semi-custom, or programmable types of ASICs. PLDs and FPGAs are popular examples of (programmable) ASICs.

Architecture See *Embedded Systems Architecture* or *Instruction Set Architecture*.

Arithmetic Logic Unit (ALU) The component within a processor's CPU which executes logical and mathematical operations.

Aspect Ratio A ratio of width to height (in memory the number of bits per address to the total number of memory addresses, the size or resolution of a display, etc.).

Assembler A compiler that translates assembly language into machine code.

Astable Multivibrator A sequential circuit in which there is no state it can hold stable in.

Asynchronous A signal or event that is independent of, unrelated to, and uncoordinated with a clock signal.

Attenuator A device that reduces (attenuates) a signal (the opposite of what an amplifier does).

Automatic Binding When an RPC client automatically locates and selects a specific server.

Autovectoring The process of managing interrupts via priority levels rather than relying on an external vector source.

B

Background Debug Mode (BDM) Components used in debugging an embedded system. BDM components include BDM hardware on the board (a BDM port and an integrated debug monitor in the master CPU), and debugger on the host (connected via a serial cable to BDM port). BDM debugging is sometimes referred to as On-Chip Debugging (OCD).

Bandwidth On any given transmission medium, bus, or circuitry – the frequency range of an analog signal (in hertz, the number of cycles of change per second) or digital signal (in bps, the number of bits per second) traveling through it (as in the case of a bus or transmission medium) or being processed by it (as in the case of a processor).

Basic Input/Output System (BIOS) Originally the boot-up firmware on x86-based PCs, now available for many off-the-shelf embedded x86-based boards and a variety of embedded OSs.

Battery A voltage source where voltage is created through a chemical reaction within it. A battery is made up of two metals submerged within a chemical solution, called an electrolyte, that is in liquid (wet cell) or paste (dry cell) form. Basically, the two metals respond with different ionic state after they are exposed to the electrolyte. Wet cells are used in automobiles (car batteries), and dry cells are used in many different types of portable embedded systems (radios, toys, etc.).

Baud Rate The total number of bits per some unit of time (kbits/sec, Mbits/sec, etc.) that can be transmitted over some serial transmission link.

Bias An offset (such as voltage or current) applied to a circuit or electrical element to modify the behavior of the circuit or element.

Big Endian A method of formatting data in which the lowest-order bytes (or bits) are stored in the highest bytes (or bits). For example, if the highest-order bits are from left to right in descending order in a particular 8-bit ISA, big endian mode in this ISA would mean that bit 0 of the data would be stored from left to right in ascending order (the value of 'B3h/10110011b' would be stored as '11001101b'). In a 32-bit ISA, for instance, where the highest-order bytes are stored from left to right in descending order, big endian mode in this ISA would mean that byte 0 of the data is stored from left to right in the word in ascending order (i.e., the value of 'B3A0FF11h' would be stored as '11FFA0B3h').

Binary A base-2 number system used in computer systems, meaning the only two symbols are a '0' or '1'. These symbols are used in a variety of combinations to represent all data.

Bit Error Rate (BER) The rate at which a serial communication stream loses and/or transfers incorrect data bits.

Bit Rate The (number of actual data bits transmitted / total number of bits that can be transmitted) × the baud rate of the communications channel.

Black-Box Testing Testing that occurs with a tester that has no visibility into the internal workings of the system (no schematics, no source code, etc.) and is basing testing on general product requirements documentation.

Blocking Communication When communication between receiver and transmitter is blocked until response is received in synchronous messaging communication scheme.

Block Started by Symbol (BSS) BSS is several different things depending on the context and who is asked, including 'Block Started by Symbol', 'Block Storage Segment', and 'Blank Storage Space'. The term 'BSS' originated from the 1960s, and while not everyone agrees on what the BSS abbreviation stands for, it is generally agreed upon that BSS is a statically allocated memory space containing the source code's uninitialized variables (data).

Board Support Package (BSP) A software provided by many embedded off-the-shelf OS vendors that allow their OSs to be ported more easily over various boards and architectures. BSPs contain the board and architecture-specific libraries required by the OS, and allow for the device drivers to be integrated more easily for use by the OS through BSP APIs.

Bootloader Firmware in an embedded system that initializes the system's hardware and system software components.

Breakpoint A debugging mechanism (hardware or software) which stops the CPU from executing code.

Bridge A component on an embedded board that interconnects and interfaces two different buses.

Buffered Queue RAM resident message queue.

Bus A collection of wires that interconnect components on an embedded board.

Byte A byte is defined as being some 8-bit value.

Byte Code Byte (8-bit) size opcodes that have been created as a result of high-level source code (such as Java or C#) being compiled by a compiler (a Java or some Intermediate Language (IL) compiler) on a host development machine. It is byte code that is translated by a Virtual Machine (VM), such as: the Java Virtual Machine (JVM) or an.NETCE Compact Framework virtual machine, for example.

Byte Order How data bits and/or bytes are represented and stored in a particular component of a computer system.

C

Cache Very fast memory that holds copies of a subset of main memory, to allow for faster CPU access to data and instructions typically stored in main memory.

Capacitor Used to store electrostatic energy, a capacitor is basically made up of conductors (two parallel metal plates), separated by an insulator (a dielectric such as: air, ceramic, polyester, mica, etc.). The energy itself is stored in an electric field created between the two plates given the right environment.

Cathode (1) The positively charged pole (terminal) of a voltage source. (2) The negatively charged electrode of a device (diode) that acts as an electron source.

Cation A positive ion, meaning an atom that has lost electrons.

Cavity Resonator A component that contains and maintains an oscillating electromagnetic field.

Central Processing Unit (CPU) (1) The master/main processor on the board. (2) The processing unit within a processor that is responsible for executing the indefinite cycle of fetching, decoding, and executing instructions while the processor has power.

Checksum A numerical value calculated from some set of data to verify the integrity of that data, commonly used for data transmitted via a network.

Chip See *Integrated Circuit (IC)*.

Circuit A closed system of electronic components in which a current can flow.

Circuit Breaker An electrical component that insures that a current load doesn't get too large by shutting down the circuit when its overheat sensor senses there is too much current.

Class Used in object-oriented schemes and languages to create objects, a class is a prototype (type description) that is made up of some combination of interfaces, functions (methods), and variables.

Clear-Box Testing See *White-Box Testing*.

Clock An oscillator that generates signals resulting in some type of waveform. Most embedded boards include a digital clock that generates a square waveform.

Coaxial Cable A type of cabling made up of two layers of physical wire, one center wire and one grounded wire shielding. Coaxial cables also include two layers of insulation, one between the wire shielding and center wire, and one layer above the wire shielding. The shielding allows for a decrease in interference (electrical, RF, etc.).

Compiler A software tool that translates source code into assembly code, an intermediary language opcode, or into a processor's machine code directly.

Complex Instruction Set Computer (CISC) A general-purpose ISA which typically is made up of many, more complex operations and instructions than other general-purpose ISAs.

Computer Aided Design (CAD) Tools Tools used to create technical drawings and documentation of the hardware, such as schematic diagrams.

Computer Aided Software Engineering (CASE) Tools Design and development tools that aid in creating an architecture and implementing a system, such as UML tools and code generators.

Conductor A material that has fewer impediments to an electric current (meaning it more easily loses/gains valence electrons) allowing for an electrical current to flow more easily through them than other types of materials. Conductors typically have ≤ 3 valence electrons.

Connector An electrical component that interconnects different types of subsystems.

Context The current state of some component within the system (registers, variables, flags, etc.).

Context Switch The process in which a system component (interrupts, an OS task, etc.) switches from one state to another.

Coprocessor A slave processor that supports the master CPU by providing additional functionality, and that has the same ISA as the master processor.

Coulomb In electronics, the charge of one electron is too small to be of practical use, so in electronics, the unit for measuring charges is called a coulomb (named after Charles Coulomb who founded Coulomb's law), and is equal to that of 6.28×10^{18} electrons.

Critical Section A set of instructions that are flagged to be executed without interruption.

Cross Compiler A compiler that generates machine code for hardware platforms that differs from the hardware platform the compiler is actually residing and running on.

Crystal An electrical component that determines an oscillator's frequency. A crystal is typically made up of two metal plates separated by quartz, with two terminals attached to each plate. The quartz within a crystal vibrates when current is applied to the terminals, and it is this frequency that impacts the frequency at which the oscillator operates.

Current A directed flow of moving electrons.

D

Daisy Chain A type of digital circuit in which components are connected in series (in a 'chain-like' structure), and where signals pass through each of the components down through the entire chain. Components at the top of the chain essentially can impact (slow down, block, etc.) a signal for being received by components further down in the chain.

Data Communications Equipment (DCE) The device that the DTE wants to serially communicate with, such as an I/O device connected to the embedded board.

Datagram What the networking data received and processed by the networking layer of the OSI model or corresponding layer in other networking models (the Internet layer in the TCP/IP model) is called.

Data Terminal Equipment (DTE) The initiator of a serial communication, such as a PC or embedded board.

Deadlock An undesired result related to the use of an operating system, in which a set of tasks are blocked, awaiting an event to unblock that is controlled by one of the tasks in the blocked set.

Debugger A software tool used to test for, track down, and fix bugs.

Decimal A base-10 number system, meaning there are 10 symbols (0–9), used in a variety of combinations to represent data.

Decoder A circuit or software that translates encoded data into the original format of the data.

Delay Line An electrical component that delays the transmission of a signal.

Demodulation Extracting data from a signal that was modified upon transmission to include a carrier signal and the added transmitted data signal.

Demultiplexor (Demux) A circuit which connects one input to more than one output, where the value of the input determines which output is selected.

Device Driver Software that directly interfaces with and controls hardware.

Dhrystone A benchmarking application which simulates generic systems programming applications on processors, used to derive the MIPS (Millions of Instructions per Second) value of a processor.

Die The portion of an integrated circuit that is made of silicon, that can either be enclosed in some type of packaging or connected directly to a board.

Dielectric An insulative layer of material found in some electrical components, such as capacitors.

Differentiator A circuit that calculates a mathematical (calculus) derivative output based on a given input.

Digital A signal that is expressed as some combination of one of two states, a '0' or '1'.

Digital Signal Processor (DSP) A type of processor that implements a datapath ISA, and is typically used for repeatedly performing fixed computations on different sets of data.

Digital Subscriber Line (DSL) A broadband networking protocol that allows for the direct digital transmission of data over twisted pair wired (POTS) mediums.

Digital-to-Analog Converter (DAC) A device that converts digital signals to analog signals.

Diode A two-terminal semiconductor device that allows current flow in one direction, and blocks current which flows in the opposite direction.

Direct Current (DC) Current that flows constantly in the same direction in a circuit. DC current is defined by two variables: *polarity* (the direction of the circuit) and *magnitude* (the amount of current).

Direct Memory Access (DMA) A scheme in which data is exchanged between I/O and memory components on a board with minimal interference from and use of the master processor.

Disassembler Software that reverse-compiles the code, meaning machine language is translated into assembly language.

Domain Name Service (DNS) An OSI model session layer networking protocol that converts domain names into internet (network layer) addresses.

Dual Inline Memory Module (DIMM) A type of packaging in which memory ICs can come in, specifically a mini-module (PCB) that can hold several ICs. A DIMM has protruding pins from one side (both on the front and back) of the module that connect into a main embedded motherboard, and where opposing pins (on the front and back of the DIMM) are each independent contacts.

Dual Inline Package (DIP) A type of packaging that encloses a memory IC, made up of ceramic or plastic material, with pins protruding from two opposing sides of the package.

Dual Port Random Access Memory (DPRAM) RAM that can connect to two buses allowing for two different components to access this memory simultaneously.

Dynamic Host Configuration Protocol (DHCP) A networking layer networking protocol that provides a framework for passing configuration information to hosts on a TCP/IP-based network.

Dynamic Random Access Memory (DRAM) RAM whose memory cells are circuits with *capacitors* that hold a charge in place (the charges or lack thereof reflecting the data).

E

Earliest Deadline First (EDF) A real-time, preemptive OS scheduling scheme in which tasks are scheduled according to their deadline, duration, and frequency.

Effective Address The memory address generated by the software. This is the address that is then translated into the physical address of the actual hardware.

Electrically Erasable Programmable Read Only Memory (EEPROM) A type of ROM which can be erased and reprogrammed more than once, the number of times of erasure and re-use depending on the EEPROM. The contents of EEPROM can be written and erased 'in bytes' without using any special devices. This means the EEPROM can stay on its residing board, and the user can connect to the board interface to access and modify EEPROM.

Electricity Energy generated by the flow of electrons through a conductor.

Electron A negatively charged subatomic particle.

Emitter One of three terminals of a bipolar transistor.

Encoder A device that encodes (translates) a set of data into another set of data.

Endianness See *Byte Order*.

Energy The amount of work performed that can be measured in units of joules (J) or watts × time.

Erasable Programmable Read Only Memory (EPROM) A type of ROM that can be erased more than one time using other devices that output intense short wavelength, ultraviolet light into the EPROM package's built-in transparent window.

Ethernet One of the most common LAN protocols, implemented at physical and data-link layers of the OSI model.

Extended Data Out Random Access Memory (EDO RAM) A type of RAM commonly used as main and/or video memory; it is a faster type of RAM that can send a block of data and fetch the next block of data simultaneously.

F

Farad The unit of measurement in which capacitance is measured.

Field Programmable Gate Array (FPGA) A type of programmable ASIC implementing the application-specific ISA model.

Firmware Any software stored on ROM.

Flash Memory A CMOS-based faster and cheaper variation of EEPROM. Flash can be written and erased in blocks or sectors (a group of bytes). Flash can also be erased electrically, while still residing in the embedded device.

Flip-Flop One of the most commonly used types of latches in processors and memory circuitry. Flip-flops are sequential circuits that are called such because they function by alternating (flip-flopping) between both states (0 and 1), and the output is then switched (such as from 0 to 1 or from 1 to 0, for example). There are several types of flip-flops, but all essentially fall under either the asynchronous or synchronous categories.

Fuse An electrical component that protects a circuit from too much current by breaking the circuit when a high enough current passes through it. Fuses can also be used in some types of ROMs as the mechanism to store data.

G

Galvanometer A measurement device that measures smaller amounts of current in a circuit.

Garbage Collector A language-related mechanism that is responsible for deallocating unused memory at runtime.

Gate A more complex type of electronic switching circuit designed to perform logical binary operations, such as AND, OR, NOT, NOR, NAND, XOR, and so on.

Glass-Box Testing See *White-Box Testing*.

Ground In a circuit, the negative reference point for all signals.

H

Half Duplex An I/O communications scheme in which a data stream can be transmitted and received in either direction, but in only one direction at any one time.

Handshaking The process in which protocols are adhered to by components on a board or devices over a network that want to initiate and/or terminate communication.

Hard Real Time Describes a situation in which timing deadlines are always met.

Hardware All of the physical components of an embedded system.

Harvard Architecture A variation of the von Neumann model of computer systems, which differs from von Neumann in that it defines separate memory spaces for data and instructions.

Heap A portion of memory used by software for dynamic allocation of memory space.

Heat Sink A component on a board that extracts and dissipates heat generated by other board components.

Henry The unit of measurement for inductance.

Hertz The unit of measurement for frequency in terms of cycles per second.

High-Level Language A programming language that is semantically further away from machine language, more resembles human language, and is typically independent of the hardware.

Hit Rate A cache memory term indicating how often desired data are located in cache relative to the total number of times cache is searched for data.

Host The computer system used by embedded developers to design and develop embedded software; it can be connected to the embedded device and/or other intermediary devices for downloading and debugging the embedded system.

Hysteresis The amount of delay in a device's response to some change in input.

I

In-Circuit Emulator (ICE) A device used in the development and debugging of an embedded system which emulates the master processor on an embedded board.

Inductance The storage of electrical energy within a magnetic field.

Inductor An electrical component made up of coiled wire surrounding some type of core (air, iron, etc.). When a current is applied to a conductor, energy is stored in the magnetic field surrounding the coil allowing for a energy storing and filtering effect.

Infrared (IR) Light in the THz ($1000\,\text{GHz}$, $2 \times 10^{11}\,\text{Hz} - 2 \times 10^{14}\,\text{Hz}$) range of frequencies.

Instruction Set Architecture (ISA) The features that are built into an architecture's instruction set, including the types of operations, types of operands, and addressing modes, to name a few.

Insulator A type of component or material which impedes the movement of an electric current.

Integrated Circuit (IC) An electrical device made up of several other discrete electrical active elements, passive elements, and devices (transistors, resistors, etc.) – all fabricated and interconnected on a continuous substrate (chip).

Interpreter A mechanism that translates higher-level source code into machine code, one line or one byte code at a time.

Interrupt An asynchronous electrical signal.

Interrupt Handler The software that handles (processes) the interrupt, and is executed after the context switch from the main instruction stream as a response to the interrupt.

Interrupt Service Routine (ISR) See *Interrupt Handler.*

Interrupt Vector An address of an interrupt handler.

Inverter A NOT gate that inverts a logical level input, such as from HIGH to a LOW or vice versa.

J

Jack An electrical device designed to accept a plug. There are many types of jacks, including coaxial, two-plug, three-plug, and phono, just to name a few.

Joint Test Access Group (JTAG) A serial port standard that defines an external interface to ICs for debugging and testing.

Just-In-Time (JIT) Compiler A higher-level language compiler that translates code via interpretation in the first pass, and then compiles into machine code that same code to be executed for additional passes.

K

Kernel The component within all operating systems that contains the main functionality of the OS, such as process management, memory management, and I/O system management.

L

Lamp An electrical device that produces light. There are many types of lamps used on different types of embedded devices, including neon (via neon gas), incandescent (producing light via heat), and xenon flash lamps (via some combination that includes high voltage, gas, and electrodes), to name a few.

Large Scale Integration (LSI) A reference to the number of electronic components in an IC. An LSI chip is an IC containing 3000–100,000 electronic components per chip.

Latch A bistable multivibrator that has signals from its output fed back into its inputs, and can hold stable at only one of two possible output states: 0 or 1. Latches come in several different subtypes, including S-R, Gated S-R, and D.

Latency The length of elapsed time it takes to respond to some event.

Least Significant Bit (LSb) The bit furthest to the right of any binary version of a number.

Least Significant Byte (LSB) The 8 bits furthest to the right of any binary version of a number; for example, the two digits furthest to the right of any hexadecimal version of a number larger than a byte.

Light Emitting Diode (LED) Diodes that are designed to emit visible or infrared (IR) light when in forward bias in a circuit.

Lightweight Process See *Thread*.

Linker A software development tool used to convert object files into executable files.

Little Endian Data represented or stored in such a way that the LSB and/or the LSb is stored in the lowest memory address.

Loader A software tool that relocates developed software into some location in memory.

Local Area Network (LAN) A network in which all devices are within close proximity to each other, such as in the same building or room.

Logical Memory Physical memory as referenced from the software's point of view, as a one-dimensional array. The most basic unit of logical memory is the byte. Logical memory is made up of all the physical memory (registers, ROM, and RAM) in the entire embedded system.

Loudspeaker See *Speaker*.

Low-Level Language A programming language which more closely resembles machine language. Unlike high-level languages, low-level languages are hardware dependent, meaning there is typically a unique instruction set for processors with different architectures.

M

MAC Address The networking address located on networking hardware. MAC addresses are internationally unique due to the management of allocation of the upper 24 bits of these addresses by the IEEE organization.

Machine Language A basic language consisting of ones and zeros that hardware components within an embedded system directly transmit, store, and/or execute.

Medium Scale Integration (MSI) A reference to the number of electronic components in an IC. An MSI chip is an IC containing 100–3000 electronic components per chip.

Memory Cell Physical memory circuit that can store one bit of memory.

Memory Management Unit (MMU) A circuit used to translate logical addresses into physical addresses (**memory mapping**), as well as handling memory security, controlling cache, handling bus arbitration between the CPU and memory, and generating appropriate exceptions.

Meter A measurement device that measures some form of electrical energy, such as voltage, current, or power.

Microcontroller Processors that have most of the system memory and peripherals integrated on the chip.

Microphone A type of transducer that converts sound waves into electrical current. There are many types of microphones used on embedded boards, including condenser microphones which use changes in capacitance in proportion to changes in sound waves to produce their conversions, dynamic microphones which use a coil that vibrates to sound waves, and a magnetic field to generate a voltage that varies in proportion to sound variations, to name a few.

Microprocessor Processors that contain a minimal set of integrated memory and I/O peripherals.

Most Significant Bit (MSb) The bit furthest to the left of any binary version of a number.

Most Significant Byte (MSB) The 8 bits furthest to the left of any binary version of a number; for example, the two digits furthest to the left of any hexadecimal version of a number larger than a byte.

Multitasking The execution of multiple tasks in parallel.

Multivibrator A type of sequential logical circuit designed so that one or more of its outputs are fed back as input.

N

NAND Gate A gate whose output is 0 when both inputs are 1.

Noise Any unwanted signal alteration from an input source, or any part of the input signal generated from something other then a sensor.

Non-Volatile Memory (NVM) Memory that contains data or instructions that remain even when there is no power in the system.

NOR Gate A gate whose output is 0 if either of the inputs are 1.

NOT Gate See *Inverter*.

O

On-Chip Debugging (OCD) Refers to debugging schemes in which debugging capabilities are built into the board and master processor.

One Time Programmable (OTP) A type of ROM that can only be programmed (permanently) one time outside the manufacturing factory, using a ROM burner. OTPs are based upon bipolar transistors, in which the ROM burner burned out fuses of cells to program them to '1' using high voltage/current pulses.

Operating System (OS) A set of software libraries that serve two main purposes in an embedded system: providing an abstraction layer for software on top of the OS to be less dependent on hardware (making the development of middleware and applications that sit on top of the OS easier), and managing the various system hardware and software resources to ensure the entire system operates efficiently and reliably.

OR Gate A gate whose output is 1 if either of the inputs are 1.

P

Packet A unit to describe some set of data being transmitted over a network at one time.

Parallel Port An I/O channel that can transmit or receive multiple bits simultaneously.

Plug An electrical component used to connect one subsystem into the jack of another subsystem. There are many types of plugs, such as two-conductor, three-conductor, and phono/RCA.

Polling Repeatedly reading a mechanism (such as a register, flag, or port) to determine if some event has occurred.

Printed Circuit Board (PCB) Thin sheets of fiberglass in which all the electronics within the circuit sit on. The electric path of the circuit is printed in copper, which carries the electrical signals between the various components connected on the board.

Process A creation of the OS that encapsulates all the information that is involved in the execution of a program, such as a stack, PC, the source code and data.

R

Random Access Memory (RAM) Volatile memory in which any location within it can be accessed directly (randomly, rather than sequentially from some starting point), and whose content can be changed more than once (the number depending on the hardware).

Read Only Memory (ROM) A type of non-volatile memory that can be used to store data on an embedded system permanently.

Real Time Operating System (RTOS) An OS in which tasks meet their deadlines, and related execution times are predictable (deterministic).

Rectifier An electronic component that allows current to flow in only one direction.

Reduced Instruction Set Computer (RISC) An ISA that usually defines simpler operations made up of fewer instructions.

Register A combination of various flip-flops that can be used to temporarily store data or delay signals.

Relay An electromagnetic switch. There are many types of relays, including the DPDT (Double Pole Double Throw) relay which contains two contacts that can be toggled both ways (on and off), a DPST (Double Pole Single Throw) relay which contains two contacts that can only be switched on or off, an SPDT (Single Pole Double Throw) relay which contains one contact that can be toggled both ways (on and off), and an SPST (Single Pole Single Throw) relay which contains one set of contacts and can only be switched one way (on or off).

Resistor An electronic device made up of conductive materials that have had their conductivity altered in some fashion in order to allow for an increase in resistance.

Romizer A device used to write data to EPROMs.

S

Scheduler A mechanism within the OS that is responsible for determining the order and the duration of tasks to run on the CPU.

Semaphore A mechanism within the OS which can be used to lock access to shared memory (mutual exclusion), as well as can be used to coordinate running processes with outside events (synchronization).

Semiconductor Material or electrical component whose base elements have a conductive nature that can be altered by introducing other elements into their structure, meaning it has the ability to behave both as a conductor (conducting part of the time) and as an insulator (blocking current part of the time).

Serial Port An I/O channel that can transmit or receive one bit at any given time.

Speaker A type of transducer that converts variations of electrical current into sound waves.

Switch An electrical device used to turn an electrical current flow on or off.

T

Target The embedded system platform, connected to the host, being developed.

Task See *Process*.

Thermistor A resistor with a resistance changes on-the-fly depending on the temperature the thermistor is exposed to. A thermistor's resistor typically decreases as temperature increases.

Thermocouple An electronic circuit that relays temperature differences via current flowing through two wires joined at either end. Each wire is made of different materials with one junction of the connected wires at the stable lower temperature, while the other junction is connected at the temperature to be measured.

Thread A sequential execution stream within a task. Threads are created within the context of a task, meaning a thread is bound to a task. Depending on the OS, a task can also own one or more threads. Unlike tasks, threads of a task share the same resources, such as working directories, files, I/O devices, global data, address space, and program code.

Throughput The amount of work completed in a given period of time.

Tolerance Represents at any one time how much more or less precise the parameters of an electrical component are at any given time based on its actual labeled parameter value. The actual values should not exceed plus (+) or minus (–) the labeled tolerance.

Transceiver A physical device which receives and transmits data bits over a networking transmission medium.

Transducer An electrical device that transforms one type of energy into another type of energy.

Transformer A type of inductor that can increase or decrease the voltage of an AC signal.

Transistor Some combination of P-type and N-type semiconductor material, typically with three terminals connecting to one type of each material. Depending on the type of transistor, they can be used for a variety of purposes, such as current amplifiers (amplification), in oscillators (oscillation), in high-speed integrated circuits, and/or in switching circuits (DIP switches and push buttons commonly found on off-the-shelf reference boards).

Translation Lookaside Buffer (TLB) A portion of cache used by an MMU for allocating buffers that store address translations.

Trap Software and internal hardware interrupts that are raised by some internal event to the master processor.

Truth Table A table that outlines the possible input(s) of a logic circuit or Boolean equation, and the relative output(s) to the input(s).

Twisted Pair A pair of tightly interwrapped wires used for digital and analog data transmission.

U

Ultra Large Scale Integration (ULSI) A reference to the number of electronic components in an IC. A ULSI chip is an IC containing over 1,000,000 electronic components per chip.

Universal Asynchronous Receiver Transmitter (UART) A serial interface that supports asynchronous serial transmission.

Universal Synchronous Asynchronous Receiver Transmitter (USART) A serial interface that supports both synchronous and asynchronous serial transmission.

Untwisted Pair (UTP) A pair of parallel wires used for digital and analog data transmission.

V

Very Large Scale Integration (VLSI) A reference to the number of electronic components in an IC. A VLSI chip is an IC containing 100,000–1,000,000 electronic components per chip.

Virtual Address A memory location based upon a logical address that allows for the expansion of the physical memory space.

Voltage Divider An electrical circuit made up of a few or more resistors that can decrease the input voltage of a signal.

Voltmeter A measurement device that measures voltage.

W

Wattmeter A measurement device that measures power.

Way-Ahead-of-Time (WAT) Compiler A compiler that translates higher-level code directly into machine code.

White-Box Testing Testing that occurs with a tester that has visibility into the system's interworkings, such as having access to source code and schematics information.

Wire A component made up of conductive material that carries signals between components on a board (i.e., bus wires) or between devices (i.e., wired transmission mediums).

X

XOR Gate A gate whose output is 1 (or on, or high) if only one input (but not both) is 1.

Index

Printed and bound by CPI Group (UK) Ltd, Croydon, CR0 4YY

03/10/2024

01040335-0001